T0326378

Case Studies in

NANOTOXICOLOGY AND PARTICLE TOXICOLOGY

Case Studies in NANOTOXICOLOGY AND PARTICLE TOXICOLOGY

Edited by

ANTONIETTA M. GATTI, PhD
Associate Professor, National Research Council, Rome, Italy;
Visiting Professor, Institute for Advanced Sciences Convergence, US Department of State, Washington DC, USA;
Founder, Nanodiagnostics Srl, Modena, Italy

STEFANO MONTANARI, PhD
Director, Nanodiagnostics Srl, Modena, Italy

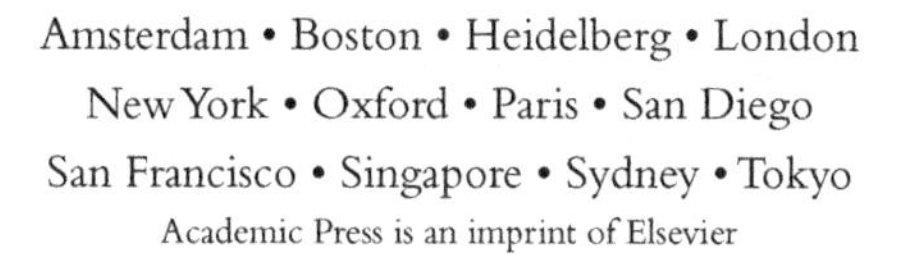

Amsterdam • Boston • Heidelberg • London
New York • Oxford • Paris • San Diego
San Francisco • Singapore • Sydney • Tokyo
Academic Press is an imprint of Elsevier

Academic Press is an imprint of Elsevier
125, London Wall, EC2Y 5AS, UK
525 B Street, Suite 1800, San Diego, CA 92101-4495, USA
225 Wyman Street, Waltham, MA 02451, USA
The Boulevard, Langford Lane, Kidlington, Oxford OX5 1GB, UK

Cover Image © 2014: *Red cells in a vessel.* Photograph by Antonietta M. Gatti and Laura Valentini; elaborated by Daniele Contini.

British Library Cataloguing-in-Publication Data
A catalogue record for this book is available from the British Library

Library of Congress Cataloging-in-Publication Data
A catalog record for this book is available from the Library of Congress

ISBN: 978-0-12-801215-4

For information on all Academic Press publications
visit our website at http://store.elsevier.com/

Publisher: Christine Minihane
Acquisition Editor: Kristine Jones
Editorial Project Manager: Molly McLaughlin
Production Project Manager: Lucía Pérez
Designer: Miles Hitchen

Typeset by Thomson Digital

DEDICATION

Everything has a cause and diseases are no exception. This book does not pretend to be the depository of truth, but is dedicated with due humility to those who want to be put on the trail of truth, even if it implies stepping beyond the frontiers of the present, received knowledge, into no man's land.

This book is also dedicated to humanity and its survival. Acts of faith may not regard science, i.e, knowledge, and believing may not be a synonym of knowing. What can be found in these pages comes from direct observation of natural phenomena and should be worthy of serious consideration. Remaining prisoners of convictions of convenience, often not supported by any kind of demonstration besides a sort of more or less vague *ipse dixit*, threatens the very survival of *Homo sapiens*, and pretending *a priori* that this is not true is perhaps the most serious among the dangers we are facing. This is not a millenarian prophecy, but a mathematical, logic extrapolation from what is already happening in the world.

CONTENTS

Foreword *xi*
Preface *xiii*
Acknowledgments *xv*

1. Introduction **1**

1.1 The history 1
1.2 What is Nanopathology? 2
References 5

2. A Very Brief History of Particulate Pollution **7**

2.1 Origin 7
References 10

3. Nanotoxicity **13**

3.1 Introduction 13
3.2 New scenario of the nano-bio-interactions 15
3.3 The nanotoxicological debate 21
References 27

4. Clinical Cases **29**

4.1 Introduction 29
4.2 Mesothelioma: a nanofiber-induced disease 32
4.3 Hashimoto thyroiditis 39
4.4 Ameloblastoma 45
4.5 Leukemia and lymphoma 50
4.6 Congenital malformations 54
4.7 Cryoglobulinemia 56
4.8 Breast cancer 62
References 62

5. Sentinel Cases **65**

5.1 Introduction: nanopathology and toxicology 65
5.2 Sentinel cases 67
5.3 Hepatic granulomas: same disease, different pathogens 69

5.4 Archeology and war 71
5.5 The case of the worker in a ceramic-tile industry 73
5.6 Precious alloys in a prostatic neoformation 74
5.7 The case of a child with prostate cancer 75
5.8 A malformed child born with leukemia 78
5.9 Malformed children 81
5.10 The child with bone cancer 85
5.11 The case of the patient killed by repeated enemas 88
5.12 The boy who played five-a-side football 90
5.13 The boy who went into a sudden coma 93
References 95

6. Environmental Cases and Nanoecotoxicology 99

6.1 The case of a power plant 99
6.2 Contamination around urban incinerators 109
6.3 The case of the incinerator of Terni 116
6.4 Contamination by engineered nanoparticles 122
References 127

7. War Cases and Terrorist Attacks 129

7.1 Introduction 129
7.2 The war environmental dust 131
7.3 The Italian case: diseases among soldiers after the Balkan war 135
7.4 The case of Soldier 1 139
7.5 The case of Soldier 2 140
7.6 A case of a soldier with contaminated semen and artificial insemination 142
7.7 A case of a soldier with aspergillosis complications 144
7.8 The case of a civilian who worked in Sarajevo during the siege and war 145
7.9 The cases of two reporters who worked in the Gulf and Balkan war theaters 147
7.10 Quirra and the Quirra Syndrome 149
7.11 The cases of rescue workers during the terrorist attacks on the Twin Towers 152
7.12 Have Hiroshima and Nagasaki been misinterpreted? 155
7.13 A brief conclusion 159
References 160

8. Food, Drugs and Nanoparticles 163

8.1 Introduction 163
8.2 Intentional and accidental contamination of food 170

8.3 Bovine spongiform encephalopathy and food 180
8.4 Vaccine contamination 187
References 193

9. Occupational Cases 195

9.1 Introduction 195
9.2 Printers and nanoink 196
9.3 Cases of spontaneous pneumothorax 200
9.4 Working pollution in nanotechnology laboratories 204
References 208

10. Miscellaneous Cases 209

10.1 Impact of smoking 209
10.2 Diabetes, chronic-fatigue syndrome and other pathologies that could be explained from a different point of view 216
10.3 Other possible effects of nanoparticle exposure 228
References 229

11. The Future of Nanotechnologies 231

11.1 Introduction 231
11.2 Can nanomedicine solve the unsolved problems of medicine? 233
11.3 The success of nanosilver 238
11.4 Analysis of the end of the life cycle of nanoproducts 240
11.5 Prevention and systems of prevention 243
11.6 Present and future 243
References 246

12. Conclusions 249

Index 251

FOREWORD

Drs. Gatti and Montanari are members of that rare species, scientists whose work is so novel and far-reaching that society is reluctant to believe it in the first instance. However, with time, the ideas and notions put forward make so much sense, they move into the realm of accepted wisdom.

The work of these particular authors relates to the origins of hitherto inexplicable but widespread diseases of the modern industrialised world.

Evolution has fitted humans to be "at one" with their environment. Disease often occurs when that environment is affected in unnatural ways; for example, the flu pandemic of 1919 was traced back to an abnormal proximity of people and animals, as are present outbreaks of bird flu, and many previous and subsequent plagues can be related to overcrowding and poor sanitation in towns and cities, unnatural situations which humankind was not evolved to cope with.

There is now a crop of new diseases that are besetting society. These diseases and their causes are the focus of this book. Many of these diseases – and their causes – are extremely controversial. Society will often deny what it does not want to admit. But Gatti and Montanari have applied their expertise, experience and professional rigor to uncovering the "key" to this multitude of modern afflictions. And like every new theory, the more it appears to provide the solution to each new problem, the more acceptable it becomes. And the sooner we accept what the authors are telling us, the sooner we can restrict or ban altogether the root causes. Though what they say makes for very uncomfortable reading.

So, what are these modern diseases? And what are their causes? The diseases, outlined in graphically riveting detail, range from clinical cancers such as mesothelioma and leukemia, to Gulf War syndrome and similar diseases, to diseases associated with proximity to certain industrial plants to workplace diseases; for example, those manifest by some hairdressers, welders, ceramic-tile workers and so on.

The authors have made a lifetime's study of many diseases whose causes have baffled other scientists, and their conclusions form the basis for this book. This publication is not a dry academic treatise; it is an earth-shattering story, written in the finest traditions of detective work, based on uncovering the facts, and then identifying the culprit. For scientists and laypeople alike, this book is a fascinating read, not only because it shows science at its best, but also because it communicates its message clearly, and having health as its theme, it touches the fears and hopes of each and every one of us.

The critical key to understanding the root cause of these "modern" diseases, according to Gatti and Montanari, revolves around the unrecognized exposure to man-made nanoparticles and their subsequent introduction into the body, resulting in damage to surrounding cells, tissues and organs. These novel particles are generally metallic, usually the result of high-temperature combustion, and are often completely new compounds.

It is likely this book will engender a great rage that innocent and unknowing individuals in the past have fallen mortally ill from causes that can be traced to industrial and military activity. My wish is that now there will be no excuse for any individual in the future to suffer a similar fate. And no hiding place for governments that do not implement the necessary legislation to stop these preventable diseases in their tracks.

Ottilia Saxl
Founder, Institute of Nanotechnology and *Nano* Magazine

PREFACE

More than a book, this work is a testimony to the uncertain future awaiting humanity if we do not learn how to wisely manage the nanoparticles and the nanoworld of which they are the fundamental constituents.

This book, written through the magnifying lens of an electron microscope, is meant to present the effects of micro- and, more in particular, nanosized products and by-products on human and animal life and on the environment already affected by them. These incidental particles are generated as unwanted consequences of many human activities involving combustion processes, and by those particles we consciously produce through nanotechnologies and through the incineration of nanomaterials at the end of their life cycle.

We report our personal investigations and their results, illustrating the fate of those particles in the environment, and more specifically, inside the animal, human body, and their offspring. Through these investigations, we try to offer information that, if duly used, can help to understand the source of some diseases of unknown origin, to have a better view on others that might not have been fully understood, and to protect us and the planet we live in against the dangers inherent in nanoparticles.

Quite often, the investigations we present are not systematic and can rightly be considered episodic, but they introduce the new concept of personalized medicine.

The work done was supported not only by a few institutions, but also by personal funds. Since our lives are not eternal, we would like to share this knowledge with open-minded scientists, in the hope that they continue our work, understand what we did not, correct our mistakes and achieve positive results. Politicians, industrialists, decision-makers and the general public may also benefit from reading this book. The dream goal of these pages is to save human lives. We are conscious of the fact that not everyone will agree with us, but we intend to continue this work and encourage others to do the same until this dream comes true.

ACKNOWLEDGMENTS

We are indebted to many collaborators, friends and institutions.

We are most grateful to Dr. Federico Capitani and Ms. Lavinia Nitu, patient collaborators of our daily work, for the technical contribution they gave us in writing this book, but also for their loyalty and competence.

Our research was supported by the European Commission with two grants (Nanopathology, DIPNA) that gave us the possibility to discover the impact of nanoparticles in living bodies. It was also supported by the Italian Institute of Technology, which funded the nanotoxicology project INESE, and by the Italian Ministry of Defense, which funded our projects BATNAN and VENAM.

We want to also thank other occasional collaborators: Professor Pietro Gobbi, Dr. Laura Valentini, Dr. Paola Boi and Dr. Anita Manti of the University of Urbino.

We are also grateful to Dr. Claudio Pizzagalli of ARPAM (Pesaro, Italy) for his availability and understanding with regard to our research needs.

We thank Professor Nobuo Kazashi of Harvard University for introducing us to the Director of the Hiroshima Peace Memorial Museum, Ms. Anne-Marie Principe, survivor of the collapse of the Twin Towers in New York on September 11, 2001.

CHAPTER 1

Introduction

Contents

1.1 The history 1
1.2 What is Nanopathology? 2
References 5

1.1 THE HISTORY

Back in early 2008 we published the book *Nanopathology: The Health Impact of Nanoparticles* [1]. In those pages we told the circumstances that led us to start our research about how solid, inorganic, and non-biodegradable micro- and nanosized particles, whatever their origin, can interfere with living beings, and described the basic principles we abide by.

But, on a larger scale, that particular dust also has an impact on the environment of which all living beings are inevitably guests and protagonists at the same time, by which they are as inevitably influenced and, in some cases, affected. And that topic was also dealt with in the book mentioned above.

Nanopathology is a word that we invented in 2002, which became the title of a European project that Dr. Gatti directed and coordinated (Nanopathology: The role of micro and nanoparticles in inducing human health effect (FP5-QOL-147-2002-05)). At that time it was not much more than an empty box. Today, after little more than a decade, nanopathology is a new philosophy, a novel approach to medicine.

As happens with many works in progress, particularly when exploring unknown or little-known territories is the aim, much has been done in the last few years and that work is summed up in this new title. In the lapse of time between the two books, Antonietta M. Gatti had a chance to coordinate a further European project and to be part of a few more research projects sponsored by different agencies; the two of us came across more clinical cases (now amounting to more than 2,000), checked more food, drugs and cosmetics, analyzed more samples collected in polluted environments and, altogether, we were lucky enough to learn a great deal. A selection of our cases, some of which will be considered "sentinel" cases, is illustrated in Chapter 4. The technique we used was described in detail in our former book.

Case Studies in Nanotoxicology and Particle Toxicology
http://dx.doi.org/10.1016/B978-0-12-801215-4.00001-7

Though it is not always possible, here we will try as much as we can to avoid repeating what has already been written and keep referring our readers to *Nanopathology: The Health Impact of Nanoparticles* for what they will not find in these chapters. The book describes also the methods and protocols used to detect nanoparticles in biological matrices performed by means of an Environmental Scanning Electron Microscope (ESEM) and of Field Emission Gun Environmental Scanning Electron Microscope (FEG-ESEM); the equipment is also described.

1.2 WHAT IS NANOPATHOLOGY?

It is understood that the particles discussed here are both engineered, i.e., fabricated in a laboratory, and non-engineered, i.e., incidental. Among the incidental, are those produced by natural phenomena, those coming from high-temperature industrial processes and those generated as by-products of a fair number of activities carried out at low temperature. The others are those intentionally manufactured by nanotechnology industries and, in this case, their fate at their usable-life end will be discussed.

Natural inorganic nanoparticles do exist in nature: sea-water aerosols, some structures of snow and dust erupted by volcanoes are a few of them and with them man has lived all along.

The nanosized by-products are the nanoparticles mostly discussed in this book, since, in our opinion, they are responsible for a non-negligible number of diseases affecting people. They are mostly generated by accidental combustions and are freely dispersed in air, in water and in soil, thus exposing their effects to people who inhale or ingest that particulate matter (PM), sequestering it, at least in part, in their organism.

In general, we take into consideration solid, inorganic and non-biodegradable, i.e., biopersistent, PM.

In this book we do not consider polymeric particles or carbon nanotubes.

Even if the ISO standard [2] defines nanoparticles – but their interest is limited to the engineered ones, those obtained with nanotechnological processes and ranging from just above 0 to 100 nm – in our opinion, for the focus on possible risks to human life, we would rather refer to a functional definition of nanoparticles, i.e., we consider nanoparticles all the particles able to escape the physiological barriers including the cell and its nuclear membrane, so capable of inducing a nanoeffect.

This is a definition made of nanoobjects, but what is nano when biology is what matters? What is the minimum size range of foreign bodies that can induce a nanoeffect, i.e., an effect that the same material, when in bulk form, is unable to induce?

A more specific definition of nanoeffect can be: Whatever damages, changes or causes a modification to a cell that can alter its physiological metabolism. For this reason, we consider "nano" to be what can be internalized by a cell: to give an order of magnitude, that which is below one micron. We know some "bureaucrats" of science will turn up their nose at the way we use the word, but words are nothing more than a means to communicate concepts in the easiest possible way, and we claim the right to use the word nano as it is more convenient for us. The adjective "submicronic" will also be used throughout the book as a synonym, meaning a particle whose size is smaller than one micron.

Until now, there has been no real definition of nanoeffect, but by that term we mean any interaction of nanoparticles with cellular organelles, proteins, enzymes, DNA, etc., interactions that can give origin to stable compounds that are not recognized or only partially recognized by the biological environment where they have been somehow introduced and can no longer participate in the local physiological metabolism. It is a "partial" foreign body that the cell probably tolerates, but this unwanted coexistence can physically alter/condition the normal cellular functions. It must be considered that a foreign body occupies a volume and, at the cell level, can block the in-and-out-flow of nutrients/metabolites or can disturb the chromatin strands during mitosis (see Chapter 10).

But it is not only nanoparticles, whatever their definition, that are responsible for health effects. Particles with a size up to some tens of microns, when they manage to enter the organism, interfere negatively with tissues and organs.

In the last few years many studies have been performed by a great number of authors on engineered nanoparticles matched with cell cultures or organs. In contrast, the reader will find in this book, a focus mostly on human cases, something largely missing in today's literature and very often seen as premature. But, premature or not, what we found in our research was a very interesting new world.

Micro- and nanoparticles are produced in quantities that are rapidly increasing and, in spite of pointless discussions, man is actually reacting to them in a very visible and, to be sure, worrisome way.

Unfortunately, we may not infer the behavior in humans of particles generated by the multitude of sources characterizing our world from how a cell reacts when matched for a short time and in laboratory conditions with very simple (and, for the time being, very uncommon in the environment) particles made by a technician. Man is far more complicated and, to be honest, we still know very little about his physiology.

The particles we find in real-life conditions are not only complicated but often unpredictable at least as to chemical composition, besides coming in a large variety of shapes, sizes, mixes, and exposure, something strictly personal, is in most cases equally unpredictable.

Particles can enter the organism through a variety of doors, and as much as what we have done allows, the different origins and carriers of those particles will be described.

Nanotechnology, treated in Chapter 11, represents the biggest investment ever made in the history of the industry. As a matter of course, investors in the industry have – or think they have – an interest in making the world believe that no risk is involved in their technologies and in their products. Though unproven, that may be the case, but we think it worth considering how the insurance companies deem nanotech one of the riskiest enterprises as seen from their pragmatic point of view.

The cases discussed in this book are mostly Italian since we live in Italy and the investigations performed need a technical background that ranges from the knowledge of the territory to repeated analyses. Of course, physiology and pathology do not change with geography.

What we describe is based on objective observation and repeatable evidence, never contradicting traditional medicine as taught in universities throughout the world. Nanotechnology's contribution is to show further content of pathological tissues that, so far, have escaped due consideration. We introduce an added value for medicine: an interdisciplinary approach that includes physics, chemistry and the study of the environment. Only with an integrated approach do some new and/or idiopathic, i.e., as a matter of fact, mysterious, diseases become understandable.

What we propose is a novel tool, i.e., "customized or personalized medicine," an approach tailored to individual patients that has the drawback of looking more expensive, but has the advantage of being far more effective. Then, all things considered, its being represented as more expensive should probably be reconsidered. In any case, it is obvious that, when the general rules that govern the behavior of particles interacting with living organisms, complex as they are likely to be, are discovered, the economic costs will be greatly reduced.

For the time being, nanopathology is particularly efficacious in finding what the origin is of some pathologies – many of which still considered cryptogenic – provided that the origin is either environmental or comes from the more or less conscious use of nanoparticles or from their presence in food, drugs, cosmetics and other products. That is of the utmost importance, since it gives the patient the chance to avoid or eliminate the exposure, and those responsible for that pollution the possibility to make up for the problem, in addition to putting lawgivers in a position to legislate correctly. Setting aside generic and often vague rules, it is easy to see that in many cases no effective laws exist to protect people against pollution and, when quantitative limits are set, they vary from country to country or groups of countries, being the results of negotiations between local lawmakers and local polluters. But, what is at least as important is the possibility that nanopathology offers to devise appropriate therapies for pathologies that now are often faced by simply trying to hide their symptoms or make the patient able to bear them more or less successfully. In the future, we hope that drugs or devices to eliminate nanoparticles when they are intimately trapped in tissues can be devised, perhaps by taking advantage of nanotechnologies.

Unfortunately, many of the sources of pollution have quickly become part of the psychological DNA of our civilization and getting rid of them looks extremely hard if not actually impossible. Indeed, often we do not even realize that most of what we deem indispensable has not existed for thousands of centuries. However, rejecting extremism and just being pragmatic, the most feasible thing we can do to balance the needs and whims of our society is to use science and technology in all its honest aspects and capacities, hopefully tipping the scales in favor of health. And nanopathology can lend a powerful hand.

In late 2013 a report was released by the International Agency for Research on Cancer (IARC), the specialized cancer agency of the World Health Organization (WHO), on "Outdoor air pollution a leading environmental cause of cancer deaths" (report No.161/2013). The agency classified air pollution as carcinogenic to humans (Group 1), but they did not put forward any hypothesis as to its patho-mechanism. The research we developed, starting from another point of view, demonstrates the invasiveness of submicronic particles and their deep interactions at the cellular level. This invasiveness, beyond the possibility of any physiological filtration, is the key point of various pathologies.

REFERENCES

[1] Gatti A, Montanari S. Nanopathology: The Health Impact of Nanoparticles. Singapore: PanStanford Pub; 2008. 1–298.
[2] ISO/TS 80004-1:2010 Nanotechnologies – Definition 2.1.

CHAPTER 2

A Very Brief History of Particulate Pollution

Contents

2.1 Origin 7
References 10

2.1 ORIGIN

Man has lived in a dusty environment all along, an environment that grows dustier everyday with pollution directly related to the increase in human activity (industrial processes, traffic, disasters, etc.). As briefly mentioned in the previous chapter, for hundreds of millennia dust was exclusively from natural sources: rock and soil erosion, sand carried by the wind, volcanic fumes and the smoke of occasional wood fires. It was the discovery of how to light, keep and use fire – something that occurred long after man was present on the Earth in a more or less now recognizable form, and probably his most important technological breakthrough ever – that introduced artificial pollution into the environment.

Along with gases, every combustion generates particles whose composition depends on what is being burned and whose shape and size depend mainly, though not only, on the temperature at which that combustion occurs. Some of them are formed immediately at the combustion site, and some are formed by condensation, as soon as the chemical elements or molecules are set free in the fire and find a cold enough temperature away from the fire in which they come from. As a general rule, the mostly-metallic particles generated directly by combustion (the ones that are part of the focus of our treatment), are spherical and hollow with a size that decreases with the raising of the formation temperature, ranging from a few tens of microns down to some tens of nanometers. Their crust, often very thin, is usually crystalline and particularly fragile, so that it breaks easily into minute fragments. Other particles, called secondary as opposed to the primary ones just briefly described, are produced at a distance from the combustion site and after some time through photochemical condensation of gases such as nitrogen oxides, sulphur dioxide, ammonia and volatile organic compounds

Case Studies in Nanotoxicology and Particle Toxicology
http://dx.doi.org/10.1016/B978-0-12-801215-4.00002-9

with water vapor, ozone and free radicals more or less naturally present in the atmosphere.

So, environmental pollution is composed of micro- and nanosized particulate matter (PM) generated by natural processes [1], added to which today's environment contains large quantities of other pollutants derived from industrial activities, high-tech nanotechnological productions [2,3], combustions in general [1,4], waste incineration [5,6], engines [7], etc.

Dust is a natural component of the Earth and has always been a part of human life, but it is also something that could be the cause, or one of the causes, of its end. As it is written in the Bible: "Remember, thou art dust and to dust thou shalt return" ("Memento homo, quia pulvis es et in pulverem reverteris") (Genesis 3:19).

Of course, we are not religious exegetes and it is very likely that this phrase was not intended as a sort of planetary prophecy, although catastrophic effects have already occurred in the world's history. In fact, life on Earth came to a sort of eye-catching standstill when dinosaurs disappeared in a very short time about 65 million years ago. There are many theories that try to explain this event. Walter Alvarez of the University of California at Berkeley is well-known for having hypothesized that a large extraterrestrial object collided with the Earth, and its impact threw up enough dust to cause a climatic change. Another theory considers dust, in this instance coming from a mass volcanism phenomenon, as triggering global changes not compatible with the life of great reptiles, for many millions of years the actual lords of the planet. No matter what happened then, dust in the atmosphere above a certain density is incompatible with the type of life we know [8]. It is obvious then that, whatever happens, the Earth will find another equilibrium. The only problem for us is if and how we can adapt to a new scenario.

As a matter of fact, today global industrial activity is generating a never-experienced-before mass of dust and what is more noteworthy is that that mass is increasing rapidly, particularly in emerging, developing countries where tumultuous economic development takes place without concern for health or the environment. On the other hand, the rest of the developed world is happy to buy goods at extremely cheap prices, apparently without realizing that such a policy is bound to be disastrous for the whole planet, since cheap means also uncontrolled. In addition to that, the Earth is much smaller than it looks, so its poisoning is not limited to confined areas and will sooner or later spread everywhere. As long as that problem remains unsolved, the pathologies caused by pollution will continue to rise steadily and inexorably.

Man lived for thousands of centuries like any other primate, without impacting on the world he inhabited. It was only when he discovered how to light a fire that man, at first imperceptibly, started to stain the planet. The quantity and concentration, especially in urban areas, of PM, though, could be considered negligible until the so-called Industrial Revolution between the eighteenth and nineteenth century. Then it started to increase in the environment at an accelerating pace and is now at objectively alarming values. Internal-combustion engines (i.e., mainly vehicular traffic, factories, foundries, cement works, power plants burning heavy oil, coal or biomass, waste incinerators) are all agents unknown until not long ago and responsible for particulate pollution, a kind of pollution that is partly irreversible since a considerable fraction of the matter generated as an unwanted by-product is neither degradable nor technically possible to isolate once it has been released into air, water and soil.

The Great Smog of 1952, or Big Smoke, was a severe air pollution event that affected London during December 1952 [9]. Medical reports in the following weeks estimated that no less than 4,000 people died prematurely and 100,000 more were made ill because the smog's effects on the human respiratory tract. More recent research suggests that the number of fatalities was considerably greater at about 12,000 [10].

As stated above, all combustive processes are a source of particles. So, the so-called "bioenergy," a form of energy obtained by burning biomasses like agriculture waste or poultry droppings, does not represent by any means an exception, and the bio prefix should be looked at as rather suspicious not as much for the origin of the sources of energy, but for the effects of the industrial process on the environment.

Particles can also be produced unintentionally at low temperatures by mutual friction of hard objects. An example may be that of car brakes with pads rubbing on disks. In those cases the PM shapes are irregular, and the size ranges from nanometric to rather coarse. Metal milling, turning and polishing are further cold sources of PM as well as the artefacts from aging buildings.

But, as already mentioned, not all particles, especially nanosized ones, have unintentional origin. With the discovery of the extraordinary physical characteristics of nanoparticles and the advancement of Nanotechnology, higher and higher quantities and more and more various compositions of particles both in the micro and nano range are constantly being introduced into the market and, as an obvious consequence, into the environment by laboratories set up for the purpose. If their presence is dwarfed by what comes from traffic, industry and incineration, it is not hard to see that this will indeed soon change. However, engineered micro- and nanoparticles currently being added to some food, drugs and cosmetics are a component of an increasing number of high-tech yarns, sports equipment, paints, glasses, etc.

The quantity of submicronic particles released into the air is constantly on the increase. While until a few decades ago the nanosized fraction of PM was comparatively low or, in any case, restricted to some industrial areas (steel melting industries, laser ablation technique, laser soldering, etc.), today other activities like those briefly listed above release an increasing quantity of dust into the air we inhale. Because of that, the exposure to every breathing creature grows every day from a safe level and that exposure increases the probability of a noxious interaction with our body and our internal organs. Human life is based essentially on the O_2/CO_2 exchange, and dust can conflict with this basic mechanism. The accumulation of these foreign bodies creates the conditions for the beginning of adverse biological reactions.

Compared to the past, today we synthesize in great quantity so-called "engineered nanoparticles": these "magic bullets" have the ability to cross and outdo all physiological barriers. Nanosized by-products also have the same ability, as will be shown.

The following chapters will illustrate what we consider to be only the tip of the iceberg. The spot identification of particles helps in understanding the exposure a patient suffered and the particles' invasiveness, but other studies are necessary to identify the pathological mechanism sometimes at the level of a single cell damaged by the physical-chemical presence of these foreign bodies.

REFERENCES

[1] Kemppainen S, Tervahattu H, Kikuchi R. Distribution of airborne particles from multi-emission source. Environ Monit Assess 2003;85:99–113.

[2] Colvin VL. The potential environmental impact of engineered nanomaterials. Nature Biotechnology 2003;21:1166–71.

[3] Nel A, Xia T, Mädler L, Li N. Toxic Potential of Materials at the Nanolevel. Science 2006;311:622–7.

[4] Bosco ML, Varrica D, Dongarra G. Case study: inorganic pollutants associated with PM from an area near a petrochemical plant. Environ Res 2005;99:18–30.

[5] Bethanis S, Cheeseman CR, Sollars CJ. Effect of sintering temperature on the properties and leaching of incinerator bottom ash. Waste Manag Res 2004;22:255–64.

[6] Moon MH, Kang D, Lim H, Oh JE, Chang YS. Continuous fractionation of fly ash particles by SPUTT for the investigation of PCDD/Fs levels in different sizes of insoluble particles. Environ Sci Technol 2002;36:4416–23.

[7] Murr LE, Soto KF, Garza KM, Guerrero PA, Martinez F, Esquivel EV, Ramirez DA, Shi Y, Bang JJ, Venzor J III. Combustion-generated nanoparticulates in the El Paso, TX, USA / Juarez, Mexico Metroplex: their comparative characterization and potential for adverse health effects. Int J Environ Res Public Health 2006;3:48–66.
[8] Luis W, Alvarez, Walter Alvarez, Frank Asaro, Helen V. Michel Extraterrestrial Cause for the Cretaceous-Tertiary Extinction. Science 6 June 1980;208(4448):1095–108.
[9] http://en.wikipedia.org/wiki/Great_Smog#Effect_on_London.
[10] Bell ML, Davis DL, Fletcher T. A Retrospective Assessment of Mortality from the London Smog Episode of 1952: The Role of Influenza and Pollution. Environ Health Perspect 2004;112(1):6–8.

CHAPTER 3

Nanotoxicity

Contents

3.1 Introduction 13
3.2 New scenario of the nano-bio-interactions 15
3.3 The nanotoxicological debate 21
References 27

3.1 INTRODUCTION

Nanotechnology represents one of the major areas of scientific and economic interest of the twenty-first century. New nanotechnological commercial products and applications are increasingly present on the market, but none are approved or certified by any official authority as to the safety of their nanocontent.

The major concern is about the lack of knowledge about the effects of exposure to particles, particularly the nano ones. Because of their comparatively small, though fast growing, presence in our society (the adverb comparatively means in comparison with other sources of particles), their impact on health is still limited, even if this book presents pieces of evidence of the deep invasiveness of micro- and nanoparticles and a few in-vivo researchers show that nanoparticles introduced in living tissues can trigger cancers [1–2].

Nanoparticles are a sort of double-edged sword: if on one side they can enhance the quality and performance of many products and can even be effective as drug-delivery agents particularly in cancer therapy, on the other hand they can also induce cancer in addition to a number of other pathologies [3].

Nanoparticles, the basic "bricks" of a growing number of new materials, are used more and more frequently in daily products because they have very interesting, in many respects unexpected, characteristics that give the final product properties that make it very profitable. The impulse given to new types of research, and the hasty applications of their results to develop products that can be sold in considerable quantities and with a great added value, allow no time to realize what the actual consequences might be in terms of environment and health. Until now the lack of regulation for nanoscaled materials allowed manufactures to sell products without carrying out preliminary, specific tests concerning the full life-cycle of their products

Case Studies in Nanotoxicology and Particle Toxicology
http://dx.doi.org/10.1016/B978-0-12-801215-4.00003-0

or to inform their customers of the nanocontents of their products. Organization such as the International Organization for Standardization and the Organisation for Economic Co-operation and Development are working to create standards (from vocabulary to toxicity tests, etc. [4,5]) and to impose them on the industry. Until now, defining and declaring nanocontent was left up to producers who, in most cases, chose not to in part because of the expensive tests required to characterize the nanocontent. However, it is interesting to note that more and more products today include on the packaging the advertisement, "no nanoparticles" (e.g., Kissmyface.com sunscreen). But the truth is, most conscious consumers do not trust these novel products or indications because of a objective lack of reliable information. This further proves the urgency of creating and mandating specific rules and standards for the industry.

The European Commission has already supported a number of nanotoxicity projects in order to fulfill this need: DIPNA, CellNanoTox, ENNSATOX, ENANOPARTICLERA, ENRHES, HINAMOX, InLiveTox, NanEAU, NanEx, NANODEVICE, NanoFATE, NANOFILM, NanoHouse, NanoImpactNet, NanoInteract, NANOKEM, NANOMMUNE, NANOPLAST, NanoPolyTox, ENPRA, EuroNanoTox, InLiveTox, MARINA, ModNanoTox, NanEx, NanoHouse, NanoImpactNet, Nanolyse, NANOMMUNE, NanoPolyTox, NanoReTox, Nanosafe2, NanoSustain NanoTransKinetics, NanoValid, NEPHH, NeuroNano and QNano [6].

There have been many attempts to create major guidelines for in-vitro research parameters, ultimately suggesting that current cytotoxicity parameters as well as standard procedures may not be suitable for most of these tasks. The well-known cytotoxicity standard tests (ISO/EN 10993-5) delivered for biomedical materials and implants proved to be applicable to nanoparticles as well, but the results appear contradictory since they vary from laboratory to laboratory. Several specific reasons can explain why those results are so inhomogeneous and, de facto, unreliable.

As a matter of fact, the basic open question seems to be related to whether physical properties at the nanoscale level (such as nanoparticle aggregation and non-homogeneous dispersion in the medium as well as internalization in the cells that may be responsible for different cytotoxic results) are indeed physiologically relevant. Additionally, nanoparticle coating, size and dispersion have also proved to be crucial for their in-vitro and in-vivo interaction, a possible internalization and cytotoxic effect [7–12].

In-vivo use of nanoparticles requires a very accurate understanding of their kinetics and toxicology due to their ability to interact with various fluids, organs and cellular complex pathways [13–17]. But, there is also the "human factor" that can make a difference.

Usually in-vitro and in-vivo biological tests are carried out by biologists or, in most cases, by operators with biological backgrounds. Through those tests, physicists or chemists are asked to give a full characterization of the particles they add to the cells they are experimenting with (e.g., size, morphology, composition, concentration, Z-potential, coating, etc.), and to check the biological parameters of the cells (e.g., vitality up to gene expression, etc.); however, they do not verify the physical-chemical parameters of the materials *after* the nanointeraction. What those researchers take for granted is that the initial, theoretical, concentration and conditions of the nanoparticles are conserved throughout the testing and those values are plotted with different biological parameters. That is an assumption that can be practically demonstrated to be untrue. Nanoparticles have strong adhesive properties, so, in far-from-negligible circumstances, they form clusters and/or aggregates. In general, those formations can be one or more orders of magnitude larger than single particle's and do not coincide with the nanotoxicity consequences the tests mean to define. Being microsized, they cannot be internalized (i.e., that effect does not happen with bulk material) so their apparent "safety" is a misinterpretation. Again, because of their remarkable tendency to adhere to each other and to surfaces, part of the nanoparticle remains attached to the pipette or the Petri capsule or other supports, while some of the clusters precipitate to the bottom of the vial, thus not participating in the cell interaction. And it doesn't matter how careful the researcher is: handling nanoparticles is an awkward business and its accuracy depends on many different conditions. So, the large and variable errors introduced by their behavior and responsible for the great differences in results, must always be considered. In many cases, when the actual presence of nanoparticles inside the cells is not verified, the results of those nanotoxicity tests are indeed invalid. Only when that presence is checked can the test have a certain validity, even if the question concerns the best parameters to describe nanoparticle/cell interaction, since the normal inflammatory or immunogenic markers seem unsuitable to describe it.

3.2 NEW SCENARIO OF THE NANO-BIO-INTERACTIONS

It is only natural that, to devise toxicity tests suitable for nanoparticles, the approach must take their properties into due consideration. They are discrete, solid, sometimes crystalline, physical entities that can be internalized by cells and interact with their physiology. They represent a non-specific stimulus, i.e., something cells cannot differentiate. As in many other respects, nanoparticles are also different from chemical compounds, agents that cells can often recognize and against which some defense can often be implemented. Unfortunately, these fundamental features are only very seldom understood in all their importance.

We can distinguish a few different types of nano-bio-interactions:

1. the interaction of nanoparticles with extracellular matrix or medium;
2. the interaction with cell membrane and sensors;
3. the interaction with organelles and cytoplasmic components; and
4. the interaction inside the nucleus.

Some researchers have demonstrated that fullerene particles bind to DNA, and if that phenomenon opens the door to a new type of gene therapy [18], this could possibly lead to unvoluntary entrance of other nanosized particles into cells inducing unwanted and casual interaction with cytoplasmic "entities." Even if these complexes are reversible, since they are biodegradable, there is a sequestration of biochemical compounds that can no longer participate in cell metabolic activities. In the case where these complexes are biopsersistent, they are perceived by the cells as foreign bodies. However, the mechanisms of nanoparticles or complex excretion from cells are not yet known [19].

The cell-nanoparticle interactions start with the interaction with the cell membrane. Figure 3.1 shows different possible interactions of antimony-oxide nanoparticles with ematopoietic cells. Thanks to a collaboration with Dr. Lisa Bregoli, then PhD student at the University of Bologna (Italy), we performed scanning transmission electron microscopy (STEM) of the contact and entrapment of these nanoparticles inside the cells. It is clear that, because of good mutual "understanding," these cells can play the role of vector of nanomedicine nanoparticles for whole-body dispersion, but the same mechanism can occur for unwanted nanosized by-products as well.

Figure 3.1 shows some possible interactions between Sb_2O_3 nanoparticles and hematopoietic cells. A few of those particles, aggregated and micrometric in size, remain outside and interact only with the extracellular medium, but at the same time they can also interfere with the signaling among cells. They can also act as a barrier for endocytic trafficking, and some of them can adhere to the cell surface and can be internalized in vacuoles. The clathrin-dependent endocytosis and macropinocytosis are the primary uptake mechanisms [21].

Endocytosis is a normal mechanism for cell functions such as nutrient uptake, cell adhesion and migration, membrane-receptor recycling and

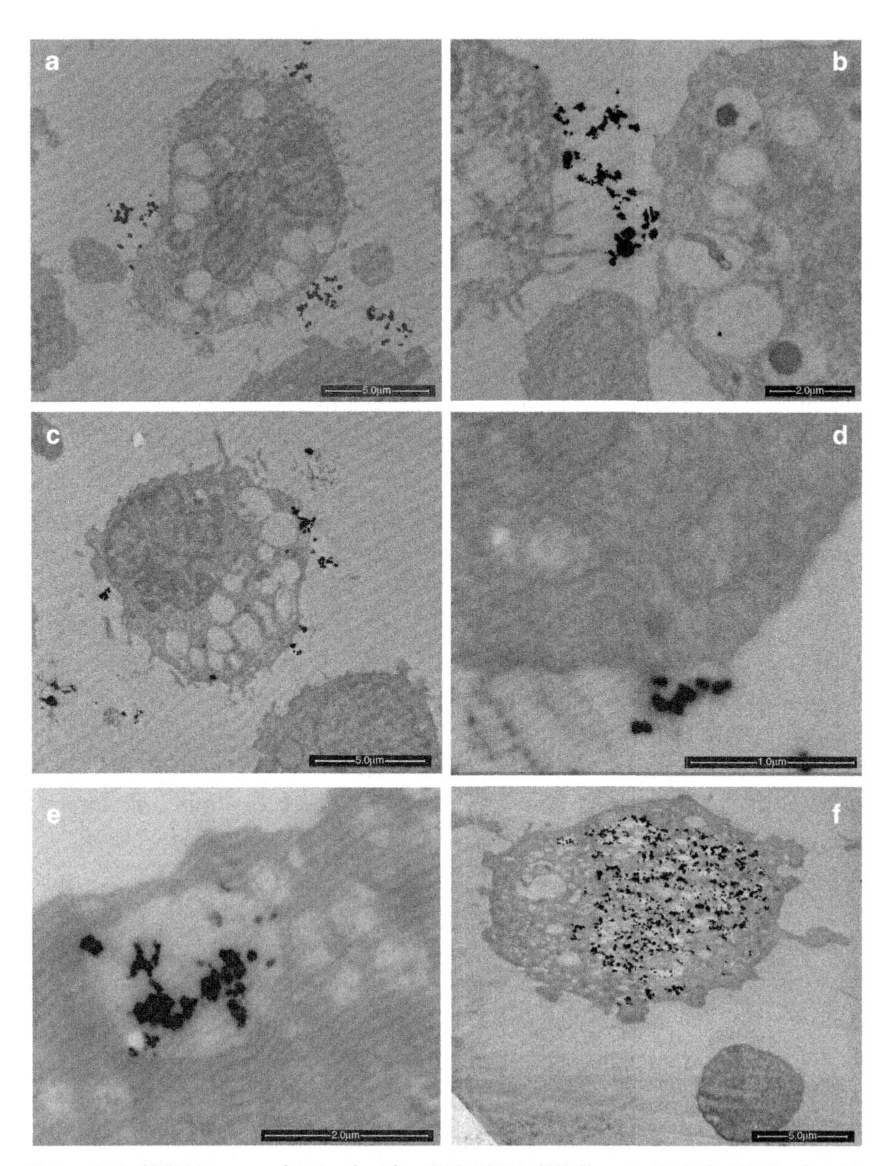

Figure 3.1 STEM images obtained with a FEG-ESEM, (FEI Quanta 250, The Netherlands), which show the interactions of Sb_2O_3 nanoparticles with erythrocytes [20]: (a, b) show cluster of nanoparticles outside hematopoietic cells; (c, d) show nanoparticles in close contact with the cell membrane; (e, f) show internalization of the nanoparticles.

downregulation, pathogen entry, antigen presentation, cell polarity, mitosis, cell growth and differentiation. All these functions can be altered by the presence of nanoparticles. But while lysosomes digest macromolecules from endocytosis, autophagy and phagocytosis, they are unable to digest biopersistent nanoparticles. Figure 3.1(f) shows an abnormal entrance of nanoparticles in the cell, but there was not a prompt following cell death [17,22].

As already noted, the cell-membrane sensors do not recognize nanoparticles, since they represent a non-specific stimulus. Once inside the cytoplasm, nanoparticles cause an obvious increase in cell volume and weight, which can be a biomarker that is never investigated. They then have the possibility of interacting with chemicals and organelles such as mitochondria, ribosomes and endoplasmic reticulum, along with their functions and their metabolites. Nanoparticles can be either biodegradable and/or corrodible or be "eternal," since they remain in place unaltered after the cell's death. Metallic nanoparticles can corrode, thus degrading and releasing ions; and corrosion is a slow phenomenon needing not only time but proper conditions of temperature, pH, etc. Corrosion causes the release of ions that can interact with specific metabolic sequences. For instance, an ion can combine with an intermediate metabolite along the Krebs cycle when it shows higher affinity than the normal metabolite. If the cycle is interrupted, there is no ATP formation. The total energetic level of the cell is decreased and there is no energy sufficient for the cell to live a "normal" life, but that does not necessarily mean that death is the consequence. Biopersistent (i.e., not biodegradable, not corrodible) nanoparticles with electric surface charges can bind local proteins that, as a consequence, are no longer available for local metabolism. This phenomenon does not kill, but induces only cell damage. A protein bound to a nanoparticle can get its three-dimensional morphology changed (chemical bounds can stretch the molecule). In that situation, the protein cannot be recognized as the body's and is thus responsible for an autoimmune response. If an enzyme is bound to a nanoparticle, specific reactions of the cell cannot be catalyzed and some metabolic passages do not occur. At this point, many questions arise. For instance, taking into account all the differences existing between vitamins, we do not know how they behave in the presence of nanoparticles, and we do not know much about the interaction between nanoparticles and mitochondria. In Figure 3.1(d) one of the two mithocondria visible is enlarged and deformed.

Furthermore, can nanoparticles induce an epigenetic effect in the RNA messenger or in the proteins synthesized by the ribosomes? [23]

A further question concerns the interaction of nanoparticles with the endoplasmic reticulum. And, as to the formation of calcium compounds so frequently observed particularly in cases of cancerous tissues where particles are present, we do not know whether the phenomenon is induced by the sole physical presence of nanoparticles or whether other conditions are involved.

Nanoparticles or nanoparticle/biological compounds can either pass through the membrane pores or occlude them and, as a consequence, alter the sodium-pump mechanism or prevent its function completely. They represent a physical obstacle to the normal uptake of nutrients and exocytosis of catabolites that can cause cell death if the phenomenon is massive enough and involves the whole of the cell membrane. When it is just partial, it can induce more or less serious cell damage. These subcellular nano-biointeractions are a key issue that must be duly considered [24].

Nanoparticles can enter not only the cytoplasm, but also the nucleus. However, not being toxic in the classical sense of the term (there is no dose-response curve), as already mentioned they do not induce cell death or, at least, not a sudden one. During mitosis, when the nucleus membrane dissolves, nanoparticles can get in close contact with the DNA strands of the mitotic spindle. Figure 3.2 shows this type of event. A part of the dividing chromatin is deformed by the physical presence of the iron-oxide clustered nanoparticles. They do not allow DNA migration along the spindle. It is likely that one out of two daughter cells will be healthy, but the other one could be "damaged."

Nanoparticles can also interact with metabolic by-products and the exocytosis mechanism causing their alteration. These interactions can give origin to entities (which are not biodegradable) that the cell does not recognize as self and perceive as foreign bodies. These inorganic-organic catabolites can accumulate and produce adverse effects.

In our laboratory we implemented a set of in-vitro tests to induce a massive nanoparticle uptake inside cells in order to detect the limit of tolerance, which was possible by using a technique of transfection of nanoparticles, a method used to deliver DNA into cells for possible gene therapy or drug delivery [26].

The protocol we used induced an important nanoparticle uptake into the cells, but beyond the physiological death correlated to the methodology

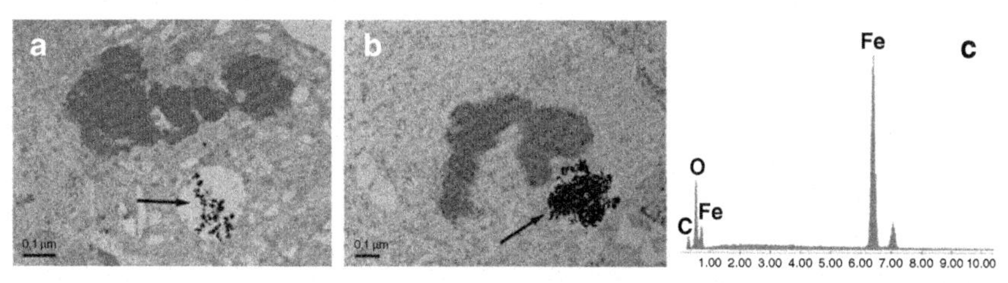

Figure 3.2 Electronic-transmission microscopy images (a, b) of a 3T3 cell during mitosis in contact with hematite (Fe_2O_3) nanoparticles (c) in a nanotoxicity test. (Image courtesy of J. of Imaging [25].)

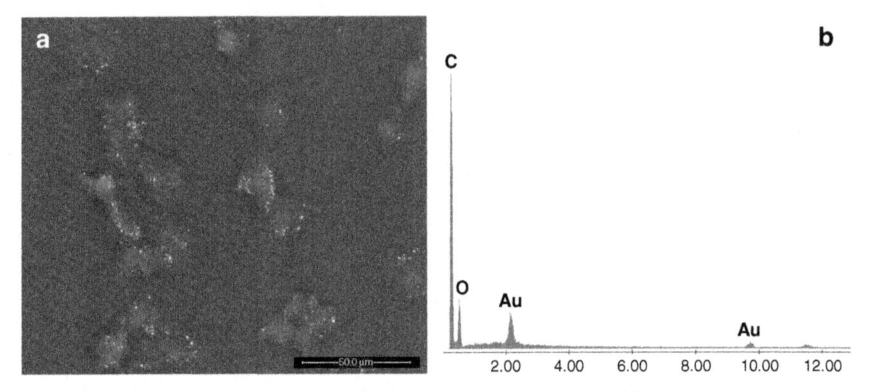

Figure 3.3 3T3 cells after transfection with gold nanoparticles (b). The white dots represent the gold nanoparticles disseminated on the cells (a).

given by the control, there was no general toxic effect. The trick was to coat the nanoparticles with a lipid layer, recognized by the cell membrane, which allowed massive nanoparticle internalization (Figures 3.3–3.5).

The results of viability (XTT, a colorimetric assay for assessing cell viability) and proliferation tests of primary cells (BrdU assays, bromodeoxyuridine cell proliferation test) of 3T3 cells in contact with cobalt, iron-oxides, titanium-oxide and cerium-oxide nanoparticles showed that after a massive uptake of nanoparticles, cells do not die within the time test (24, 48h), but an over-proliferation does indeed occur. These novel results indicate that only after repeated or chronic exposure to nanoparticles does a biological response take place, and that response is similar to the tissue proliferation in cancer cases, a parameter to take into consideration in nanotoxicity tests.

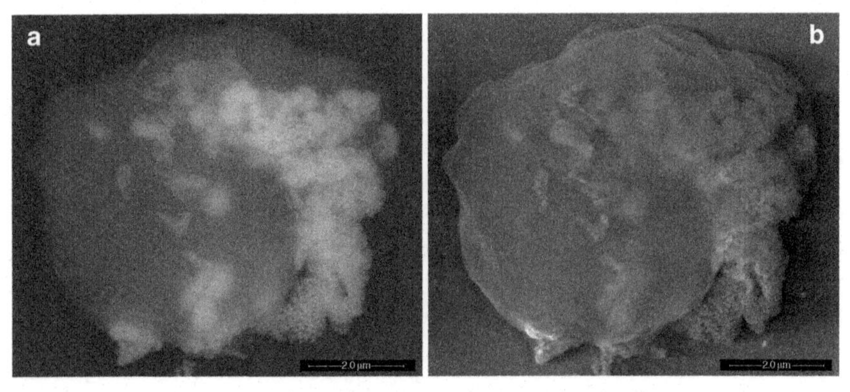

Figure 3.4 Back-scattered electron (a) and secondary electron (b) mode of observation under a FEG-ESEM of a 3T3 cell full of CeO_2 nanoparticles.

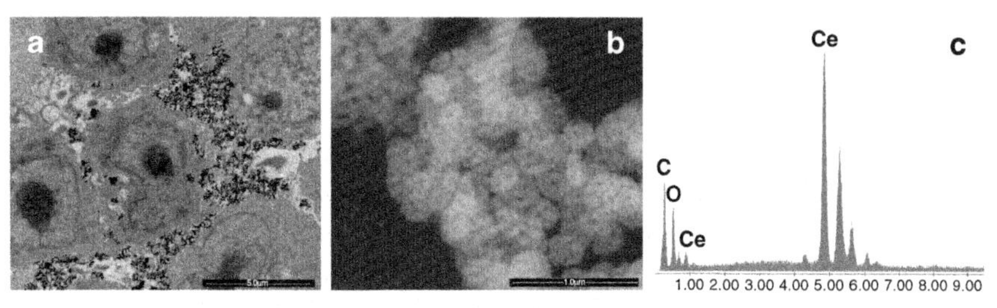

Figure 3.5 STEM (a) and SEM (b) images of CeO_2 nanoparticles in 3T3 cells; they fill the extracellular space thus preventing the normal cell signaling and trafficking.

3.3 THE NANOTOXICOLOGICAL DEBATE

The nanotoxicology debate poses some controversial questions, the answers to which are in many cases debatable:

1. Can the handling of engineered nanoparticles during tests influence the nanotoxicological results?
2. Are different protocols needed for physical and chemical characterization of dry vs. wet (suspension) nanoparticles?
3. Sterilization is mandatory for the materials to be subjected to in-vitro and in-vivo toxicological tests. Is it as necessary with nanoparticles? If so, what are the most reliable methods to sterilize dry and wet nanoparticles before they undergo biological testing? Are current sterilization methods compatible with nanosized morphologies?
4. Is it necessary to carry out preliminary investigations on contamination by lipopolysaccharides of dry and wet nanoparticles, and of their media/solvents (an indispensable procedure for toxicological tests)?
5. Are different protocols needed to characterize nanoparticles either with or without engineered coating?
6. Can the radio-labeling of nanoparticles for dosimetry studies induce changes in the nanosized shape that bias or alter at all the results?
7. What, if any, are the limitations of cells used for in-vitro nanosafety assessment? Should primary cells be preferred over immortalized cell lines?
8. Are the persistence of nanoparticles in cells and the occurrence of biotransformation reactions significant parameters?
9. What are the basic biological parameters to be measured that better define the nano-bio-interaction? Since immediate cell death cannot be used as a parameter, what are the parameters that better describe cell damage? Is the in-vitro test time (24, 48,72h) sufficient to assess possible cell damage?

10. How can we simulate repeated and chronic exposure of cells to nanoparticles? Is the short time of an in-vitro test sufficient to assess a significant nano-bio-interaction?
11. What are the most suitable animal species for in-vivo short- and long-term tests with nanoparticles? Are there elective animal species that are best suited for testing particular types of nanoparticles? What are the most relevant types of in-vivo delivery routes that must be assessed?
12. What are the most significant adverse effects to consider after in-vivo implantation?
13. How do in-vitro results compare with in-vivo results? To what extent can animal-based in-vivo results be extrapolated to humans for risk assessment?
14. Are in-vivo models of dose response needed over chronic and cumulative exposure?

Not all these questions have answers and those that do are debatable. What follows will try to cover some of these concepts.

We have already pointed out that handling nanoparticles is a problem. If we want to avoid clustering or aggregation, we need to coat them with a repulsive coating, but in that case it is obvious that we alter the chemical composition of the surface. Although the particle keeps playing a role as a foreign body, the chemical interaction of proteins and organelles occurs with the coating, thus making the results of the test unreliable.

Since nanoparticle concentration is not a true quantitative parameter, it is necessary to quantify the cell uptake in a quantitative way. As already mentioned, radio-labeling a nanoparticle can involve a morphological modification, and that undermines the reliability of the test.

Every manipulation we perform on nanoparticles changes their state, and that prevents the desired results from being obtained. This situation in a way resembles the Heisenberg's uncertainty principle: in an attempt to study a particle, we affect its characteristics and the result eludes us. In this case, the conservation of the nanoparticle's integrity does not allow us to know its nano-bio-interaction. Is there also an uncertainty principle for biology?

The nanoparticle full-life cycle starts from its synthesis. The presence of additives or the exact formulation of the sample solution has to be known and controlled during all tests. So, nanoparticles can be contaminated, for instance, by the catalyst used for their synthesis, which can remain in the final product and be responsible for a toxic effect in the biological medium. That is the case of carbon nanowires contaminated with aluminium and iron, as shown in Figure 3.6.

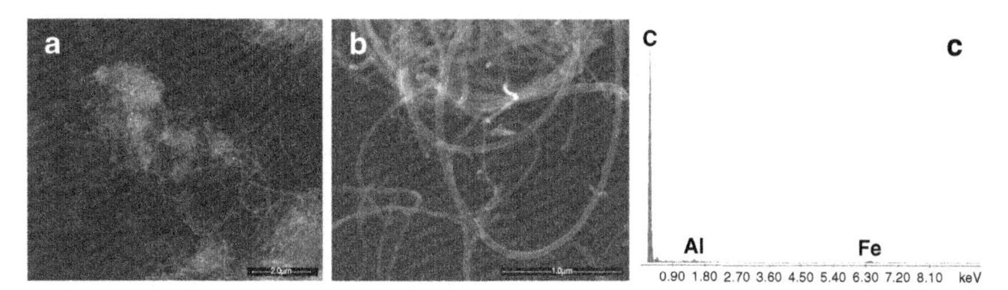

Figure 3.6 FEG-ESEM images show a cluster of carbon nanotubes (a), which at higher magnification (b) reveal the presence of a contamination of aluminium and iron (c).

Unreacted precursors or additives may have stronger biological effects than the nanoparticles themselves. Similarly, endotoxin or other biological contaminations have been found in many nanoparticle preparations. Those undue presences are never considered in cytotoxicity tests.

To understand how mutable nanoparticles are, it is useful to think of them as different states of matter rather than new compounds. For example, water, when its thermal energy is lowered and the fluid turns to solid, decreasing the surface energy of nanoparticles involves their aggregation. Becoming a micrometric entity results in an important change in their properties. Thus, it is clear that clusters are not representative of nanotoxicity.

Dispersion of nanoparticles in biological media for toxicity testing may also destabilize them. As an example, we observed that a non-toxic colloid became toxic when the same particles were aggregated without any chemical modification in the formulation of the sample. Strategies and a multiplexed set of characterizations can determine and control the aggregation state of the nanoparticles.

Also interesting is looking at how the conjugation of soluble entities produces an insoluble conjugate [6]. Conjugate instability has been observed even when soluble nanoparticles are conjugated with soluble molecules, since colloids are out-of-equilibrium systems in spite of their apparent stability. According to this, solid particles suspended in a liquid medium in a non-interacting regime, whether conjugated or not, sediment following Stokes' law describing the frictional force exerted in a continuous viscous fluid on spherical objects with very small Reynolds numbers as is the case with nanoparticles. That means that for every species of nanoparticles and size there is a critical concentration that maintains them in suspension.

In the interacting regime, the increase in nanoparticle concentration leads to an increase of the number of collision events with further destabilization of the nanoparticles. Eventually, a saturation concentration is reached and nanoparticles spontaneously agglomerate and sediment. This condition also makes the test unreliable.

On the other side, nanoparticles may undergo further oxidation or crystal symmetry transition. So, for example, magnetite may be transformed into hematite in a short time because of the very small size of the changing crystal. In this example, the magnetic signal disappears but, chemically, the iron-oxide nanoparticle is still there, and the results must be related to the altered initial boundary conditions. Additionally, the nanoparticle surface may also suffer a number of modifications.

Protein corona, i.e., the formation of a protein layer on the surface of a nanoparticle, is an example of the possible effects of nanoparticles in a biological environment, which is governed by the protein/nanoparticle binding affinity. As those particles come into contact with a biological environment, they do not remain bare; proteins and other biological molecules can be adsorbed on the nanoparticles' surface [27]. The composition, thickness and conformation of these layers are affected by various parameters: nanoparticle size, surface morphology, chemical composition, surface charge, coating and surface functionalization, hydrophilicity or hydrophobicity and type of biological environment (blood or other tissue, cytoplasm, etc.). Parameters such as temperature, pressure and concentration can also play a role in this interaction.

The interaction results in a modification of the nanoparticle surface, a modification that can exert a strong influence on the bio-compatibility and bio-distribution of these particles. The protein corona layer can be formed on top of the existing nanoparticle coating, or it can replace the previous coating, which can involve a modification of the initial electric charge or of the particle's hydrophilicity. The protein/nanoparticle interaction also has the effect of modifying proteins, unfolding them and making them no longer available for cell metabolism and unrecognizable. The presence of these complexes are not recognized by the immune system as the body itself, in-vivo, which can trigger autoimmune pathologies. Cryoglobulinemia (see Chapter 4) could be the expression of the presence of these complexes in the human body (Figure 3.7).

The fate of particles in a biological environment can be multifarious. Disintegration/dissolution of the inorganic core is possible, depending on the chemical composition of the material. If the phenomenon occurs, the particle can release ions whose biological effects should be evaluated from time to time.

In an in-vitro test, a further effect of nanoparticles in a biological medium is the possibility of producing optical interference. A critical point is

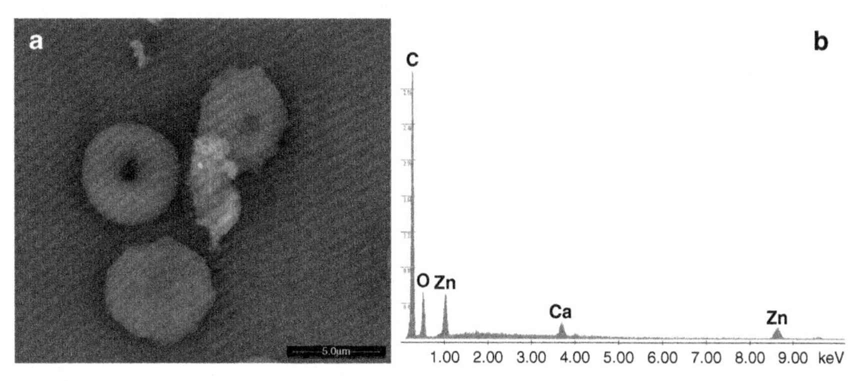

Figure 3.7 FEG-ESEM image (a) shows a red cell with a granulocyte of human blood and an unusual biological agglomerate containing an inorganic zinc-calcium particle (b).

to evaluate the potential optical interference between the dyes used and the nanoparticles, since inorganic nanoparticles can be particularly active in the visible range, where, for instance, plasmonic nanoparticles can have absorption cross-sections 10,000 times larger than organic dyes. As an example, the optical interference of gold with the colorimetric cell-viability assays yields false negatives. Be they isolated or aggregated, small or large, nanoparticles are optically active, and interaction of carbon nanotubes with formazan dyes leads to false positives.

Finally, crystallographic properties and other physical characteristics of engineered, inorganic nanoparticles can be used to detect and differentiate nanoparticles from the biological media (cell, organism or environment) where they are dispersed.

Because of that, at the end of any biological experiment with nanoparticles their final state must be checked from the morphological, chemical and physical standpoints to verify the type of cell/nanoparticle interaction and to make the results meaningful.

As to the possible need to follow different protocols when characterizing dry vs. wet nanoparticles, it must be considered that the instrumentation to characterize nanoparticles generated in liquid suspension is different from that used to work on particles obtained in dry conditions. Physical characteristics as size, aggregation/agglomeration, surface area, surface charge, surface functionalization, crystallinity and shape must be investigated as well as chemical composition before and after interaction with different media. If determining the elemental composition of dry nanoparticles is rather simple, analyzing them in suspension is certainly more complicated. The interactions of nanoparticles with components of the medium where they are suspended to set up the contact with cells can cause the formation of a coating specific for every medium, which can modify their reactivity. In some instances, that coating can be technologically predetermined (functionalization), while in other cases natural reactivity influences the nanoparticle/cell reactivity.

Concerning sterilization, for toxicological tests, in order not to blame possible bacterial or, in any case, biological, pollution for the effects observed, it is actually necessary to use sterilized nanoparticles decontaminated from lipopolysaccharides, and the method employed must not change their physical state and morphology as previously characterized. But sterilization means a superficial oxidation, and inevitably that alters morphology and chemistry. So, just as inevitably, the test will yield results not from the nanoparticles in question but from modified ones.

Among the many current methods of sterilization, the chemical method is obviously not suitable. Although the recommended method, ethylene-oxide, also causes problems because the previously weighed particles must be inserted in a bag to whose internal surface the nanoparticles adhere. So, before being used, they must be weighed again and the operation risks new pollution. The ethylene-oxide process does not cause an irreversible agglomeration and the dispersion of nanoparticles in the medium chosen for the test is homogenous enough, a characteristic indispensable for toxicological tests, although a strong, repeated sonication is necessary after the particles have been dispersed in a medium, no matter what. To assess the validity of the biological results of the test, it is necessary to verify that a real nanoparticle uptake and cell-nanoparticle contact have really occurred.

Nanoparticles must also be free from pyrogens. We put aliquoted nanoparticles in a dry oven at 180°C for 45 minutes, a process that was indeed efficient as long as depyrogenization was concerned, but the side effect was that the particles agglomerated with differences in the state of agglomeration depending on the type of nanoparticles used, but always in an irreversible way. Of course, a condition like that makes the toxicological experiment meaningless, since only a nominal quantity of nanoparticles is known and not the actual quantity matching the cells.

To overcome these problems, manufacturers should supply dimensionally calibrated, aliquoted, sterilized, pyrogen- and lipopolysaccharides-free nanoparticles to scientists. Nanoparticles should be synthesized with sterile equipment in a clean room and every batch should be certified.

As to which cells, primary or immortalized, to use, it must be considered that primary cells cannot be efficiently grown. So, they are available for nanotoxicology experiments only in limited quantities and cannot be standardized. The same kinds of primary cells, e.g., monocytes, when coming from different donors do not respond in a consistent way to the same stimulus and they may even show differences when taken as a second donation from the same subject. Immortalized cells are more resistant than primary ones, are easier to use and the results they yield are more reproducible. As already stated, for the limited life-span of cells, the time of the in-vitro tests must be short, and it is too short to observe all the possible interactions [28].

Inside the cell, nanoparticle/organelle/compound interactions occur, but they are stochastic events that are difficult to predict and that can be driven by chemical-physical binding affinity, but also by the case specifics. That leads to controversial results from laboratory to laboratory.

For all the above quoted reasons, in-vitro tests are not suitable to verify all possible nanoeffects. In-vivo tests are more reliable and supply more information to verify adverse reactions.

REFERENCES

[1] Poland CA, Duffin R, Kinloch I, Maynard A, WallaceWAH, Seaton A, Stone V, Brown S, MacNee W, Donaldson K. Carbon nanotubes introduced into the abdominal cavity of mice show asbestos-like pathogenicity in a pilot study. Nature Nanotechnology 2008;3:423–8.
[2] Hansen T, Clermont G, Alves A, Eloy R, Brochhausen C, Boutrand JP, Gatti A, Kirkpatrick J. Biological tolerance of different materials in bulk and nanoparticulate form in a rat model: Sarcoma development by nanoparticles. J R Soc Interface 2006;3:767–75.
[3] Jennifer M, Maciej W. Nanoparticles technology as a double–edged sword: cytotoxic, genotoxic and epigenetic effects on living cells. J of Biomat and nano-biotechnology 2013;4:53–63.
[4] http://www.iso.org/iso/home/store/catalogue_tc/catalogue_tc_browse.htm?commid=381983.
[5] http://www.oecd.org/science/nanosafety/47104296.pdf.
[6] NanoSafetyCluster- Compendium 2010, pag 11 ed. M Riediker, Georgios Katalagarianakis, Brussels, Belgium, (ftp://ftp.cordis.europa.eu/pub/nanotechnology/docs/compendium-nanosafety-cluster2010_en.pdf).
[7] Cui W, Li J, Zhang Y, Rong H, Lu W, Jiang L. Effects of aggregation and the surface properties of gold nanoparticles on cytotoxicity and cell growth. Nanomedicine:Nano technology, Biology and Medicine 2012;8(1):46–53.
[8] Okuda-Shimazaki J, Takaku S, Kanehira K, Sonezaki S, Taniguchi A. Effects of Titanium Dioxide Nanoparticle Aggregate Size on Gene Expression. Int J Mol Sci 2010;11(6):2383–92.
[9] Albanese A, Chan WCW. Effect of Gold Nanoparticle Aggregation on Cell Uptake and Toxicity. ACS Nano 2011;5(7):5478–89.
[10] Magdolenova Z, Bilaničová D, Pojana G, Fjellsbø LM, Hudecova A, Hasplova K, Marcomini A, Dusinska M. Impact of agglomeration and different dispersions of titanium dioxide nanoparticles on the human related in vitro cytotoxicity and genotoxicity. J Environ Monit 2012 Feb;14(2):455–64.

[11] Zook JM, MacCuspie RI, Locascio LE, Halter MW, Elliott JT. Stable nanoparticle aggregates/agglomerates of different sizes and the effect of their size on hemolytic cytotoxicity in (http://www.nist.gov/manuscript-publication-search.cfm?pub_id=906154).
[12] Zhang S, Li J, Lykotrafitis G, Bao G, Suresh S. Size-Dependent Endocytosis of Nanoparticles. Adv Mater 2009;21:419–24.
[13] Lynch I, Salvati A, Dawson KA. Protein-nanoparticle interactions: what does the cell see? Nat Nanotechnol 2009;4:546–7.
[14] Verma A, Stellacci F. Effect of surface properties on nanoparticle-cell interactions. Small 2010;6:12–21.
[15] Saptarshi S, Duschl A, Lopata A. Interaction of nanoparticles with proteins: relation to bio-reactivity of the nanoparticle. J Nano-biotechnol 2013;11:26.
[16] Peters K, Unger RE, Gatti AM, Sabbioni E, Gambarelli A, Kirkpatrick CJ. Impact of Ceramic and Metallic Nanoscaled Particles on Endotheliar Cell Functions in Vitro. Nanothechnologies for the Life Sciences 2006;5:108–29.
[17] Peters K, Unger RE, Kirkpatrick CJ, Gatti AM, Monari E. Effects of nanoscaled particles on endothelial cell function in vitro: Studies on viability, proliferation and inflammation. J Mater Sci Mater Med 2004;15:321–5.
[18] Maeda-Mamiyaa R, Noirib E, Isobec H, Nakanishic W, Okamotob K, Doib K, Sugayad T, Izumie T, Hommaa T, Nakamuraa E. In vivo gene delivery by cationic tetraamino fullerene. PNAS 2010;107(12):5339–44.
[19] Sakhtianchi R, Minchin RF, Lee KB, Alkilany AM, Serpooshan V, Mahmoudi M. Exocytosis of nanoparticles from cells: Role in cellular retention and toxicity. Advances in Colloid and Interface Science 2013;201–202:18–29.
[20] Bregoli L, Chiarini F, Gambarelli A, Sighinolfi G, Gatti AM, Santia P, Martelli AM, Cocco L. Toxicity of antimony trioxide nanoparticles on human hematopoietic progenitor cells and comparison to cell lines. Toxicology 2009;262:121–9.
[21] Harush-Frenkel O, Debotton N, Benita S, Altschuler Y. Targeting of nanoparticles to the clathrin-mediated endocytic pathway. Biochem Biophys Res Commun 2007;2(353-1): 26–32.
[22] Bouzier-Sore AK, Ribot E, Bouchaud V, Miraux S, Duguet E, Mornet S, Clofent-Sanchez G, Franconi JM, Voisin P. Nanoparticle phagocytosis and cellular stress: involvement in cellular imaging and in gene therapy against glioma. NMR Biomed 2010;23(1):88–96.
[23] Mytych J, Wnuk M. Nanoparticle Technology as a Double-Edged Sword: Cytotoxic, Genotoxic and Epigenetic Effects on Living Cells. Journal of Biomaterials and Nanobiotechnology 2013;4:53–63.
[24] Gupta SK, Baweja L, Gurbani D, Pandey AK, Dhawan A. Interaction of C60 fullerene with the proteins involved in DNA mismatch repair pathway. J Biomed Nanotechnol 2011;7(1):179–80.
[25] Gatti AM, Quaglino D, Sighinolfi GL. A Morphological Approach to Monitor the Nanoparticle-Cell Interaction. International Journal of Imaging 2009;2(S09 Spring - Editorial 1):2–21.
[26] Park W, Yang HN, Ling D, Yim H, Kim KS, Hyeon T, Na K, Park KH. Multi-modal transfection agent based on monodisperse magnetic nanoparticles for stem cell gene delivery and tracking. Biomaterials 2014;35(25):7239–47.
[27] Rahman M, et al. Protein-Nanoparticle Interactions, Springer Series in Biophysics 15. Berlin Heidelberg: Springer-Verlag; 2013. Nanoparticle and Protein Corona Chap. 2: 21-44.(http://www.springer.com/978-3-642-37554-5).
[28] Kagan VE, Bayir H, Shvedova AA. Nanomedicine and nanotoxicology: two sides of the same coin. Nanomedicine 2005;1:313–6.

CHAPTER 4

Clinical Cases

Contents

4.1 Introduction	29
4.2 Mesothelioma: a nanofiber-induced disease	32
4.3 Hashimoto thyroiditis	39
4.4 Ameloblastoma	45
4.5 Leukemia and lymphoma	50
4.6 Congenital malformations	54
4.7 Cryoglobulinemia	56
4.8 Breast cancer	62
References	62

4.1 INTRODUCTION

This chapter is intended to demonstrate through the analysis of specific real cases that the exposure to micro- and nanosized nanoparticles, be they engineered or generally incidental by-products of (mainly but not exclusively) industrial processes, can prompt undesirable reactions in humans. At the level of tissues or cells, solid nanoparticles, no matter how they got into the organism, are always perceived as foreign bodies, and, for that reason, they induce a more or less visible, inflammatory "foreign-body reaction."

In the case of macro- or microsized entities, the biological variety of responses is well-known: inflammation, edema, granuloma, fibrotic capsule. What happens is the "recognition of an injurious agent" and the body activates a specific reaction to this recognition. In the case of nanosized objects capable of crossing the cell barrier, the way cells behave is not fully clear and remains under investigation.

The recognition is mostly chemical or better, biochemical, but, if the foreign agent is a bulk material with an inert surface chemistry, there is no prompt recognition. That is the reason why implantable medical devices such as breast prostheses, cardiac valves, hip joints, knees or other orthopedic prostheses can be used without any particular danger of biological, adverse rejection.

When we started this novel research in the late 1990s, the cases we were attracted to and investigated were of diseases of unknown etiology such as cancer and also some pathologies categorized as mysterious or particularly rare and, because of that, actually neglected mainly for lack of economic

Case Studies in Nanotoxicology and Particle Toxicology
http://dx.doi.org/10.1016/B978-0-12-801215-4.00004-2

interest. This chapter shows some clusters of patients affected by mesothelioma, leukemia, ameloblastoma (a rare, benign neoplasm of the odontogenic epithelium predominantly affecting the lower jaw) and so-called genetic fetal malformations. At the end of this chapter, we present a new type of analyses carried out on firefighters who became ill after being exposed to the dust generated by the collapse of the Twin Towers in New York City on 9/11.

According to the method we still use, the pathological samples we then took into consideration were analyzed by a scanning electron microscope (SEM), an instrument used more in physics than in medicine.

We started from a simple hypothesis and a simple concept: to check if these diseases are triggered by environmental pollution. The most immediate and logical thing to do was to verify if pollution could enter the body. To check that, it was necessary to use an instrument and a procedure, that could identify this particulate matter (PM). As an instrument, we chose the SEM because among the ones theoretically capable of working for that aim, it was the one we knew best and, equally important, we had available.

When this kind of particulate pollutant is embedded in biological tissues, it forms with that tissue a sort of composite material that is not clearly visible. So we developed a new technique based on the difference of atomic density between the tissue and the foreign bodies. In fact, using energy dispersive spectroscopy (EDS), technically an X-ray microprobe, a sensor that emphasizes the atomic density of the materials under the electron beam inside the chamber of the microscope, we saw that this pollution is mostly composed of inorganic elements that are "denser" than the tissue. So, these foreign bodies can look lighter or, apparently, more "luminous," than the biological surroundings where they are trapped [1].

The technique was refined as part of a European Project called Nanopathology (FP5-QOL-2002-147) and thanks to it we can identify organic, but, indeed, mostly inorganic, solid materials also at nanosized dimensions in organic matrixes. The identification of organic particles, i.e., containing organic carbon, is not always possible since the biological tissue also contains a very significant amount of carbon and it is often hard to distinguish with certitude between pollutant and tissue. Using EDS, we checked the signals coming from the biological tissue but the ones from the carbonaceous debris contribute to the height of the same peak in the EDS spectrum so the identification of the single contributions is impossible. In some cases, if the organic pollutant contains more elements, we can discriminate between the internal and exogenous carbon. For instance, the particles of

PVC (polyvinyl chloride) can be identified inside biological tissue thanks to the presence of chlorine and the debris of PTFE (polytetrafluoroethylene) can be clearly identified because of fluorine.

All the images and spectra shown in the following chapters were obtained with three types of microscopes, FEG-ESEM Quanta, ESEM Quanta 200 and FEG-ESEM250, all manufactured by FEI Company (The Netherlands) coupled with an EDS by EDAX (United States) for the identification of the elemental composition of the pollutants. The selection of these instruments was essentially due to the high technology used in the instruments. The electron-microscopic investigations were carried out under high vacuum, but around the year 2000 FEI invented new air traps to be introduced in the electronic column. They allow us to analyze samples without any preliminary treatment (e.g., carbon or gold-palladium coating, in order to make the samples electro-conductive), at room conditions and in the presence of air and water. Those are essential conditions for our investigations, since it is mandatory to be sure that what is identified belongs to the patient and represents his exposure and is not an artifact due to the preparation of the sample. For this reason, all the preparations of the samples (10-micron-thin sections either from paraffin blocks containing the archived pathological tissues, or formalin-fixed samples or drops of cryo-preserved fluids) are made in a very clean room and inside a laminar-flow cabinet. Electro-conductive coating can introduce the pollution of the laboratory or of the coating system on the surface of the sample, thus altering the result of the analysis (Figure 4.1).

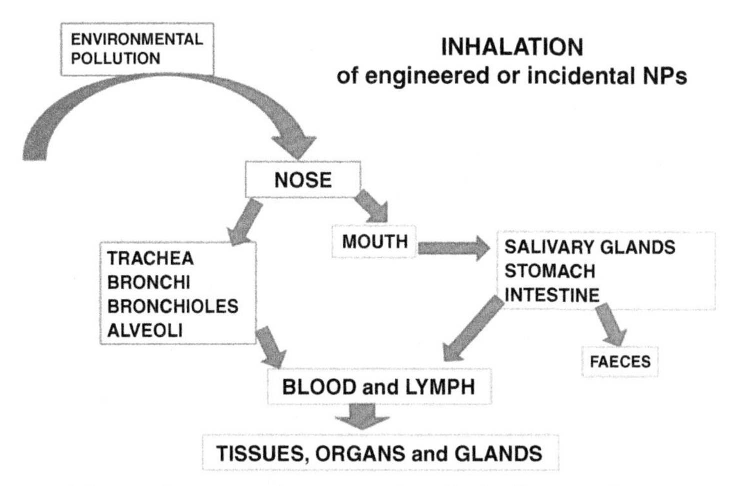

Figure 4.1 Scheme of nanoparticle entrance into the body. Any submicronic particle accumulation in a tissue can trigger a disease.

As an example, this chapter presents the results of a few investigations carried out on groups of patients suffering from the same diseases who personally asked to be checked and have their samples analyzed for the possible presence of foreign particles in their bodies. In some cases, the results of the analysis induced us to trace the origin of the debris detected in the environment frequented by the patient. This is particularly important as it allows to identify and register the possible sources of particulate pollutants and their capability of interacting with organisms.

4.2 MESOTHELIOMA: A NANOFIBER-INDUCED DISEASE

The first disease we consider here is pleural mesothelioma, a relatively rare cancer of the pleura, the two-layered serous membrane that covers the lungs and their adjoining structures. A similar disease is peritoneal mesothelioma, a malignant cancer that affects the serous membrane that lines the abdominal cavity. It is amply reported in literature that there is a close correlation between an exposure to an environment where asbestos fibers are released and mesothelioma.

According to the international references, asbestos fibers are classified as nanoparticles since at least one of their dimensions is in the range 1-100 nm (ISO ISO/TS 27687:2008). Therefore, mesothelioma is a nanopathology in all respects and, for that reason, we looked for asbestos particles in the pleural tissues of patients suffering from mesothelioma.

In his *Naturalis Historia*, the Latin scientist Pliny the Elder (I century A.D.) wrote that the slaves working in the asbestos mines died earlier than expected and he also observed that the premature death was due to the inhalation of those mineral fibers, so as to recommend the use of mouth and nose protective filtering masks for asbestos miners. All that remained without any effect until the 19th century, when the first article on a case of pleural mesothelioma appeared in a scientific journal and a medically based correlation with asbestos exposure was advanced. In 1936, the first complete epidemiological study on asbestos-exposure consequences should have induced policymakers to adopt a moratorium on the use of that material, but nothing happened because of the wide use and manifold applications of the material and its huge economic value. It wasn't until the early 1990s that asbestos was definitively banned in most countries with the notable exceptions of Russia, China and Canada. The sad truth (or, better yet, one of the sad truths) is that the effects of asbestos on health will be still visible for many years to come, since the disease has a latency time that reaches several decades, up to 40 years, from exposure. In [2] the exact time for every worker was identified:

> *Latency periods examined in 370 cases of malignant pleural mesothelioma, ranged from 14 to 72 years (mean 48.7, median 51). According to the occupational group, the mean latency periods were 29.6 among insulators, 35.4 among dock workers, 43.7 in a heterogeneous group defined as various, 46.4 in non-shipbuilding industry workers, 49.4 in shipyard workers, 51.7 among women with a history of domestic exposure to asbestos, and 56.2 in people employed in maritime trades. (Abstract)*

For centuries those natural fibers have been used mainly because of their excellent properties for fireproof insulation, and because they are abundant, easy to extract and work with.

Asbestosis and mesothelioma in its pleural and peritoneal forms are the classical diseases whose origin is imputed to asbestos, but there is no reason at least to suspect that those fibers do not behave unlike other particles of similar shape and size. Those similarities, could potentially, induce other similar pathologies. Unfortunately, virtually no research has been done on the subject or, to be more accurate, such an approach has not yet been considered by medical doctors, since it implies an interdisciplinary background based on chemistry, environmental pollution, etc. [3].

From a geological point of view asbestos is divided into two classes: serpentine with no other variety than chrysotile, and amphiboles that include anthophyllite, grunerite (amosite), riebeckite (crocidolite), tremolite and actinolite. The asbestos generally used by industries and, therefore, more dangerous for humans are the blue fibers of crocidolite: a compound of silicon, magnesium, iron and calcium.

Asbestos is probably the most dramatic [4,5] of the examples that provide evidence of a discrepancy between scientific evidence of an adverse effect and its translation into adequate preventive measures. Because of the determination of powerful economic interests to maintain the level of their profits at all costs, no international agreement yet exists to ban the processing of asbestos worldwide, and more than 2 million tons are still worked annually. While progress is slowly being made in the right direction, some rich countries continue to exploit the permissive or absent occupational legislation in poor countries to send them old ships, stuffed with asbestos, to be demolished [6].

Though asbestos-induced pathologies are now far from rare and, in fact, in spite of asbestos having been banned in many countries relatively long ago, they are on the increase because of the already-mentioned generally long time they usually need to manifest themselves.

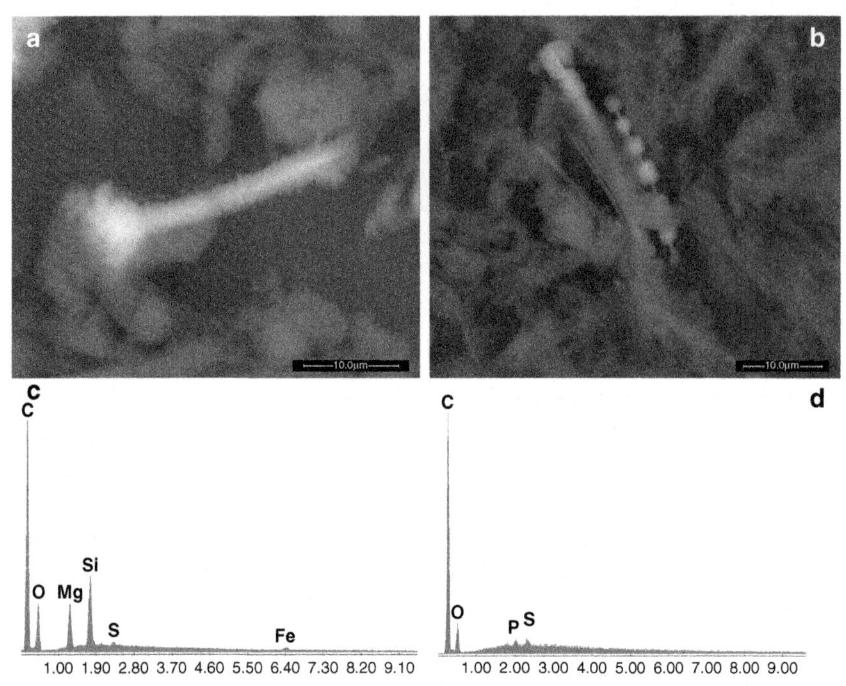

Figure 4.2 Fibers of asbestos (a, b) embedded in cancerous tissue with its chemical spectrum (c), and the EDS spectrum of the normal composition of the lung tissue (d).

We had the opportunity to analyze a total of 13 cases from a group of workers in a shipyard in Trieste (Italy) and some in another in La Spezia (Italy). Both groups of workers were engaged in an environment where asbestos was used for insulation, and they were certainly exposed to its fibers (Figure 4.2).

We examined the pleural tissue and in many cases found the fibers of asbestos with the characteristic pearls. The periodically disposed whiter precipitates around the fibers are called pearls since their shape and their spatial periodicity look like pearls in a necklace. These pearls are precipitates and are composed of iron proteinate. The periodicity of the precipitates around the fibers shows (at least to a material scientist) that the precipitation is ruled by diffusive phenomena. Diffusion processes of the ions occur from the fibers to the biological surroundings and inversely to equilibrate the ionic presence. The presence of the inhaled fibers located very far from the entrance "door" needs an explanation. When the fibers stop in the pleura (it must be remembered that they entered the organism from the nose/mouth and traveled, pushed by every respiratory act, through the bronchi, across the alveolar wall and pulmonary tissue toward the pleura), there is the possibility of an interaction with the extracellular fluids and since the silicon-based fiber is partially biodegradable, there is the possibility of diffusion of ions from and toward the fibers. Other ceramic materials, such as bioglass, show the same phenomenon [7,8].

The diffusion of ions (Si, Mg, Fe) from the material destabilizes the local metabolism. An accumulation of iron ions occurs at the surface of the fibers and, as soon as the concentration of saturation is reached, iron proteins precipitate around the asbestos fibers, forming the so-called pearls. It is likely that the patho-mechanism is due to the physical presence of the fibers and also due to the effects induced by their degradation. The reason why a worker usually becomes ill many years after exposure is because of the long time needed for the fibers to reach the pleura (the last border). While the fiber is traveling, the right condition does not exist for its degradation and interaction with the biological environment. As soon as it stops in the pleura, other phenomena have the time and find the conditions to occur.

The mesothelioma induced by asbestos fibers was the first example of physical-chemical toxicity for us. During our study of tissues affected by mesothelioma, we identified the asbestos fibers but also examining other organs of the same patients such as lung and liver, it was interesting to discover that other foreign bodies were present in the tissues and particles with similar chemical composition and size. It is obvious that a worker also inhales urban pollution just like anybody else and, if he is a smoker, he breathes his tobacco's pollution as well. So, it is obvious that in his lungs as in other organs the presence of other pollutants can be found, as an evidence of the multiple exposures he underwent, as shown in Figures 4.3–4.6.

The presence of other micro- and nanosized "foreign bodies" means that PM with similar submicronic size has the same probability to be trapped in the lung and there induce an adverse biological reaction. But if the PM's

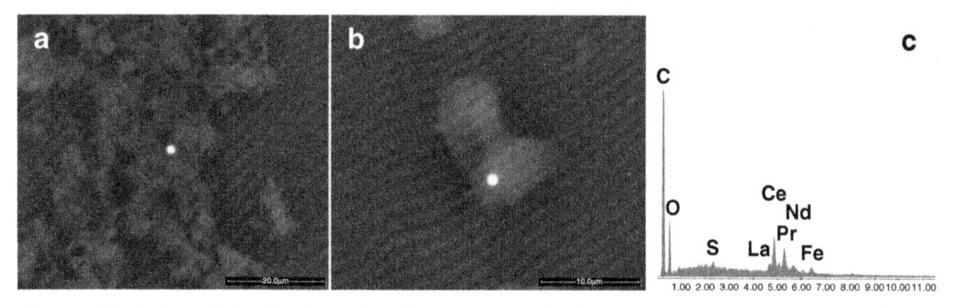

Figure 4.3 Spherules of cerium, lanthanum, neodymium and praesodimium (c) found in the lung (a) and in the pleura (b) in a case of pleural mesothelioma.

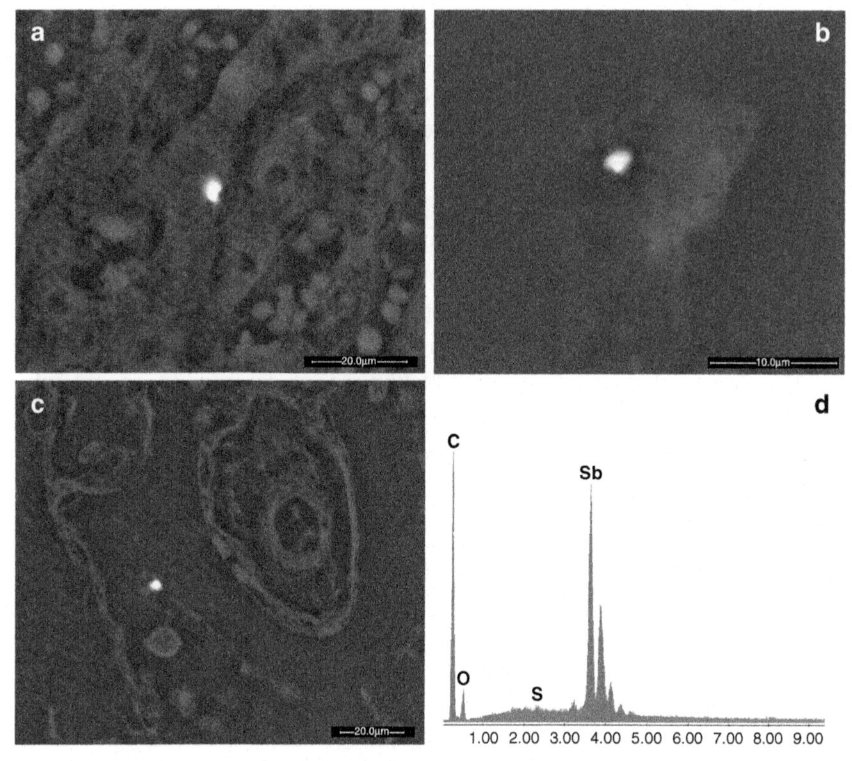

Figure 4.4 Particles of antimony (d) found in the liver (a), in the pleura (b) and in the lung (c) of a patient affected by mesothelioma.

size is small enough, it can cross the lung barrier and, being carried by the blood circulation, be also captured by other organs (Figures 4.7 and 4.8).

Investigating two cases of pleural mesothelioma or, at least, that was the diagnosis, we could not find any asbestos fibers. We do not know if the samples we analyzed were not the "right" ones for what we expected to find, but the fact was that asbestos was not there. What we found, instead, were other debris, sometimes of submicronic size, composed, for instance, of iron-chromium-nickel, i.e., stainless steel.

In a case of peritoneal mesothelioma (as was the diagnosis) we found a silicon-based compound that also contained uranium and thorium. We discovered that that particular patient had eaten wild vegetables for decades, grown under the fallout of a ceramic-tiles factory. The different colors of glaze or decoration are obtained by using a variety of chemical elements (uranium and thorium among them) that, after being treated in an oven (oxidation), take on the desired hue [1].

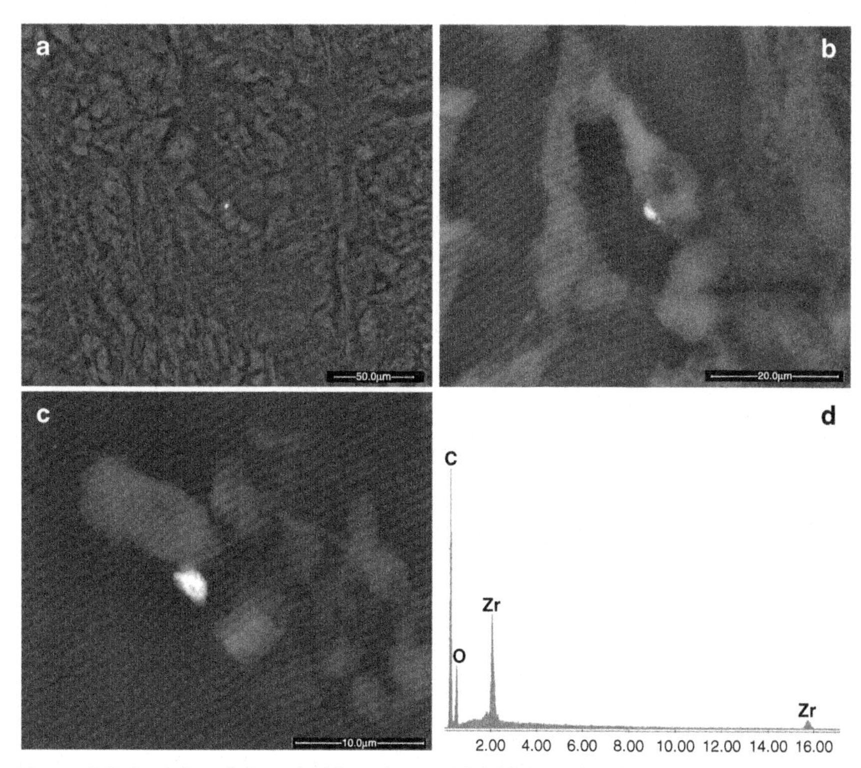

Figure 4.5 Particles of zirconia (zirconium oxide) (d) found in the liver (a), in the pleura (b) and in the lung (c) of a patient who died of pleural mesothelioma.

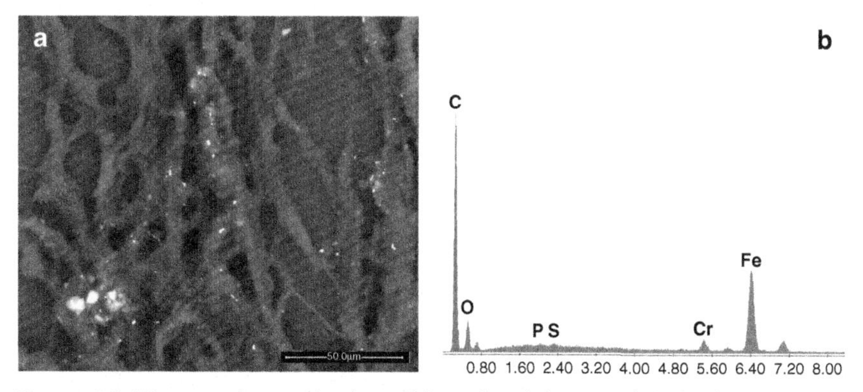

Figure 4.6 Micro- and nanosized particles of stainless steel (iron-chromium) (b) identified in the pleura (a) of a patient affected by pleural mesothelioma.

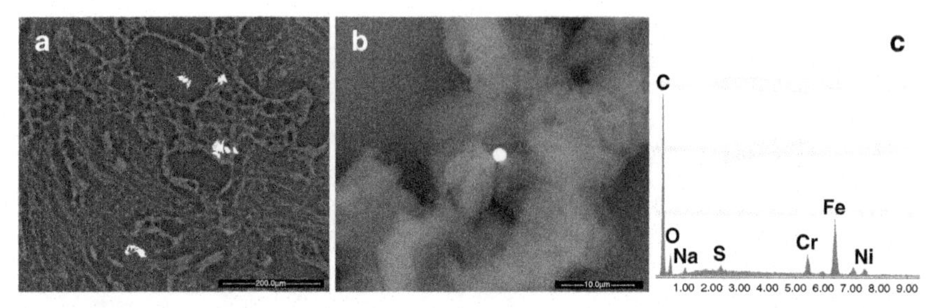

Figure 4.7 Images of the debris (a, b) of stainless steel (iron-chromium-nickel) (c) found in a peritoneal tissue affected by mesothelioma.

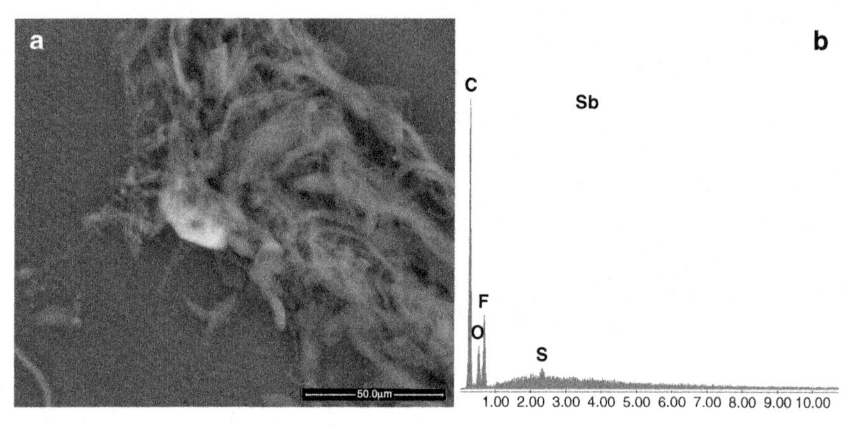

Figure 4.8 Image of another debris (a) found in the same patient as the one from Figure 4.7. The microsized particle was a carbon-fluoride (b) debris, probably Teflon, and possibly a piece of a frying-pan coating.

The term "right" needs an explanation. More than once we verified that the sampling of the biological tissue is crucial. In many cases of primary cancer the identification of the interface between the pathological and the healthy tissue is crucial for the analysis. In fact, there is a high probability that these foreign bodies will be found not in the middle of the cancer but at its interface. That is an area where the accumulation of the debris is particularly considerable and their density decreases as one moves away from the interface. If the surgeon or the histopathologist do not select this part as a sample to be examined, virtually no pollutants can be detected but only necrotic or normal tissue. Of course, there are instances in which nanoparticles cannot be found simply because they are not the cause of the pathology and, in fact, in about 4% of the cases we observed we did not find foreign bodies. That is only natural, since the cancer mechanism can also be triggered by agents that have nothing to do with particles: e.g., chemicals, radiations, etc. It is also likely that the percentage of cancers caused by particles is much lower than our 96%, a quantity probably due to a sort of pre-selection of the cases we receive, cases that often come to us because they could not get a "traditional" etiological explanation.

4.3 HASHIMOTO THYROIDITIS

The thyroid is a large endocrine gland of the human body located below the laryngeal prominence, or "Adam's apple" as it is popularly called, and thyroiditis is its inflammation. Sometimes this condition can result in cancer.

There is more than one hypothesis to explain its possible etiology, and one of them takes into consideration a possible environmental origin but, until now, there have been no systematic studies and data available in literature regarding a possible association between thyroiditis [9] or thyroid tumors and a patient's exposure to a specific environmental pollution.

We started from the hypothesis that tiny particles floating in the blood could be trapped in the gland, as they do virtually in any other organ, and that these particles could be responsible for the thyroid dysfunction. We created a survey to analyze 41 samples: 10 cases of patients suffering from thyroid malignant tumors, 13 cases of Hashimoto's thyroiditis, 14 cases of benign thyroiditis with an adenoma nodule and 4 cases of goiter. The normal thyroid tissue has three different areas that can be targets for nanoparticles: the thyroid follicle, the border of the follicle and the tissue external to the follicle. From the elemental point of view, the normal thyroid tissue is composed of carbon, oxygen, iodine, sulphur, sodium and phosphorus (Figure 4.9). So, the presence of any material composed of other elements must be explained.

The results of the analyses of the tumor cases showed the widespread presence of calcification areas, and in some cases, these disseminated presences were so numerous that we could not identify further debris. In another section of this book we will describe the calcification areas as spherules of calcium phosphate.

Inorganic debris in microsized, submicronic and nanosized form were also clearly identified in those areas classified as either goiters or benign or malignant neoplasms. The difference was in what appears to be the density

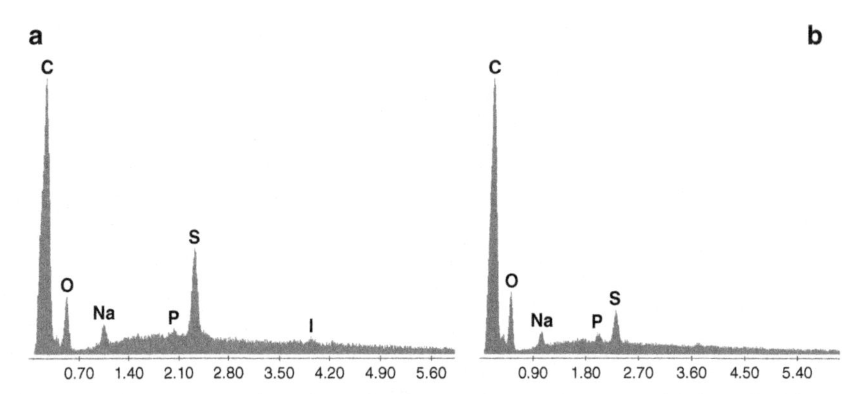

Figure 4.9 EDS spectra of the thyroid's elemental normal content: (a) the colloid's (inside the follicle) content with iodine; (b) the composition of the other tissue (carbon, oxygen, sodium, sulphur and phosphorus).

of the particles and their chemical composition. We must emphasize that very little can be said about the actual number of PM contained in an organ or even in a part of it. What we have available is no more than a thin section of a much larger volume, and particles are not distributed in an even, homogeneous way, but tend to accumulate in clusters, leaving large virtually empty areas (Figure 4.10).

In this case, sulphur-barium spherical particles were detected, while in Figure 4.11 microsized and submicronic particles of iron compounds are visible.

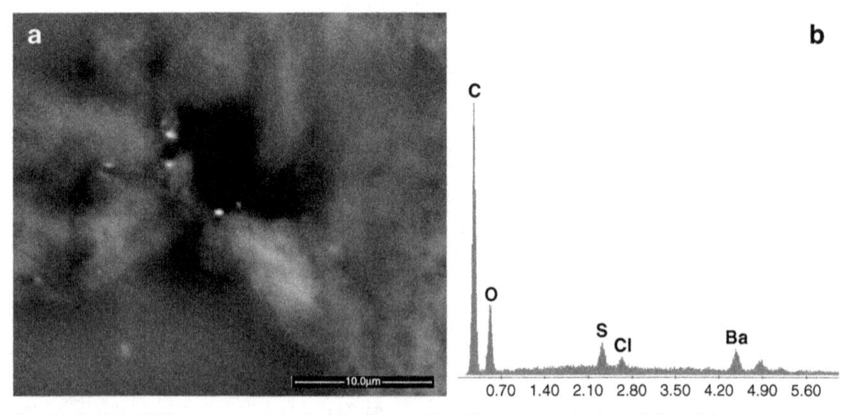

Figure 4.10 SEM image of an area affected by Hashimoto's thyroiditis (a) where some particles (sizes ranging from 0.3-0.8 μm) are accumulated. The EDS spectrum identifies them as composed of sulfur and barium (b).

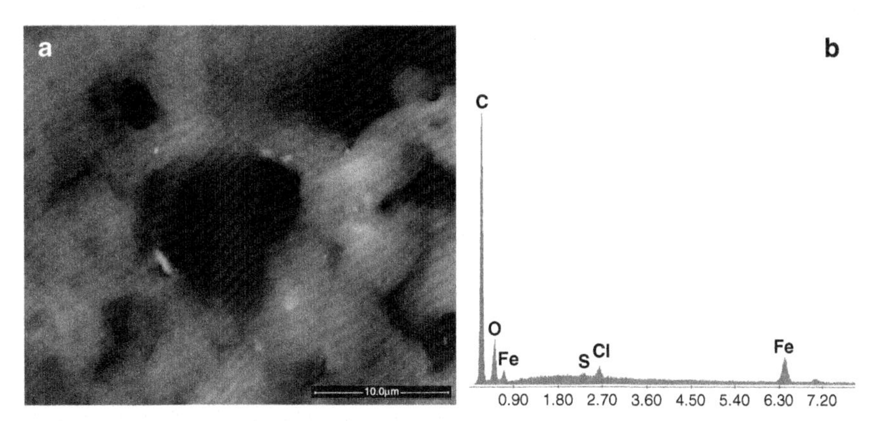

Figure 4.11 SEM image of a Hashimoto's thyroiditis sample with submicronic and nanosized particles (outside the follicle) (a) composed of iron, carbon, sodium, sulphur and chlorine (b) belonging to biological tissue.

Sometimes, single particles of iron or lead-chlorine are detected (Figures 4.11–4.12), and in some cases groups of particles with non-homogeneous chemistry are also identified.

We classified the particles we identified into three classes (Table 4.1):

– calcium compounds of probable endogenous origin (pale gray color)
– ceramic (white color)
– metallic materials (dark gray color).

The particles detected in standardized areas (12-14 mm^2) of the sections of the specimen examined were counted and the analyses revealed that the Hashimoto's thyroiditis samples contained a higher number of foreign bodies than the reference samples (Table 4.2).

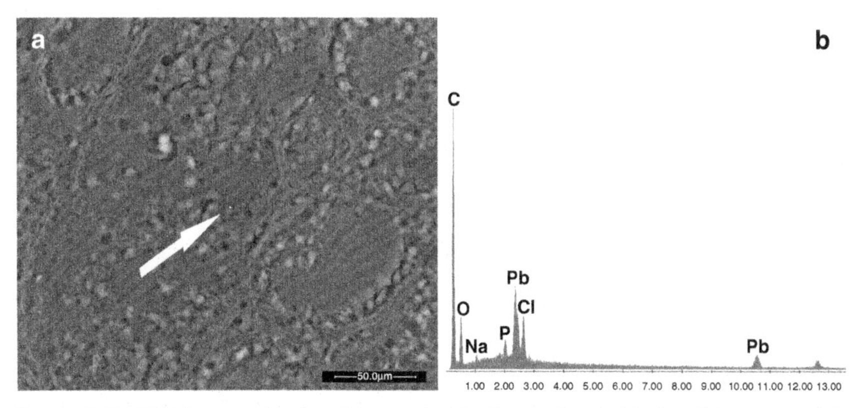

Figure 4.12 SEM image of a thyroid sample (a) where a lead-chlorine (b) nanoparticle is visible.

Table 4.1 List of particles detected in a sample with compositions identified according to the most abundant element in the spectrum

Elements	Size (μm)
Ca	1.5
Ca Mg	0.05
Si Al	1
Si Al Fe	1.5-2-2
Fe	0.7-1
Fe Cr Ni	4
Fe Cu	0.5-1
Al Ba	0.8
Al Fe Ti	2
Sb	1
Ti	1
Pb	1
Mg	2
Bi	1-1-2
Cr	1
Ba S	2-2

The size of the particles identified ranged from 0.050 (50 nm) microns to 10 microns.

It must be noted that some reference samples were specimens of the contralateral thyroid tissue apparently not affected by the disease. So, the higher presence of PM in the reference tissue than in the goiter is probably due to an erroneous hypothesis that we put forward. We supposed that the same tissue not affected by the disease was healthy. This is not true. When a person is exposed to a particulate pollution, the pollutants are probably distributed all over the body until there is a specific uptake by some organs. The thyroid can be one of the organs attracting PM, and the higher the concentration, the higher the probability that an adverse reaction has been triggered. At present we do not know if there is a threshold concentration

Table 4.2 Number of foreign bodies identified in standardized areas

	Ca-Com	Metals	Ceramics
Hashimoto's	77	75	46
Goiter	6	15	3
Reference	43	22	22

that triggers the reaction or if it also depends on the ability of single submicronic particles to enter cells and interact with the local metabolism.

The high concentration of calcium precipitates identified may be explained by the altered production of calcitonin by the parafollicular cells of the thyroid, a hormone that works to reduce blood calcium, counteracting the effects of the parathyroid hormone. The presence of the foreign bodies can induce an inflammation that changes the normal metabolism of the thyroid and calcium deposition could be one of the response effects.

The EDS spectrum of the calcium spherules detected contains the peaks of sodium, magnesium, silicon, sulphur, chlorine and potassium. All the elements except silicon can be of endogenous origin. Silicon is not always present. These ions are of vital importance in many physiological phenomena: muscle contraction, cell secretion and cell permeability. Nevertheless, they should not be contained in particles but must be bio-available, which is not the case with PM (Figure 4.13).

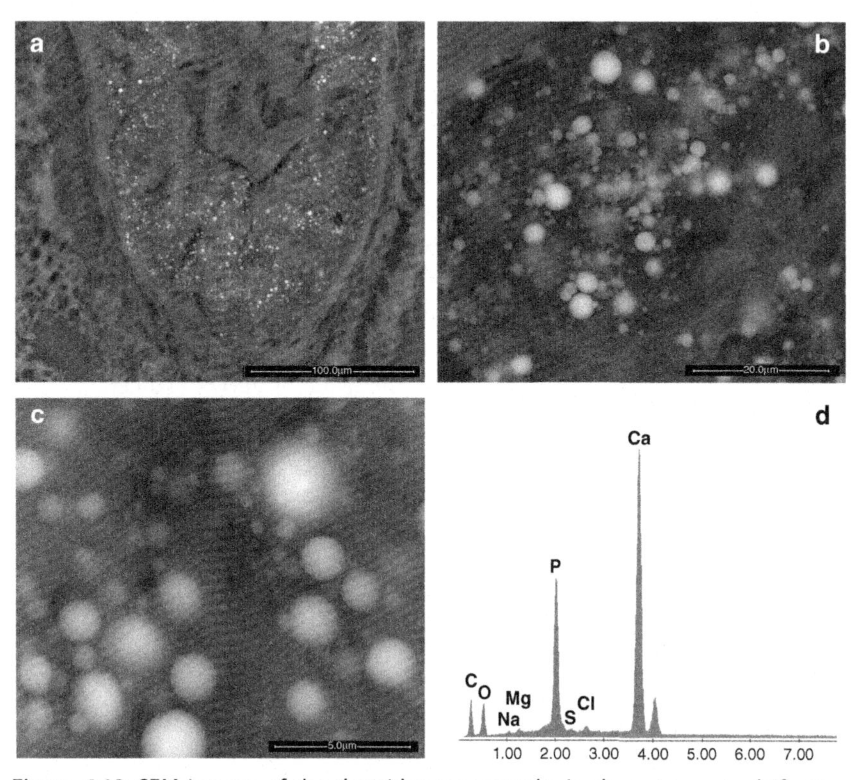

Figure 4.13 SEM images of the thyroid tumor sample. In these images calcification areas (Ca, P) are shown at different magnifications (a,b,c). Spectra of these particles are visible in (d). (Markers 300 micron and 20 micron, respectively)

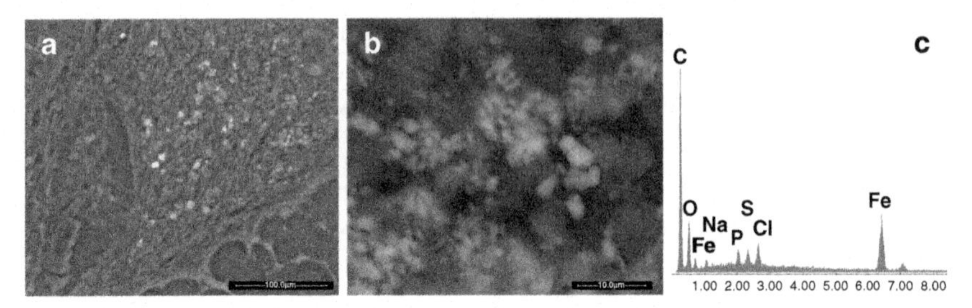

Figure 4.14 SEM images of a thyroid-tumor sample at different magnifications (a, b). In these images the calcification areas (Ca, P) also contain iron (Fe) (c).

Calcification is normal in biological tissue after an inflammatory reaction, but, until now, it was not clear why these calcified bodies assume a globular size. From a chemical point of view, the globular formation of hydroxyapatite (the mineral form of this kind of calcification) could not be formed at a temperature of 37°C and at an atmospheric pressure. The mechanism of this formation is a mystery we could not solve for now.

The tissue also contained wide areas rich in Fe debris (Figure 4.14) that probably originate from iron proteinates.

We assume that both these types of particles are of endogenous origin, i.e., they originated from the precipitation of endogenous elements. The precipitation causes the formation of non-resorbable particles that became and act as foreign bodies for the organism. We suspect that the chemical formation of these precipitates is part of the cancer mechanism. These elements probably come from the extracellular matrix, released by the dead cells, but there is also the possibility of a precipitation inside the cell itself. In this case, after the cell death those precipitates are released into the tissue, as ghosts of a cell.

The hypothesis can find its basis in some recognized diseases such as siderosis (Figure 4.15).

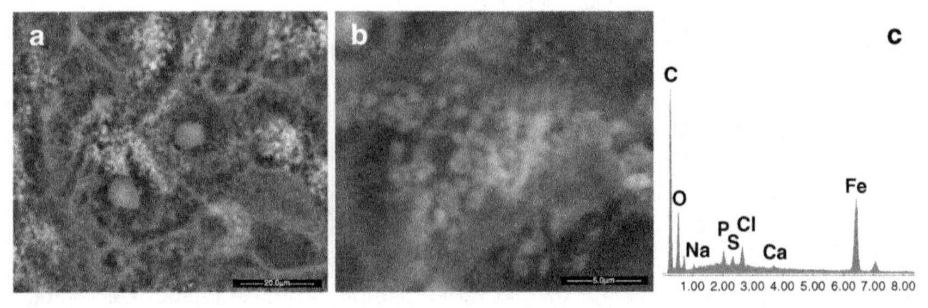

Figure 4.15 SEM images of liver tissue affected by siderosis (a, b). The white dots inside the cell cytoplasm are the iron precipitates (c).

In order to understand the spherular calcification we consider another pathology where there is the precipitation of iron for the genetic absence of a protein: siderosis.

The images show submicronic and nanosized precipitates of iron in a cytoplasmic area. The absence of suitable metabolism for iron causes its accumulation and, when the saturation concentration is reached, it precipitates according to the laws of chemistry. These precipitates are not biodegradable, so they are just foreign bodies to the tissue.

The formation of similar precipitates in globular form in a cancerous tissue can make us presume that in that case a saturation concentration is also reached and that the environmental pollution identified has disturbed the metabolism.

Can the nanosized formation of those spherules be due to the presence of nanosized foreign bodies? Is their presence expression of a nanoeffect? These are questions we need answers to.

4.4 AMELOBLASTOMA

Ameloblastoma is a tumor of the jaw and accounts for approximately 1% of all oral and maxilla-mandibular tumors. It is a rare, benign tumor of the odontogenic epithelium. (Ameloblasts are cells that disappear after tooth development and whose function is to deposit tooth enamel.) It affects the lower jaw much more commonly than the upper jaw [11]. The pathology was recognized in 1827 by Thomas Cusack [12]. Broca first described this tumor in 1868, whereas the term "ameloblastoma" was coined by Churchill in 1933 [13].

Ameloblastoma is notorious for its slow growth, histologically benign appearance and high incidence (50-72%) of local recurrence. This cancer can also metastasize to the cervical nodes and the lungs. Its etiology is also controversial because it is benign but over time can show malignant features· Typically, it occurs in intraosseous tooth-bearing areas of the jaws, especially in the molar region, and on X-ray appears as a cystic, multilocular or unilocular, lesion.

In 1955, Small and Waldron [14] collected and discussed the different hypotheses about ameloblastoma's origin: cell rests of the enamel organ, odontogenic rests, surface epithelium, epithelium of dentigerous cysts and epithelial rests in follicles of unerupted teeth. Other authors cited these possibilities in literature, although others didn't believe odontogenic cysts should be included. In 1933, Cahn et al. [15] first proposed that ameloblastoma could arise in a dentigerous cyst and several authors tried to find and describe the factors that could be the initiating event in stimulating the transformation of the cyst's lining. In the 1960s, several authors investigated the possibility of a viral role (Papova family) in the origin of ameloblastoma [13–16]. In recent years, there has been a surge of interest in human papilloma viruses (HPVs) and they have been extensively studied using recent advances in immunohistochemical and molecular biology techniques [16,20].

Migaldi et al. [21] examined 18 formalin-fixed and paraffin-embedded specimens of ameloblastoma of the jaws treated between 1991-2001 in order to verify the possible etiopathogenetic role of HPV in ameloblastoma by means of in-situ hybridization and polymerase chain reaction techniques as Kahn, Sand and Namin used [18–20].

Both analyses indicated that there was no presence of the so-called low-risk HPV types (6/11) or high-risk HPV types (16/18) both in cancer cells and in normal mucosal controls.

We analyzed the same samples starting from another point of view: the basic hypothesis for the onset of this disease was that it is not due to organic stimuli as HPV, but to inorganic ones such as particles.

In 14 cases we examined (77.7%) the ameloblastoma was located in the mandible and in four cases (22.3%) the tumor was in the maxilla. Seven patients (38.9%) had a recurrence and one patient had two, with the second recurrence occurring 41 months after primary surgery.

The dimensions of the ameloblastomas varied between 0.7 and 6 cm. In 11 patients (61.1%), primary surgery consisted of radical resection of the lesion, whereas 7 patients (38.9%) underwent non-radical surgery. The patients' follow-up was carried out over 13-145 months and two of them died because of cardiovascular or pulmonary diseases, i.e., non-tumor-related causes.

As negative controls in this study we used the specimens of both the maxillary and mandibular molar regions of six patients, who gave their consent to have a biopsy performed during the extraction of their upper or lower third molar.

The results indicated that while there was no presence of foreign bodies in all the negative controls analyzed in each ameloblastoma sample, exogenous micro- and nanoparticles of different size and composition were found (Table 4.1).

As Table 4.3 shows, the sizes ranged from 40 micrometers down to 100 nanometers and the composition of the particles was mainly metallic.

Table 4.3 List of the chemical composition of the particles found

	Case	Size	Shape	Chemistry
1	1	3–8	debris	Fe,C,Cr,O,S,Si,Na
2		0.3	debris	C,Fe,O,Cr,S,P,Ni
3			calcification	Ca,P,C,O,Zn,S
4		8	debris	Fe,C,Cr,O,Al,Si
5		1	debris	W,C,O
6	2	0.7–4	debris	C,Fe,Cr,O,S,Si
7		0.2–5	debris	C,Fe,O,Cl
8		2	debris	C,Hg,Cl,O,P
9	3	2	debris	C,Fe,O,S
10		5	debris	C,Zn,O,S
11		2.5	debris	C,Fe,Cr,O,Ni,S,Ca,Si
12		0.1–0.5	debris	C,O,Pb,Ca,Cr,Na,Si,Al
13		0.1–0.5	debris	C,O,PbSi,Al,Ca,Ti,Cr,Na
14		0.8–12	debris	Zn,C,O,Ca,Ce,La,S,Fe,Cl,Al
15	4	0.8	debris	W,C,O,S,P
16		8	debris	Fe,C,O,Si
17		2.5	debris	C,W,Fe,O,Cr,S
18		0.3–1	debris	C,S,Ba,O
19	5	10	debris	C,Au,O
20		0.1	debris	C,W,O
21			debris	C,Ca,P,O,Mg,S,Na
22	6	0.5–40	debris	Ti,C,O
23		10	debris	Fe,C,Cr,O,S
24		6	debris	Fe,C,O,Zn,Cu,S
25	7	2	debris	C,Fe,Cr,O,S
26		0.8	debris	C,W,O,S
27	8	2.5	debris	C,Fe,O,Si,S
28	9	6	debris	Si,C,O
29		8	debris	C,Ag,S,O
30		10	debris	C,Fe,O,S,P,Si,Na
31		18	debris	C,Fe,O,Ag,S,Al,P,Si,Mg,Ca,Cr,Na
32		0.2–3	debris	C,Ag,S,O,Fe,P,Cr,Ca
33		3	debris	C,Cu,S,O,Ca,P
34		0.8	debris	C,Fe,O,S,P,Si,Na
35		5	debris	Fe,C,O,Ca,Zn,Mn
36		4	debris	Si,C,O,Al,Cl,S,Ca
37		2	debris	C,Fe,O,S,Mn
38		0.3–1	debris	C,Bi,O,Cl,Si
39	10	2.5	debris	C,Si,O,Na,Al,Ca,K,Mg,Ba
40	11	2	debris	C,Fe,O,Si,Ca,S
41	12		debris	C,O,F,S,P
42		1.5	debris	C,Fe,O,S,P,Cr
43		0.1	cluster	C,Ti,O,Si,Al,P,Ca,Na
44			precipitates	C,Fe,O,P,S,Ca

The analyses of the debris are very interesting. First of all because we identified among them nanoparticles (analyses 20, 12, 13) and submicronic ones ranging from 0.1-0.8 microns, and also because we identified materials related to dentistry: stainless steel (Fe-Cr-Ni) (analyses 1,2,4,6,11,17,23,25,42), tungsten and tungsten carbide (5,15,17,19,26), the materials of the dental burs used by dentists for excavation of decay, finishing a cavity or restoration surface, finishing or cutting of castings and/or prostheses. We can recognize the calcium-phosphate cements present with zinc (3,24). The calcium phosphate with copper of analysis 33 could also be a component of dental cement, while the calcium phosphate of analysis 21 is the calcification reaction to the presence of foreign bodies. There are debris related to an amalgam drilling probably occasioned by a removal (8,32,33). Titanium of analyses 22 and 43 could be related to elastic burs or endodontic tools.

So, out of the 44 chemical compositions identified, only 10 could not have a dental material origin. It is also likely that analyses 12, 13 and 14, with lead and zinc, are related to something exogenous. We hypothesize that they may be residues of tobacco smoking. It is well known that tobacco leaves contain much of the environmental pollution residues of the place where the plant grew and, burning them, the pollutants can be released into the mouth (Figure 4.16).

The presence of these debris is almost certainly related to the release of submicronic and nanosized debris due to dentistry malpractice. The dentist might have neglected to use due protections such as, for instance, a dental dam. As a support to this hypothesis, we remember a case of a woman who had many drillings made with a high-speed handpiece to remove dental

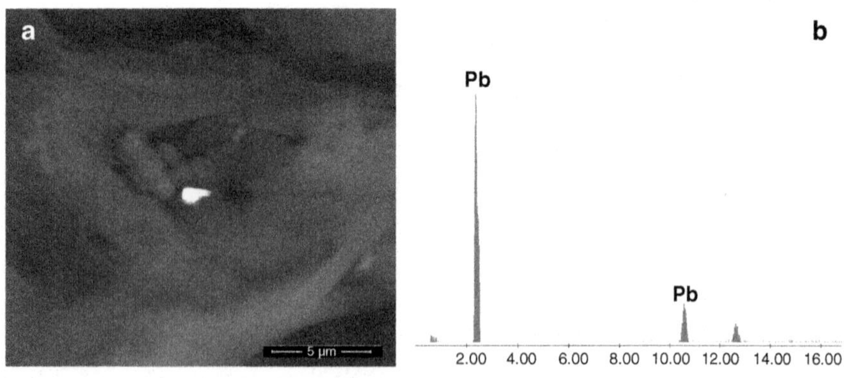

Figure 4.16 SEM image (a) of a lead debris (b) found on a tobacco strip, which could be related to an environmental pollution.

bridges and a crown and, after that, she developed burning-mouth syndrome, a frustrating condition often described as a scalding sensation in the tongue, lips, palate or throughout the mouth. A simple dabbing of the gum with an adhesive carbon stub revealed the presence of debris related to a metallic prostheses drilling. In that case, we suggested repeated washings of the mouth with relatively hot water. Vesicles were formed that broke, releasing the debris trapped as a consequence of the use of the high-speed handpiece. The debris of the prostheses and of the burs scattered in all directions in the mouth. Some were ingested and some were trapped in the oral mucosa. As expected, the latter induced a punctual inflammatory reaction, i.e., burning-mouth disease. Luckily, the hot washings helped to form the vesicles that burst. When debris are trapped deeply enough, they cannot be eliminated and their presence can induce adverse effects (Figure 4.17).

The ameloblastoma findings show a relationship between the pathology and the physical presence of nanoparticles trapped by the tissue. So, the

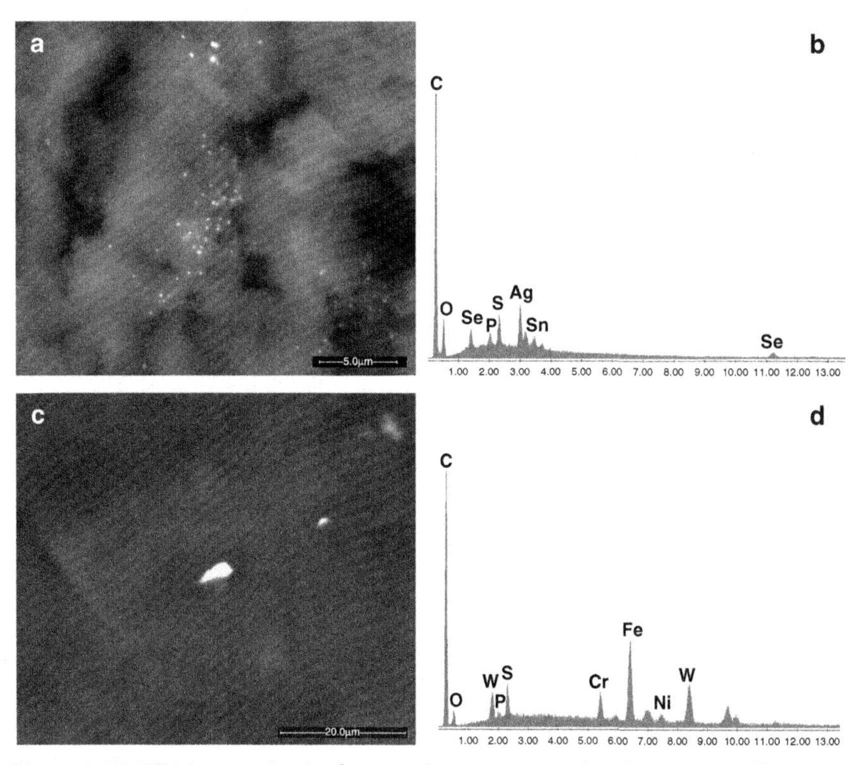

Figure 4.17 SEM images (a, c) of a mouth mucosa sample of a patient affected by "burning mouth disease." Silver-tin-selenium-sulfur-phosphorus (b) and iron-tungsten-chromium-nickel-sulfur-phosphorus (d) debris are embedded in the mucosa.

presence of HPV in some samples found by a few authors could be a secondary effect, related to the fact that the presence of the nanosized foreign bodies can impair the cell-defense mechanisms [22] and every other biological stimulus can find a fertile substrate, since the immune response is underactive.

4.5 LEUKEMIA AND LYMPHOMA

Our study of blood diseases started with the cases we were requested to investigate of some Italian soldiers who served in peacekeeping missions in the Balkan war. It is mandatory to emphasize that the Italian soldiers went to the former-Yugoslavia territories when the bombing by the American-British alliance was already over, but during their activity they patrolled war-polluted areas (in some cases contaminated by depleted uranium bombs) and also lived in partially-destroyed buildings and in bombed factories. A camp-hospital was located inside the area of the bombed Zastava factory, a car, firearm and artillery manufacturer based in the city of Kragujevac in central Serbia. So, medical doctors, caregivers, patients and soldiers spent their whole days in contact with the pollution generated by bombs and targets, in this case a huge quantity of metallic pollutants. Much of the particles produced floated in the air, but part of them penetrated through the air-conditioning system, thus entering the rooms.

We received samples taken from the soldiers, young men who left Italy in healthy condition, were sent on peacekeeping missions, and, after a few months, returned ill. Some had developed blood disorders, others leukemia and, in some cases, they died in a short time. After some years, some of the subjects suffering from blood disorders showed a myeloma.

Some time later we also received samples of soldiers affected by lymphoma, and the same occurred with samples of civilians coming from Sarajevo (Bosnia & Herzegovina). In total, we analyzed nine cases of leukemia and 87 cases of Hodgkin's and non-Hodgkin's lymphomas. We thought that there was a sort of homogeneity of exposure among the 24 cases of civilians exposed to the "war pollution," but it was not so, since, during the Balkan war, there were many migrations among the population, so they were subjected to a variety of exposures.

Some of the war cases will be discussed in Chapter 7, where similar situations occurred at Quirra, a small village close to a firing range in Sardinia (Italy). Also in that location, cases of lymphoma or other forms of cancer were registered both among the population and soldiers working in the firing range.

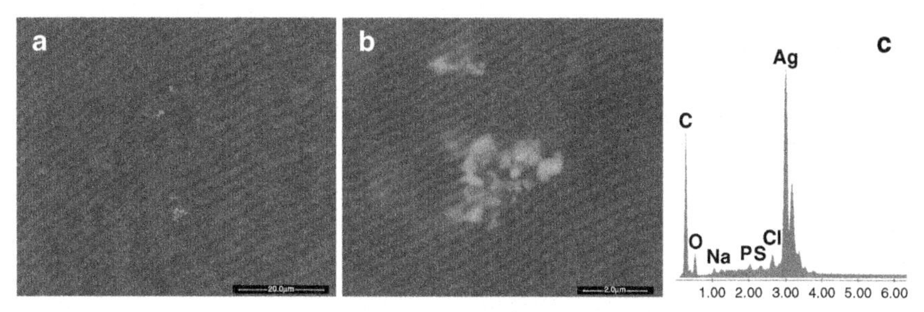

Figure 4.18 SEM images of a bone biopsy (BOM) (a, b) with embedded silver debris (c).

Among the cases we had the opportunity to analyze, was the case of an 18-year-old girl who was admitted to a hospital with a diagnosis of leukemia. We analyzed a sample of her bone marrow where we identified silver submicronic particle (Figure 4.18). Then the presence of a noble metal as PM was something unheard of in that district of the organism.

The toxicity of silver has been known since ancient times because of argyria, a condition due to chronic ingestion or inhalation of silver dust or compounds of that metal. The most visible symptom is the bluish color of the skin in more or less extended regions of the body, the mucous membranes or the conjunctiva. The deposit of silver particles is irreversible and explains the color. Today nanosilver is the main component of many nanoproducts available on the market, from wound bandages to sports garments, from filters for water to toothpastes, from spermicide vaginal foam to pesticides, etc., and many international organizations are wondering if these particles, used because of their antibacterial activity, can have adverse side effects on human health.

The translocation of the inhaled or ingested particles to the bone marrow, their permanence there and their bio-persistence was demonstrated

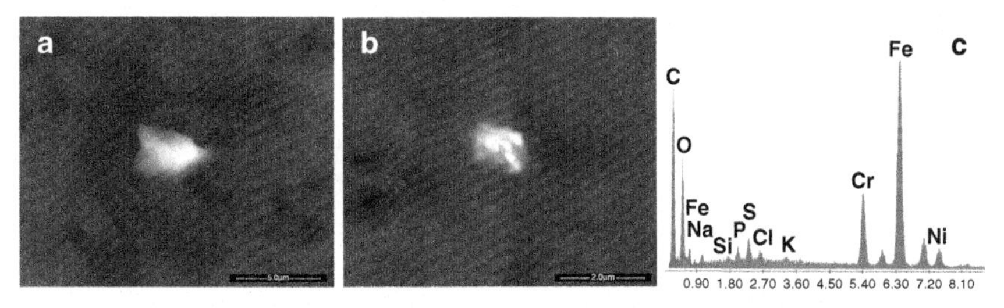

Figure 4.19 SEM images of a seminal fluid drop (a, b) where stainless-steel particles (c) are clearly visible.

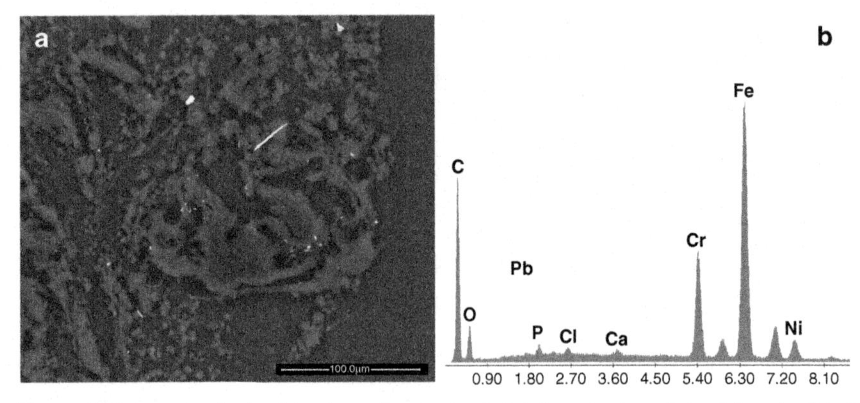

Figure 4.20 Image of submicronic and nanosized stainless steel particles (b) embedded in an osteomedullary biopsy (a).

thanks to an Italian attorney general. He was in charge of the prosecution office of Lanusei, a Sardinian town, and investigated to identify possible responsibilities for the inhabitants' and soldiers' diseases due to the military activity carried out in the already mentioned firing range of Salto di Quirra (Figures 4.19 and 4.20).

He ordered the exhumation of a shepherd who died of leukemia and gave us a tibia to check if inside the medullary cavity where once was the bone marrow foreign bodies were present. After sectioning the bone with a transversal cut a few-mm section was obtained and, after curetting the internal surface, some residues were extracted and deposited on the usual aluminium stub coated with carbon, then inserted into a FEG-ESEM microscope for the elemental analysis. Even after 10 years from the subject's death we could clearly identify metallic debris of iron. The presence of calcium and phosphorus were due to the bone surrounding the spherical debris.

The 6-micron-sized spherical debris, where the grains of the lattice structure are visible, has a clear combustive origin, and demonstrates that the shepherd was exposed to particulate pollutants and that non-biodegradable debris can persist in the body long after death (Figures 4.21–4.24).

The investigation demonstrated an exposure to an environmental pollution. His profession of shepherd was performed in an environment free from industrial pollution. With his flock he stayed in a wild environment but where experimental war activities were also carried out. The most logical explanation for the strange chemical compositions of the debris found inside the bone is that he and his animals were exposed to war pollution.

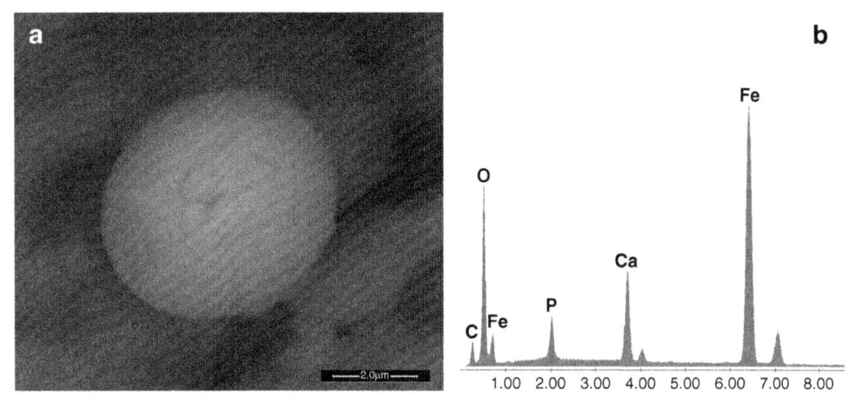

Figure 4.21 SEM image of an iron-based (b) spherule (a) identified inside the bone marrow of a tibia of a shepherd who had died 10 years previously of leukemia.

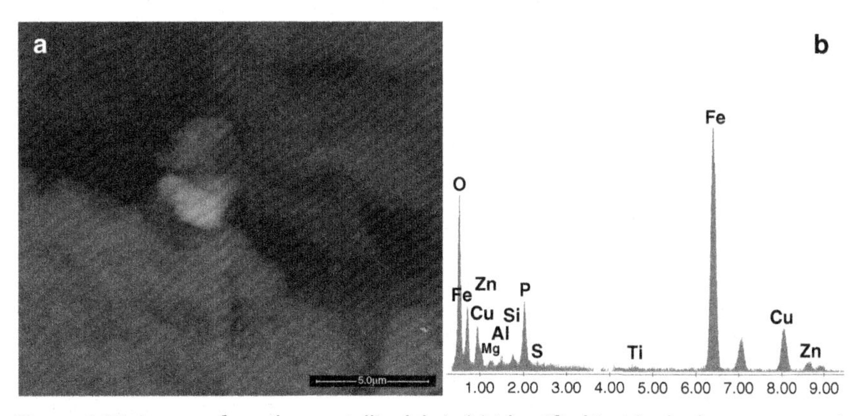

Figure 4.22 Image of another metallic debris (a) identified inside the bone marrow. It is composed of iron, copper, zinc, phosphorus, magnesium, aluminium, silicon, sulphur and titanium (b). Other nanometric debris is hardly visible.

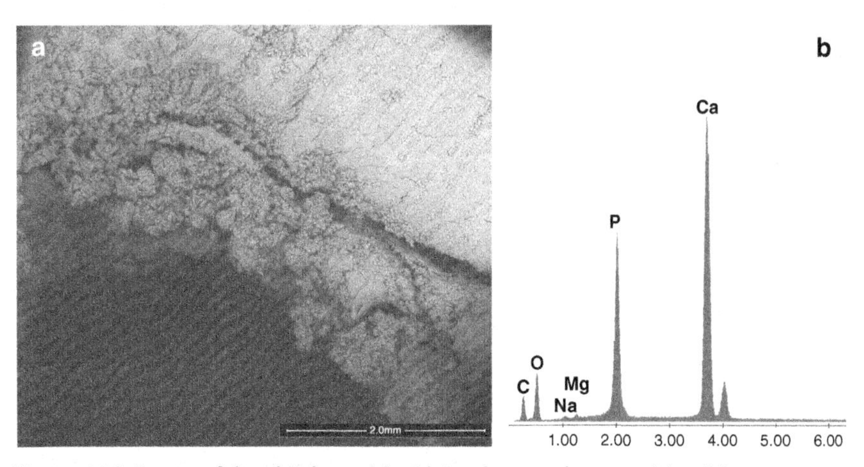

Figure 4.23 Image of the tibia bone (a) with its elemental composition (b).

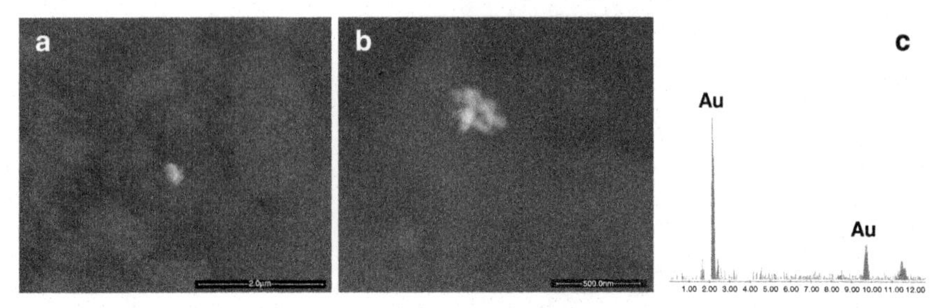

Figure 4.24 SEM microphotographs (a, b) of gold nanoparticles (c) at two different magnifications found inside the tibia's bone marrow.

4.6 CONGENITAL MALFORMATIONS

At the beginning, fetal malformations were not included in the groups of pathologies we planned to investigate, since they were considered genetic diseases and we are not experts of genes and genomics. However, as a matter of fact, malformation cases are increasing in polluted areas.

The approach to malformed babies was due to an elementary problem of "reference samples." In scientific experiments inclusion is requested of the control samples, i.e., samples not influenced by the conditions of the test. Although it is obvious that biological tissues contain mainly just carbon, oxygen, hydrogen and nitrogen besides other elements less represented in the tissues and not, for instance, spherules of stainless steel, it is necessary to demonstrate that the particles we find in the progress of our analyses do not belong in non-exposed living bodies. Life on Earth is based on the cycle of carbon. As just mentioned, cells, tissues and organs contain carbon, oxygen, nitrogen, hydrogen, but the living matter also contains Na, Mg, P, Fe, K, Cu, Zn plus a few other elements in trace, but those are not detectable with our technique. In any case, none of those elements is present either singularly or in combination as particles in a non-contaminated organism. So, it is only obvious that the reference healthy tissue only contains the physiological chemical elements combined in molecules or alone as ions, but never condensed as solid PM.

In many cases we used as reference, biological samples taken in the proximity of the pathological area (in the case of the thyroid, for example, the contralateral part) that did not contain the PM. But, in the case of nanopathologies, we thought that the best controls should be the biological tissues that were never exposed to environmental pollution. So, since we assumed that fetuses are not in communication with the external environment or, at least, they are effectively protected, the most logical choice seemed to be tissues from miscarried or voluntarily aborted fetuses. According to what we thought, those tissues should be free from exogenous pollutants and could ideally be used for reference.

The analyses of the internal organs of fetuses showed us that their tissues were actually free from foreign bodies, but in one case we detected the presence of submicronic solid pollutants. We discovered this because the technician who prepared our samples also included a case of a malformed fetus. That was the first case of spina bifida, the developmental congenital disorder caused by the incomplete closing of the embryonic neural tube, where we identified PM. The mother who had conceived the baby lived in a highly industrialized, heavily polluted area. The presence of those particles demonstrated that the placental barrier was not efficient enough to stop the submicronic and nanosized particles, which could reach the fetus, and which lead us to suspect that they could negatively influence its development. A logical consequence was that the PM could also generate genetic malformations.

A specific study on eight malformed babies born in an industrialized area of Sicily (South of Italy) where oil refineries are located was carried out. The malformed babies were affected by neural tube defects such as spina bifida, and eight controls with no overt malformation syndromes whose mothers were authorized for abortion between 21 and 23 weeks of gestation were also checked. The study demonstrated that in the internal organs (liver and kidney) of the fetuses metallic particles were present in direct relation to the environmental pollution released by the refinery effluents and, in general, related to the combustive processes involved in that kind of industrial activity [23].

The number of total particles was significantly higher in the neural-tube-defect group, while the mean size and number of the particles detected in kidney and liver tissues did not differ between the groups. In neural-tube defects we noticed that the number of iron, silicon and magnesium particles was significantly higher ($P < 0.05$, for all).

It is possible that the presence of a foreign pollution at an early stage of development can trigger a cascade of events multifactorially or genetically mediated. Further investigations are necessary to elucidate the relationship between central nervous system malformations and pollution at a nanoscale stage.

We found also very unusual particles; for instance, crystals of bismuth, chromium, aluminium and copper. Particularly the form they were found in, these elements are not biocompatible/biodegradable and are chemically toxic. Therefore, the possibility that neural-tube-defects fetuses were exposed to a kind of environmental pollution aggressive to the central-nervous system development is substantial.

In a case of a baby born with hand, foot and multi-organ malformations and acute myeloid leukemia in the city of Mantua (Italy), we found in every organ the presence of multiple submicronic pollutants. Particles similar in size, morphology and chemical composition were identified also in the air filters of the atmospheric control units of the local agency for the environmental protection used during the previous winter months. It must be emphasized that the child's mother was perfectly healthy.

This finding suggests that fetuses act as a sort of filter for the mother's blood, keeping, at least in part, pollutants from reaching the mother's organs. So, while the mother is somehow protected, the growing embryo accumulates exogenous foreign bodies that can interfere with cell duplication.

In that fetus, we found debris containing the following elements: iron, silicon, chromium, zinc, aluminium, magnesium, nickel, copper, titanium, lead, barium and tin. Elements not present in a healthy, normal tissue.

It is logical to think that:

1. If small debris, transported by the blood, reach the embryo they can interfere with its normal development, causing deviation from normal growth.
2. If foreign metallic particles, in contact or internalized in the embryo, are found, they can corrode and release ions that can interfere negatively with the cell metabolism and the normal fetal evolution.

4.7 CRYOGLOBULINEMIA

For a number of reasons we had a chance to talk to a rheumatologist who introduced us to some of the problems associated with rheumatic pathologies and mentioned a condition called cryoglobulinemia. A rather uncommon disease, even if it seems to grow and occur more frequently than in the past, and there are no suitable drugs available to eliminate the disease, but only pharmaceutics to silence the symptoms. It is a medical condition caused by proteins called cryoglobulins, abnormal proteins, mostly immunoglobulins, present in the blood that have the unusual, reversible, property of precipitating from the serum specimen at low temperatures.

After cooling the serum at -4°C, they precipitate, forming a white substrate at the bottom of the vial and then they can be dissolved again into the serum upon rewarming at 37°C [24].

Brouet [25] classifies cryoglobulinemia according to cryoglobulin composition: Type I is the result of a monoclonal immunoglobulin, often immunoglobulin M (IgM) or, more rarely, immunoglobulin G (IgG), immunoglobulin A (IgA), or light chains. Types II and III cryoglobulinemia contain rheumatoid factors, which are usually IgM and, more rarely, IgG or IgA.

These factors form complexes with the crystallizable portion of polyclonal IgG. The actual rheumatoid factor may be monoclonal (in type II cryoglobulinemia) or polyclonal (in type III cryoglobulinemia) immunoglobulin. Types II and III represent the majority of all reported cases. Hepatitis C and some autoimmune diseases may be associated with type III cryoglobulinemia.

The symptoms of the disease include hyperviscosity of the blood, arterial and venous thrombosis, acrocyanosis, retinal hemorrhage, severe Raynaud phenomenon [26] with digital ulceration, livedo reticularis [27,28], purpura [29], membranoproliferative glomerulonefritis and multiple myeloma [30].

We investigated this disease starting from the hypothesis that cryoglobulin formation could be triggered by an environmental, not biocompatible PM present in the blood and that, having a different thermal conductivity from the blood components, at -4C° can act as nucleating agents for the precipitation of the globulins. So, we wanted to verify if there were foreign bodies inside the precipitates. In this hypothesis, the autoimmune symptoms could be induced by the proteinous coating of these foreign bodies, which adhering to the debris can change their three-dimensional conformation and cannot be recognized by the body.

A cohort of four healthy volunteers and 19 patients (age 60-77) affected by membranoproliferative glomerulonephritis (MPGN) (Table 4.4) were admitted to the study divided into positive and negative to cryoglobulinemia and to hepatitis C virus (HCV) [31].

They were divided into:

10 patients MPGN and HCV positive;
3 patients with MPGN, but without infection;
5 patients without cryoglobulinemia and HCV negative;
1 patient with cryoglobulinemia but not affected by MPGN; and
4 healthy subjects as control).

Samples of blood and kidney biopsies, since the patients suffered from other systemic diseases, were investigated by means of a FEG-ESEM and EDS and analyzed according to the protocol already described.

The results indicate that all samples had the systematic presence of micro- and nanometric-sized (0.2-5 microns) ceramic and metallic particles of exogenous origin, but no debris were identified in the same volume of serum taken from the healthy volunteers. The frequency of the particles identified was four times greater in MPGN patients with cryoglobulinemia than in patients with MPGN but without cryoprecipitates.

In a 20-microl sample of cryoglobulins we found up to 28 debris, ranging from some hundredths of nanometers to 7 micron: the latter the same size as a red cell. The composition varied from subject to subject.

Table 4.4 List of patients analyzed and number of particles found in the blood serum, inside the cryoprecipitates and the kidney biopsies of the same patients

Patients	MPGN	Cryoglobulinemia	HCV	NPs in serum	NPs in cryo	NPs in biopsies
10	+	+	+	40	174	122
3	+	+	-	4	51	40
5	+	-	-	17	25	24
4 (reference)	-	-	-	0	0	-
1	-	+	-	4	16	-

The debris found had compositions that were classified as ceramic and metallic materials. The ceramics were silicates or aluminosilicates. Silicon-based materials were composed mainly of silicon and aluminium along with other elements such as calcium, iron, sodium, magnesium, etc. Other ceramic debris found was composed of aluminium oxide, zirconium oxide, calcium oxide and barium sulfate.

Metals were mostly represented by iron-based compounds also containing zinc and manganese. Many particles of stainless steel (iron, chromium and nickel), iron and titanium were identified.

Inside the cryoglobulins we also found particles with complex compositions such as C-Si-Ca-O-Pb-P-Na-C; Fe-C-Cr-O-S-Cl-Br-Si-P-Na; C-Cu-Cl-Na-Al-O-P-Ca-S-Fe-K;C-O-Sb-Cl-Al-P-S-Na-Ti;C-Sn-O-Cl-Pb-P-Si-Al; C-Si-Al-O-S-Ca-P-K-Mg-Cl-Fe-Na-Ti-Zn; W-C-Fe-S-Cr; C-O-Fe-Cl-Cr-Si-Na-S-P-Ca-V.

The most represented elements were: Si 22%, Al 19%, Fe 18%, Ca 13%, Mg 12%, Ti 4% and Cr 2%, and the rest was composed of Mn, Zn, Cu, Ba, Pb, Zr, V, W, Sb and Au.

Silicates and iron-based debris were found embedded both in the globulins and in the kidney biopsies.

From a materialistic point of view, the PM identified can have different origins. The ceramic debris (silicon or aluminium compounds) can be from ceramic factories. In fact, around Modena, the Italian town where the study was carried out, there are more than 200 ceramic factories that pollute the area. The manufacturing of their products implies the use of high-temperature furnaces with smokestacks that emit huge quantities of pollutants, among which is PM.

As in many other circumstances, most of the compositions we found in the particles are not mentioned in any handbook of materials, since they would not have any practical application. The odd element combinations are often due to a chance meeting of chemicals released by combustions such as, for instance, the ones occurring in waste incinerators, in which what is burned has a chemistry that rapidly varies over time.

The chemical coincidence of the debris found in the cryoglobulins and in the biopsies of the kidneys proves that the particles, be they inhaled or ingested, were, at least partially, trapped in the plasma proteins while others escaped and were captured by the liver and the kidneys. That can explain the systemic effects of the condition.

The elements reported in Figure 4.25 were listed starting from the element with the highest peak in the EDS spectrum. Some of the particles could be chemically toxic because of the presence of elements such as lead, tin, chromium, tungsten, zinc, vanadium and barium (Figures 4.26 and 4.27).

In one case, we verified that the gold particles identified in the cryoglobulins and in the kidney biopsy could have come from a pharmacological therapy to treat arthritic pains with colloidal gold.

In other cases, it was not possible to identify the source of the exposure only through an anamnestic study, but what is important to underline is that this contamination can remain in the blood and from a theoretical point of view it should be possible to eliminate from the blood and so improve the patient's health. When the contamination is deeply embedded in a tissue/organ only surgery can eliminate it.

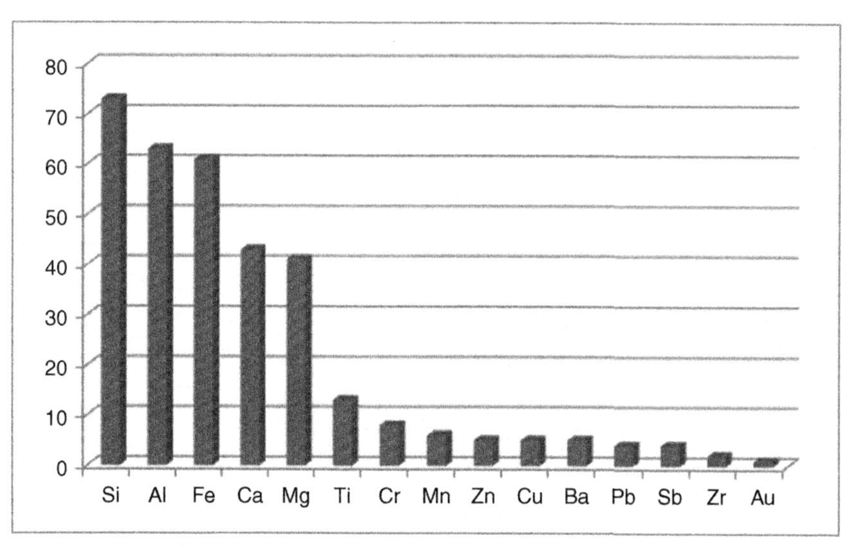

Figure 4.25 Frequency of the elements found in subjects affected by cryoglobulinemia.

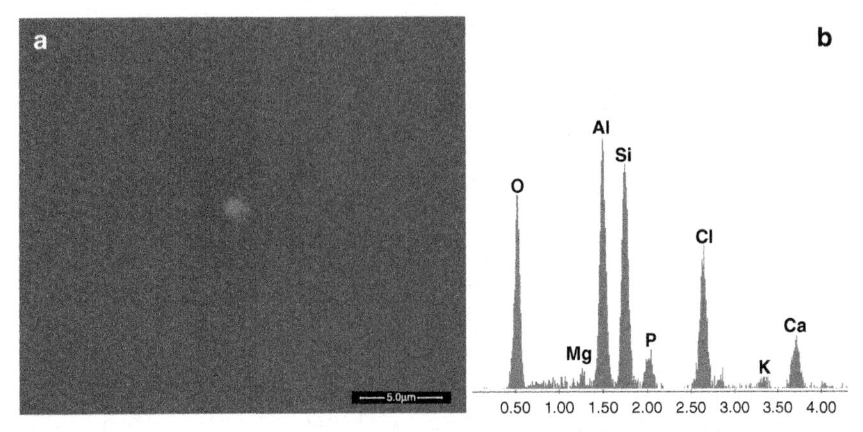

Figure 4.26 SEM image of a 3-micron-sized debris (a) of aluminium-silicon, oxygen, chlorine, calcium, phosphorus, magnesium and potassium (b).

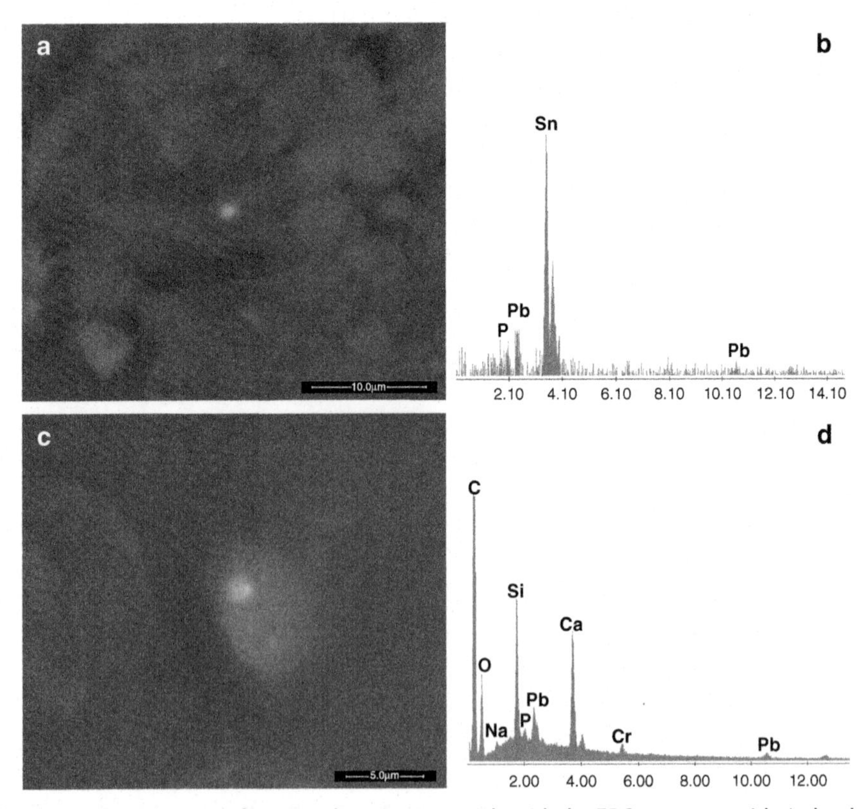

Figure 4.27 Image (a) shows a submicronic particle with the EDS spectrum with tin-lead peaks (b) entrapped in a kidney biopsy. Image (c) shows a particle of silicon-calcium-lead-chromium-sodium (d) found in the cryoprecipitate of the same patient.

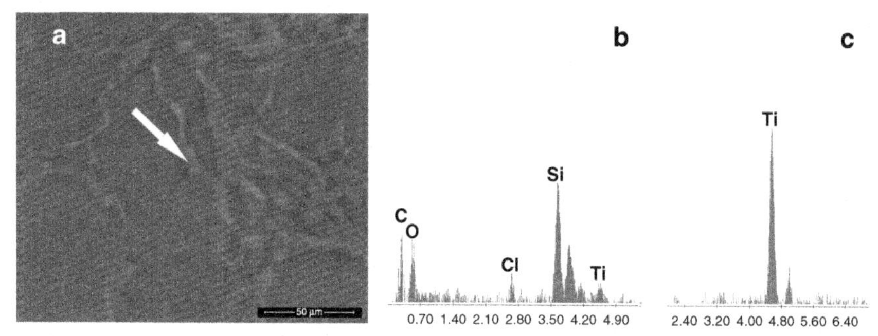

Figure 4.28 SEM image of a kidney biopsy (a) with silicon-titanium-chlorine and titanium nanoparticles (b, c) in a patient affected by cryoglobulinemia.

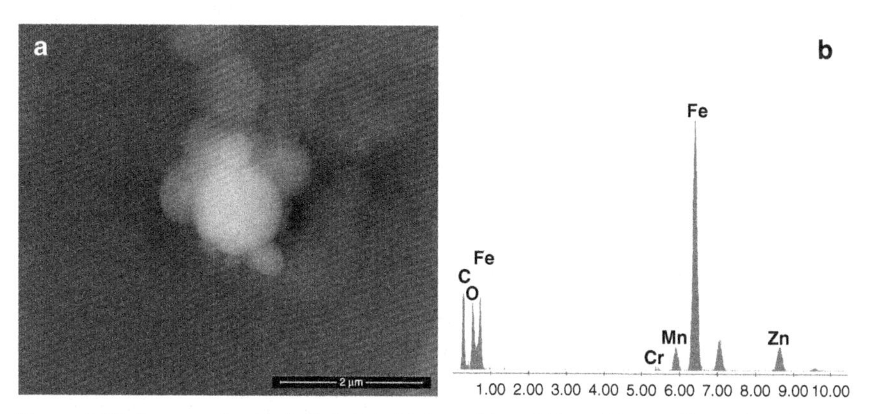

Figure 4.29 Image of spherical debris (a) found in a kidney biopsy. The spherules are composed of iron-zinc-manganese and chromium (b).

Inside all the precipitates of globulins there are many debris of exogenous origin. Debris that are present in the blood, where they interact with proteins, are also carried by the blood circulation and can be entrapped in other organs, as the kidney biopsies investigations showed (Figures 4.28 and 4.29).

This demonstrates that infections, which sometimes are present in patients, can be a consequence of a depressed immunosystem, system that was impaired by the presence of the foreign bodies and their interaction with the proteins. This nanobiointeraction creates organo-inorganic complexes that put strain on the immunosystem.

The presence of those pollutants in the blood can be eliminated with the consequent improvement of the patients' health.

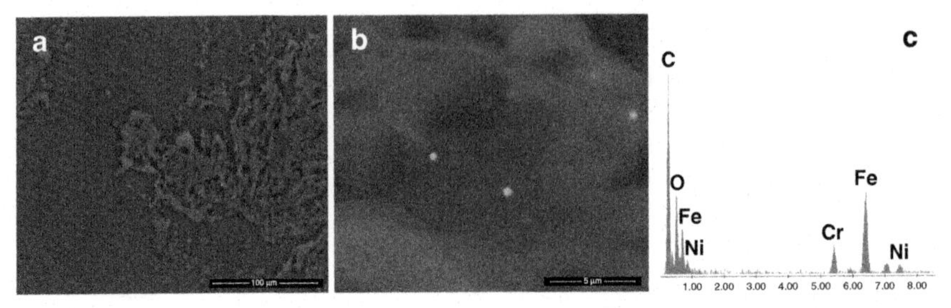

Figure 4.30 Images of debris (a,b) found inside a breast cancer. The spherical morphology of the stainless-steel (c) debris identifies them as generated by combustive processes.

4.8 BREAST CANCER

Preliminary work was performed on cases affected by breast cancer. A small cluster of women affected by ductal breast carcinoma (20 cases) and women with benign breast lesions (18 cases), plus a man with breast cancer were studied. The first results indicated that the cancerous tissues contain foreign-origin particles, and the wide areas of calcification, mostly shaped as spherules, could be ascribable to the biological inflammatory reaction as a consequence of the particles' presence.

In all cases we found debris concentrated in one part of the breast and not dispersed in all the breast area. This localization is very important because it matches the surgical indications that Dr. Umberto Veronesi discovered in the 1980s. In fact, he put forward that the cancerogenic mechanism is interested in only a quadrant of the breast and not the whole structure, and hence he identified quadrantectomy as the best approach to the breast cancer, instead of mastectomy. After more than 25 years, that particular surgery is still performed world-wide [32].

The first results indicate that the invasion of the tiny environmental pollution also affects the breast area. Figure 4.30 shows stainless-steel spherules embedded in the breast-cancer area. Additionally, for breast cancer the mostly identified pollution consists of metallic debris.

REFERENCES

[1] Gatti A, Montanari S. Appendix in Nanopathology: the health impact of nanoparticles. Pan Stanford Pub (Singapore) 2008;27–290.
[2] Bianchi C, Giarelli L, Grandi G, Brollo A, Ramani L, Zuch C. Latency periods in asbestos-related mesothelioma of the pleura. Eur J Cancer Prev 1997;6(2):162–6.
[3] Champness PE, Cliff G, Lorimer GW. The identification of asbestos. J of Microscopy 1976;108(3):231–46.

[4] Castleman B. Asbestos: medical and legal aspects. Clifton, NJ: Law and Business Inc; 1983.
[5] Ladou J. The asbestos cancer epidemics. Environ Health Perspectives 2004;112:265–90.
[6] Harris LV, Kahva IA. Asbestos: old foe in 21st century developing countries. Sci Total Environ 2003;307:1197–9.
[7] Gatti A, Ducheyne P, Piattelli A, Schepers E, Trisi P, Chiarini L, Monari E. Glass corrosion layers on bioactive glass granules of uniform size affect cellular function. In: Ducheyne P, Christiansen D, editors. Bioceramics, Vol. 6. Philadelphia: Butterworth-Heinemann Ltd.; 1993. p. 395–400.
[8] Gatti A, Hench L, Yamamuro T, Andersson OH. In-vivo reactions in some bioactive glasses and glass ceramics. Cells and Materials, USA 1993;3(3):283–91.
[9] Hashimotos Disease. Health Encyclopedia Diseases and Conditions. 2008. http://www.healthscout.com/ency/68/277/main.html.
[10] Muller H, Slootweg PJ. The ameloblastoma, the controversial approach to therapy. J Maxillofac Surg 1985;13:79–84.
[11] Lucas RB. Pathology of tumors of the oral tissues. Edburgh: Churchill Livingstone; 1976. p. 30.
[12] Broca P. Recherches sur un nouveau groupe de tumeurs désigne sous le nom d'odontome. Gaz Hebd Méd Chir 1868;5:19.
[13] Churchill HR. Histological differentiation between certain dentigerous cysts and ameloblastoma. Dent Cosmos 1934;74:1173.
[14] Small IA, Waldron CA. Ameloblastoma of the jaws. Oral Surg 1955;8:281–5.
[15] Cahn LR. The dentigerous cyst is a potential adamantinoma. Dental Cosmos 1933;75:889–93.
[16] Fleming HS. Polyoma virus tumor and the teeth. J Dent Res 1963;42:1405–15.
[17] Kahn MA. Ameloblastoma in young persons: a clinicopathologic analysis and etiologic investigation. Oral Surg Oral Med Oral Pathol 1989;67(6):706–15.
[18] Kahn MA. Demonstration of Human Papilloma Virus DNA in a Peripheral Ameloblastoma by in situ hybridization. Human Pathol 1992;23(2):188–92.
[19] Sand L, Jalouli J, Larsson PA, Magnusson B, Hirsch JM. Presence of Human Papilloma Virus in Intraosseus Ameloblastoma. J Oral Maxillofac Surg 2000;58(10):1129–36.
[20] Namin AK, Talat MA, Eslami B, Sarkarat F, Shahrokhi M, Kashian F. A study of the relationship between Ameloblastoma and Human Papilloma Virus. J Oral Maxillofac Surg 2003;467–70.
[21] Migaldi M, et al. Does HPV play a role in the etiopathogenesis of ameloblastoma? An immunohistochemical, *in situ* hybridization and polymerase chain reaction study of 18 cases using laser capture microdissection. Modern Pathology 2004;18:283–9.
[22] Lucarelli M, Monari E, Gatti AM, Boraschi D. Modulation of defence cell function by nanoparticles in vitro. Key Engineering Materials 2004;254-56:907–10.
[23] Gatti AM, Bosco P, Rivasi F, Bianca S, Ettore G, Gaetti L, Montanari S, Bartoloni G, Gazzolo D, Heavy metals nanoparticles in fetal kidney and liver tissues, Frontiers in Bioscience (Elite edition, E3) 2011;1 (January):221-6).
[24] Ramos-Casals M, Stone JH, Cid MC, Bosch X. The cryoglobulinaemias. Lancet 2012; Jan 28;348–60.
[25] Brouet JC, Clauvel JP, Danon F, Klein M, Seligmann M. Biologic and clinical significance of cryoglobulins. A report of 86 cases. Am J Med 1974;57(5):775–88.
[26] http://www.nhs.uk/Conditions/Raynauds-phenomenon/Pages/Introduction.aspx.
[27] http://www.mayoclinic.com/health/livedo-reticularis/AN01622).
[28] https://www.google.it/search?q(livedo(reticularis&rlz(1C1GGGE_itIT468IT469&espv(210&es_sm(93&source(lnms&tbm(isch&sa(X&ei(VTjFUuyKCcGihgeSw4HgDg&sqi(2&ved(0CAcQ_AUoAQ&biw(1366&bih(704.
[29] https://www.google.it/search?q(livedo(reticularis&rlz(1C1GGGE_itIT468IT469&espv(210&es_sm(93&source(lnms&tbm(isch&sa(X&ei(VTjFUuyKCc

GihgeSw4HgDg&sqi(2&ved(0CAcQ_AUoAQ&biw(1366&bih(704#es_sm(93&espv(210&q(purpura&tbm(isch.
[30] Symptoms: http://emedicine.medscape.com/article/329255-clinical#a0217).
[31] Master degree Thesis of Erica Artoni at the University of Modena and Reggio Emilia "Presenza e possibile ruolo di corpi estranei nanodimensionati nella crioglobulinemia" 2012.
[32] Veronesi U, Banfi A, del Vecchio M, Saccozzi R, Clemente C, Greco M, Luini A, Marubini E, Muscolino G, Rilke F, Sacchini V, Salvadori B, Zecchini A, Zucali R. Comparison of halsted mastectomy with quadrantectomy, axillary dissection, and radiotherapy in early breast cancer: long-term results. European Journal of Cancer and Clinical Oncology 1986;22(9):1085–9.

CHAPTER 5

Sentinel Cases

Contents

5.1 Introduction: nanopathology and toxicology 65
5.2 Sentinel cases 67
5.3 Hepatic granulomas: same disease, different pathogens 69
5.4 Archeology and war 71
5.5 The case of the worker in a ceramic-tile industry 73
5.6 Precious alloys in a prostatic neoformation 74
5.7 The case of a child with prostate cancer 75
5.8 A malformed child born with leukemia 78
5.9 Malformed children 81
5.10 The child with bone cancer 85
5.11 The case of the patient killed by repeated enemas 88
5.12 The boy who played five-a-side football 90
5.13 The boy who went into a sudden coma 93
References 95

5.1 INTRODUCTION: NANOPATHOLOGY AND TOXICOLOGY

This chapter presents some cases in which we could identify the origin of the pollution that had caused the disease affecting the subject we were studying. The task may prove to be anything but simple, since we live in a very polluted world affected by a far from encouraging trend, with a huge variety of different pollutants that only exceptionally come from a single source. In general, what we face is a complex mixture with unspecific identification markers, and in most circumstances our objective is to identify the main agent.

It must be understood that, consistent with the spirit of the whole book, here as well as in the rest of the treatment of the subject we deal with, we consider our contributions not as complements to an already well-known scientific discipline, but as what they actually are: evidence of works in progress taking place in a new scenario. For that reason, any criticism, any corrections and any additional information, provided they are backed by scientific rigor, are gratefully welcomed. And more than welcome are questions and doubts, as they are the goads and the spurs all researchers need.

Unlike classic toxicology, nanopathology's first aim is not the search for chemicals, but that for corpuscular foreign bodies inside tissues. No matter

Case Studies in Nanotoxicology and Particle Toxicology
http://dx.doi.org/10.1016/B978-0-12-801215-4.00005-4

what the foreign body is made of, it prompts an inflammatory reaction with the consequences already described. That is something that, although very obvious, is the source of some lack or, in any case, of some difficulty of understanding with toxicologists. The chemical components of particles are certainly important and can cause the symptoms peculiar to those elements, but the unwanted presence of solid matter has consequences that classical toxicology alone cannot explain. Paracelsus' much quoted words (not accurately, at that), "dosis sola facit venenum," i.e., it is only the dose that makes a thing poison, is one of the universally accepted foundations of pharmacology and toxicology. However, that tenet is not always and automatically applicable to nanopathology. The "poisonous" potential of dust depends on many different factors, dose being only one and to be approached with the utmost caution. Just to make one example in which dose is of minor importance is when particles, even in a very small quantity, manage to enter cell nuclei and start a chain of adverse biological reactions. In other circumstances, dose is certainly crucial and most of the nanopathologies suffered by the military are heavily dependent on the amount of particles that have been able to enter into their organisms. The phrase "For PM, no safe level has been identified" published by the European Environment Agency [1] is faultless and reflects the state of things. Our addition is that, in our opinion, identifying a safe level for PM tout court will be very hard if not impossible at all.

Taking due account of toxicology today, after having detected particulate matters in a pathological tissue, we have been questioned more than once about the toxicity of this or that chemical element, but a limited approach can lead to misdiagnosis or to no diagnosis at all, as it often does.

Classical blood and urine analyses are of little or no use to us and may yield results that have nothing to do with non-degraded particles found in an organ. Generally speaking, in fact, particles can be found in the blood for a limited time and in limited quantity, as a considerable fraction reaches very quickly other tissues; and urine, because of its production process, does not contain the particles we deal with. The same thing can be said for hair or nails where particles cannot travel to. It must be clear that, obviously, no criticism is intended for those analyses, very useful for many purposes as they are, but it must be as clear that they are not the ones needed when non-degradable or non-degraded particles are at stake. Besides many other compounds, what can be found through those analyses could be the chemicals derived from particle degradation, e.g., their oxides in the case of corrosion. Actually, one could be unable to find even the slightest trace of particles in the urine, in the hair or in the nails taken from a subject who is literally invaded by dust, since the organism, as far as we know, owns no efficient mechanisms to get rid of them. On the other hand, when toxic chemicals other than the impossible particles are found in those samples, an elimination process is evidently in progress.

Though at first sight it may look contradictory to what we have just written, at present we are working to improve a blood analysis aimed at early detection of micro- and nanosized foreign bodies. To that end, we are considering and analyzing every blood fraction (red cells, white cells, platelets, plasma) because we are persuaded that, when particles are in the blood stream, parameters such as morphology, size, chemical composition, surface electrical charges, free surface energy, etc., can play important roles, influencing a direct interaction with one or more than one blood fraction. That interaction can prompt the formation of aggregates with particles that can be momentarily there or stable. We also find particles in the blood we analyze, particles that, for some reasons, have not been captured by tissues. As declared, ours are works in progress and much of our science is just something temporary, pending confirmations or denials.

The formation in stable configurations determines the presence of foreign entities that can be responsible for both physical and biological consequences. They represent a complex that is not recognized or only partially recognized by the organism; so they can trigger an immune and probably an autoimmune response. But, at the same time, they represent a nucleus to which other molecules can adhere, contributing to the formation of a thrombus. The aggregates with a size larger than the ones of the cellular elements, carried by the blood stream, once in a tissue other than the blood can be trapped and activate a local reaction [2].

So, there is no contradiction to what we have just written about blood analyses. The procedure we are pretty advanced in trying to set up is not, or not necessarily, a direct detection of particles but the search for the traces they left. Something new that, being new, does not belong to the current medical paraphernalia and, in any case, is still under investigation.

Anyhow, all this promises to shed light upon the origin of some diseases, but can also be important to clarify the concomitance of some systemic symptoms of many pathologies.

5.2 SENTINEL CASES

Generally speaking, in our opinion, "sentinel cases" are a new aspect and a necessary part of personalized medicine whose present definition is that "Medicine that will help tailor healthcare solutions to the individual patient, instead of relying on a 'one-size-fits-all' approach" [3].

There is a growing feeling that the diagnostic and therapeutic approaches with standard protocols, however sophisticated, are not always sufficient and cannot be applied successfully to all patients indiscriminately. It is a well-known fact that the more complex an organism is, the more it becomes difficult to predict its biological reactions, and no organism is as complex as man's. That does not mean that medicine as conceived today is to be rejected or abandoned: just that, in some circumstances, better accuracy is needed.

Besides a non-negligible number of pathologies objectively dealt with unsatisfactorily, it is well-known that diseases once rare or even unknown are peeping out and are becoming, if not common, at least not unusual. In most cases, e.g., the so-called "orphan diseases," and for reasons that are not always connected with the nobility of medicine, no protocols exist at all for them. There is a need for the "right intervention to the right person at the right time of the person's life course." Personalized medicine ought to have the potential to improve the outcomes for patients and, as a very desirable bonus, cut down on the use of unnecessary (what is not necessary is often harmful) and expensive treatments. Given the right application, personalized medicine can contribute to securing more value from healthcare spending, to the benefit of healthcare providers and patients, and, in the end, of all citizens [4].

As far as we are concerned, we extend this concept and propose to evaluate case-by-case if the patient has been exposed to a pollution that could have caused the disease. Secondly, knowing the chemical composition of the pollutants and some other characteristics like size and shape of the particles, we try to trace them in the environment the patient attends. Then, we check that environment and, if the pollutants are still present, we try to eliminate them or, if that cannot be done, remove the patient from their influence before undertaking any therapy. In fact, it is nonsense to start a pharmaceutical treatment, be it personalized or not, in a patient who is still exposed to the pathogen. It is necessary to stress again that drugs are unable to destroy the kind of pollutants we deal with and the only advantage one may hope to obtain from them is the palliation of some symptoms. Anti-inflammatory drugs, though often used without the right perception of what the real cause of the disease is, are the most widespread example of such a use.

Apart from drugs, and considering what the organism has available to defend itself, as to macrophages, to our knowledge no demonstration exists that they are attracted to what are commonly called nanoparticles ($\leq$ 200 nm), while the attraction is visible to larger particulate matter. In any case, they too would be unable to destroy those entities physically, since the particles we deal with are not biodegradable and the way macrophages work involves necessarily the biodegradability of the object of their activity. The only thing macrophages could possibly do would be to move the particles captured to other anatomical positions such as, for instance, the lymph nodes without any possibility of actual elimination.

5.3 HEPATIC GRANULOMAS: SAME DISEASE, DIFFERENT PATHOGENS

Granulomatous hepatitis is a lesion found in a wide spectrum of liver diseases [5]. Tuberculosis and sarcoidosis have been most frequently incriminated, but many other causes can be responsible for that pathology, among which are viral infections, immunological disorders, drug-induced injuries, Crohn's disease and foreign-body reactions [6]. A minor percentage of cases, about 10%, remains of unknown etiology and is classified as granulomatous lesions of unknown significance (GLUS) [7].

From the histological point of view, those lesions are usually non-caseating, with a few multinucleate giant cells and some surrounding chronic inflammatory infiltrate [8].

From literature and personal experience we know that, among other causes, more generally, granulomas can be local effects brought about by wear debris from joint replacement as the result of in-situ degradation of orthopaedic implants [9]. Besides topical effects, particulate debris generated internally by a prosthetic device like hip-joint prostheses has been reported to be dispersed in lymph nodes [10, 11] and in distant organs [12].

As a general rule, the larger wear particles prompt a fibrous or giant-cell reaction, while the smaller ones (in any case, larger than 200 nm) incite a macrophage phagocytosis reaction, and submicronic particles can hide inside single cells. We had a chance to work on seven cases of hepatic granulomatosis classified as "of unknown etiology" coming from Italy and the UK.

Seven patients were chosen for their different meaningful clinical history among 14 analyzed cases with a histopathologic diagnosis of "granuloma" or "granulomatous flogosis" from liver biopsies. Some details of the seven cases are shown in Table 5.1.

The use of a polarized lens in an optical microscope showed histologically the presence of birefringent foreign material inside the granulomas of three cases (Nos. 1, 2 and 7). Pas, Grocott and Zihl-Nielsen reactions excluded an infectious origin.

Table 5.1 Clinical and histological data of the cases of liver granulomatosis analyzed

Case n	Sex	Age	Clinical data	Histological data
1	female	61	rheumatoid arthritis hepatic markers altered	mild chronic flogosis; a small granuloma near a portal tract containing black material
2	female	25	echinococcosis (liver) normal hepatic markers	(cyst wall): fibrosis and granulomatous flogosis (foreign-body type)
3	male	63	flogosis indices altered	steatosis and a granuloma with an epithelioid and a lympho-monocytoid halo
4	male	65	spleen neoplasia	hepatic granuloma with homogeneous eosinophilic material
5	male	27	fever; low CD4	Intra-parenchimal granulomas
6	male	62	renal failure; 10-year peritoneal dialysis	granulomatous flogosis and fibrosis
7	male	60	fever; weight loss	mild chronic hepatitis cholestasis; giant-cell granulomas

The physical-chemical ESEM observations revealed particulate materials inside the granulomatous-reaction tissue in all seven cases. The particles were different in size (in general < 20 μm) and in chemistry.

The first case showed nanometric (50 nm) gold particles in the liver granuloma, which originated from engineered gold nanoparticles injected in one of the patient's knees affected by arthrosis. The second case showed barium-sulfur submicronic particles, probably originating from the injection of a contrast medium (barium sulfate) that induced an immediate allergic reaction and was identified in the tissue some years later with a biopsy. The third case, concerning a mason, showed the presence of calcium sulfate and phyllosilicate debris probably inhaled at his work. The fourth case showed spherules of calcium-phosphorus inside the granuloma. The fifth case, that of a drug-addicted boy, showed some particles of different chemistry, but a few of them (calcium carbonate) were probably related to the mixture of drugs taken. The sixth case, a patient who underwent chronic peritoneal dialysis, contained particles of ceramic origin (carbon, oxygen, silicon, sulfur, calcium and titanium) that could be correlated with the pollution of a prolonged dialytic treatment. The seventh case presented a wide spectrum of particles sized 0.8–8 micron whose origin we could not locate.

The study demonstrated that every patient was subjected to an exposure of particulate matter that, carried by the blood, was filtered by the liver where it remained. It also demonstrated that ceramic, large-sized particulate matter can induce a granulomatosis reaction. The accumulation of this debris can either cause or worsen a hepatic pathology. A personalized approach, identifying the source of the specific contamination, can allow banning the exposure (if still present) and treating the symptoms with anti-inflammatory (cortisone-related) drugs in order to mitigate the flogistic reaction.

As described in the previous book [2], this approach was used for the first time in the case of a patient who developed liver and kidney granuloma due to the ingestion of porcelain debris produced by the two poorly-manufactured porcelain bridges he was carrying. The removal of those dental prostheses saved the patient from the dialysis treatment that had been predicted for him [13].

5.4 ARCHEOLOGY AND WAR

One day we received a phone call from an archeologist. Very simply and directly, he told us that he was dying of a pancreatic adenocarcinoma and told us his story. When the first Gulf War was declared and the bombing started, he went to Iraq hoping to rescue at least some of the ancient treasures from the plunder by local and foreign people. His first stay was an 8-month one and, after that stay and his return, he had a further chance to go again to Iraq. In 2004 he was diagnosed with a cancer of the pancreas: not many chances to survive. He offered us his biopsies to help us detect "depleted uranium" debris, something that was the current fad then and we were busy trying to detect in the biopsic samples of soldiers. What we found in those fragment of pancreas was, in a way, a surprise.

The particles we identified in the pancreas biopsy tissue have chemical compositions that need an explanation. The war bombing (see Chapter 7) generated a new environmental pollution due to the aerosolization of bomb and target together. For that reason the aerosol is mostly metallic. Both tungsten and steel are rather usual findings (Figure 5.1), but the presence of gold and silver (Figure 5.2(a,b)) in the archaeologist's tissue was something we had never come across and is hard to explain. We cannot be sure, but we think that it may be possible to relate such unusual presences to the places where the subject had spent most of his time: museums. As is obvious, the explosion of weapons reduced part of the structures hit to fine dust: buildings above all, but also what was inside, and inside those museums gold and

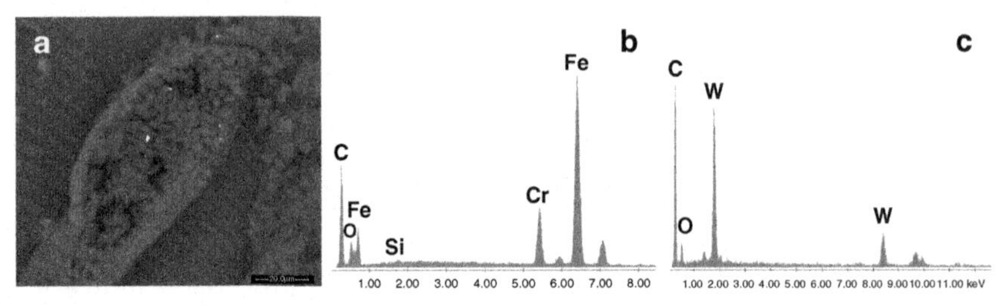

Figure 5.1 ESEM image (a) and EDS spectra (b, c) of a blood vessel where, close to red cells, (whiter) debris of iron-chromium (stainless steel) and tungsten are present.

silver were particularly abundant. The gold contained in the electronic part of modern bombs is there in really small quantities and, though there may be a better one, we could not find any other logical explanation for those unusual particles.

If the archaeologist was one of the many innocent victims of a war that did not belong to them, perhaps too romantically we like to believe that he lost his life with a part of something he had tried to save hidden in his body.

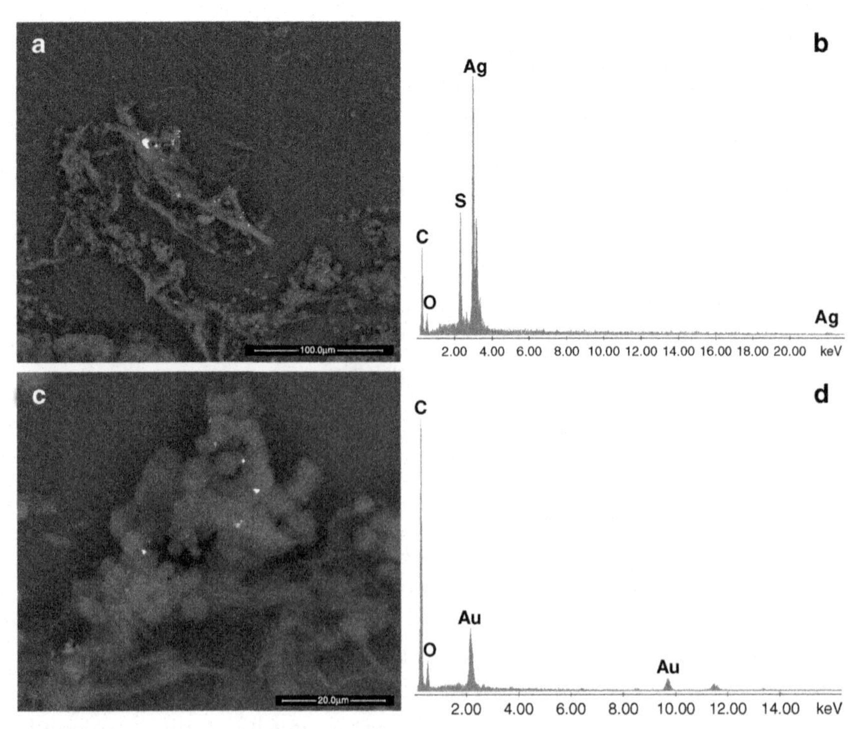

Figure 5.2 ESEM image (a, c) and EDS spectra of micron-sized and submicronic sulfur-silver (b) and gold (d) particles found in the pancreas affected by carcinoma.

5.5 THE CASE OF THE WORKER IN A CERAMIC-TILE INDUSTRY

Testicle cancer was diagnosed in a 22-year-old accountant working in a ceramic-tile industry, a variety of cancer that affects preferentially young individuals. We were requested to analyze the surgical samples and what we found demanded an explanation that we looked for in the patient's working life.

We asked the patient to describe his working place, assuming it was a small office. He told us, instead, that it was a very large room with many big instruments and machine tools. The corner where his desk was located was beside a machine that ground the earth to fine particles before mixing it to water to obtain the material with which to make tiles. On the opposite side there was another machine that cut at a 45°-angle, the edges of already formed and baked tiles. Telling us that information, the subject remembered that the machine was run by an elderly man who had contracted a lung cancer from which he eventually died.

Inside the testicle cancer tissue (Figure 5.3) we found silicon-based debris and stainless-steel particles, while in the spermatic cord we identified tungsten debris. The same kind of pollution was found in the subject's seminal fluid.

Though the boy had been fired some time ago because of his illness, he could still list the machines used in the room where he had worked, including their technical names. An online search showed that the machine used for crushing the earth had stainless-steel moving parts, while the other one was equipped with tungsten-carbide cutting elements.

To provide an explanation for what we found, it must be considered that:

1. None of the machines was protected by systems to confine the dust they produced, dust mainly composed of silicon-based material that, in those conditions, was released in the indoor environment.
2. The crushing machine also released stainless-steel wear debris.
3. The cutting machine released fine tungsten-carbide wear debris.

For years the elderly worker had inhaled the raw silicon-based dust and the tiny tungsten-carbide particles that stopped in his lungs and caused the cancer. His young colleague was not in direct contact with the heavier dust, but had inhaled fractions of the tiniest dust, those that stay suspended longer and travel farther, floating in the air. The differences between the two cases lay probably in the size of the particulate pollutants. The finest ones, being more invasive, had easily negotiated the lung barriers and reached the internal organs, including the testicles, where they had exercised their pathogenicity.

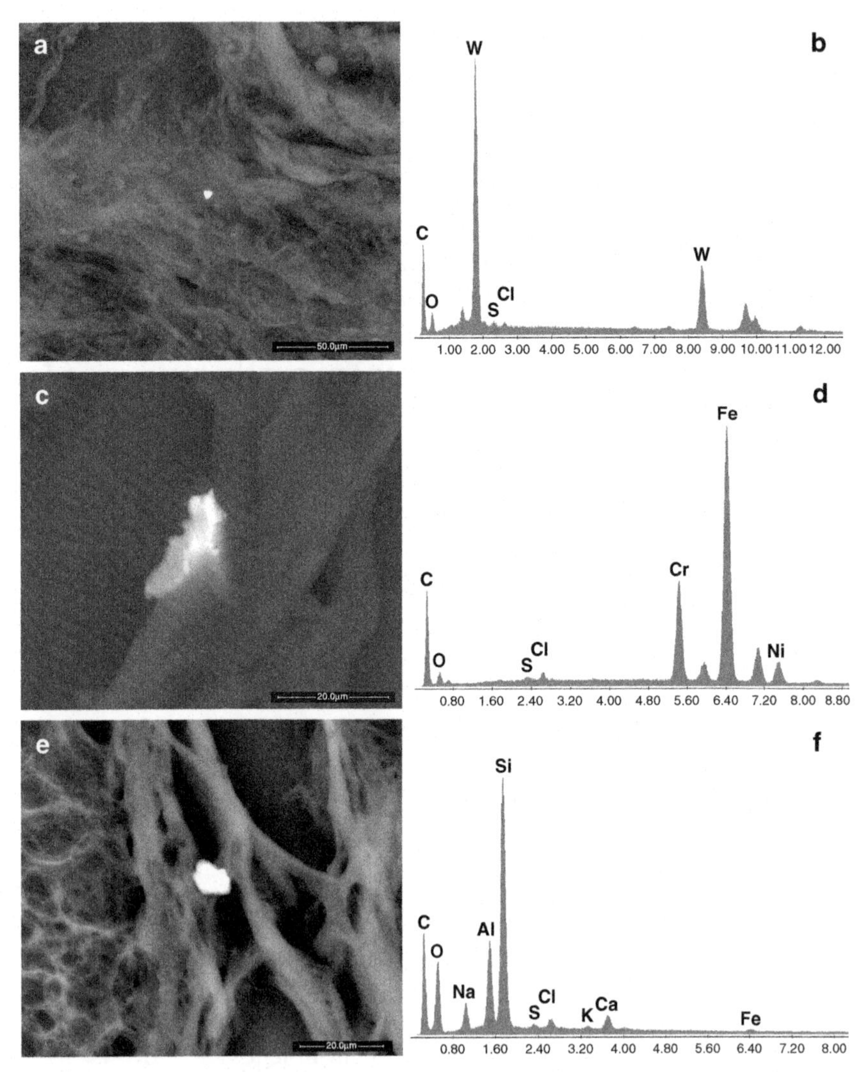

Figure 5.3 ESEM images (a, c, e) show micronic particles found in testicular tissue affected by cancer. The EDS spectra show different compositions: tungsten or tungsten carbide (b), stainless steel (d) and aluminosilicates (f) typical of ceramic tiles.

5.6 PRECIOUS ALLOYS IN A PROSTATIC NEOFORMATION

This case detected in an elderly patient during a routine screening was selected as a reference for another study on cancer, since it showed some calcifications, but no (or not yet a) cancer.

The tissue examined contained a variety of platinum-iridium spherules (Figure 5.4). We mention the case because this alloy, a very hard and chemically

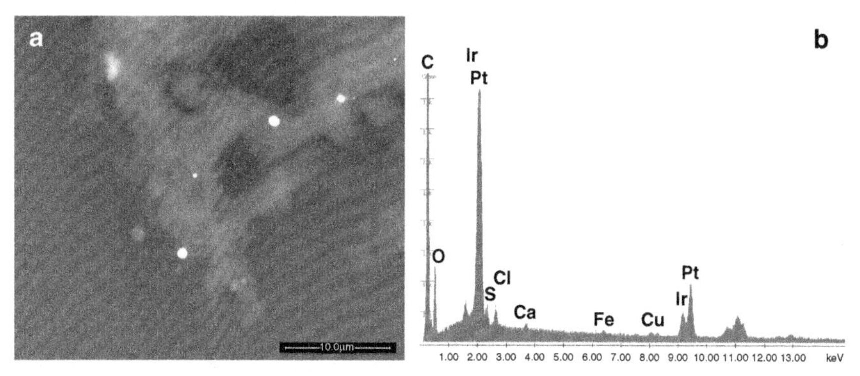

Figure 5.4 ESEM image (a) shows a prostate tissue with spherical particles composed of a platinum-iridium alloy (b).

resistant one, is not commonly found, if only because of its high cost. Besides having been used to make the international prototype kilogram and the international prototype meter, its current, limited, uses are in jewelery (e.g., expensive watch cases); in some electrical contacts, particularly microelectrodes used for electrophysiology recordings and nervous-tissue stimulation; in electrodes for pacemakers; in some high-performance spark plugs; and in sophisticated explosive devices.

Because of privacy laws, the patient's identity could not be disclosed to us, so we could not interview him and understand the kind of exposure he had been subjected to, but we are convinced that those rare indeed particles came from his working activity, probably with combustion processes involving temperatures ranging from 1,750° to 2,500°C, the range of melting-point temperatures of the Pt-Ir alloy depending on the percentage of each element [14].

5.7 THE CASE OF A CHILD WITH PROSTATE CANCER

Prostatic cancer is a far-from-common event in children. It is so rare that, to our knowledge, no statistics for the pathology incidence are available for men under the age of 35. So, when we were asked to investigate the case of a six-year-old boy affected by prostate cancer without receiving further and more accurate information, we were rather surprised. The diagnosis issued by the hospital was of embryonic rhabdomyosarcoma of the prostate, a variety of cancer different from the one that usually affects elderly males and compatible with the age of the boy. Though relatively rare [15–17], from the statistical point of view sarcomas are the second most frequent pediatric tumors after blood cancers. They are a histological form of cancer of the connective tissue wherein the malignant cells derived from the mesenchyma are similar to the primitive developing skeletal muscle of the embryo. As to possible genetic causes, in this particular case they looked particularly unlikely because of the presence of foreign bodies of obvious exogenous origin in the tissue. It must be added that both parents and the boy's sister (born after the subject's death occurred at the age of 11) enjoyed good health and, according to what the parents told us, no cases of cancer existed in their family at the time of the diagnosis.

In that instance, we also found concentrations of inorganic, non-biodegradable particles in the primary cancer, whereas very little could be seen in the metastatic tissue that had already spread in the bladder. In passing, it must be clarified that, when inorganic particles are the cause of a cancer, none of those are to be found in metastases since they owe their existence not to foreign bodies but to pre-existing cancerous cells. The particles that can be found in the metastatic tissues are there because the subject keeps being exposed to pollution.

The Italian town of Forlì has two incinerators, one dedicated to urban and one to hospital waste. It is interesting to note that the house where the boy had lived stands between those two plants and at a comparatively short distance from both (1.2 and 2 km, respectively as the crow flies) [18–22].

The particles we found in the boy's cancer specimens (prostate and bladder) were mostly metallic micro- and nanosized particles of different alloys (gold-silver, tungsten, stainless steel) (Figure 5.5), but ceramic debris were also identified (zirconium-calcium-silicon-aluminium and calcium-sulfur). Some are reported in Table 5.2.

Table 5.2 List of elements found in the biopsy

Particle size (um)	Sample	Debris chemistry
1	**Prostate**	Au, Ag
1		C,Au,O,Cl,Ag,Cu
2		C,O,S,Ba,Cl
	Bladder	
0.1-0.4	nanospherule	C,W,O,S,Cl,Na
1-3		C,Fe,O,Cr,S,Cl,Ni,Na
2		C,Zr,O,Ca,Si,Al
0.2-1	spherules	Fe,Si
0.2-1	spherules	Fe,Ni
	Normal tissue	C,O,Cl,P,S,Na

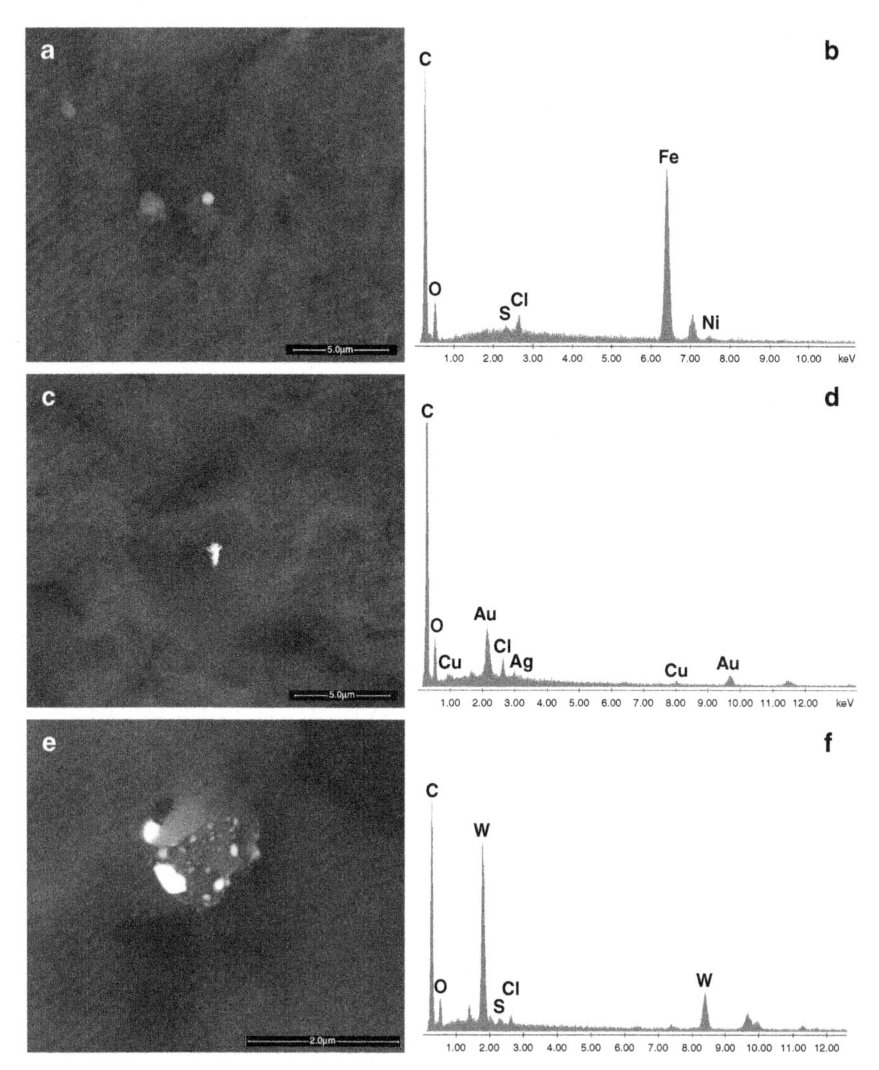

Figure 5.5 ESEM images (a, c, e) show some of the particulate matter identified in the prostate and bladder tissue affected by sarcoma; (a, e) show spehrical debris composed of iron-nickel (b) and tungsten (f); while (c) shows a submicronic debris of a golden alloy (gold-silver-copper) (d, f).

Finding metallic debris is unusual indeed, particularly spherical, i.e., with a high-temperature origin, inside the bladder and prostate tissue of a child who could not have had many possibilities to be exposed to such a contamination. The only possible exposure was just the one experienced at home, in a house located, as already mentioned, between two waste incinerators. The presence of debris of golden alloy testifies to a random combustion of waste containing small parts of that metal, and the spherical particles are a further evidence of a high-temperature origin. The exposure could have occurred by inhalation or by ingestion, probably from the vegetables grown in the adjoining kitchen garden.

Later, after the child's death, his grandfather also developed prostate cancer, a fact that can be reasonably attributed to causes different from those that killed the boy.

5.8 A MALFORMED CHILD BORN WITH LEUKEMIA

A full-term male infant delivered by emergency caesarean section as a result of diminishing fetal heartbeat was admitted to the neonatal intensive-care unit of a hospital of Mantua (Italy) three hours after birth, suffering from hypotonia and respiratory distress. The mother was healthy, had a normal medical history and during her uneventful pregnancy was not exposed to drugs or any other abusive substances such as tobacco or alcohol.

A sampling of chorionic villi resulted negative. The Apgar score (an easily replicable method to quickly and summarily evaluate the health of a newborn child immediately after birth) was 5 and 7 at 1 and 5 minutes, respectively. (Scores 7 and above are normal.) Birth weight, length and head circumference were in the 50th percentile.

Examination revealed liver-and-spleen megaly and pallor. The entire body was covered with multiple, firm, mobile nodules ("blueberry muffin" spots). The infiltrates were 1-2-cm deep and ranged in color from blue to purple.

Blood examination revealed a high white-cell count (73,020/mm^3) and severe anaemia (haemoglobin 40 g/l). The infant's condition deteriorated rapidly and he developed acute hypotension, which failed to respond to colloid and inotropic treatment.

Death occurred nine hours after birth. Soon after, an autopsy was performed and a histological and immunohistochemical diagnosis of congenital acute myeloid leukemia (AML), an unusual neoplasia in subjects that age, was issued. The autopsy revealed leukemic infiltrates in several organs including skin, liver, kidneys, spleen, heart, thymus, lungs, testicles and spine. Lymph nodes were also abnormal. Samples of all those tissues were collected at necropsy.

Dr. Luigi Gaetti, anatomo-pathologist at the Hospital "Carlo Poma" of Mantua (Italy), which we recognize as an essential contributor to this section of the chapter, in his capacity of anatomopathologist at the local hospital, gave us the post-mortem samples of the tissues taken from the corpse. What we found in all specimens analyzed was, without exceptions, metallic

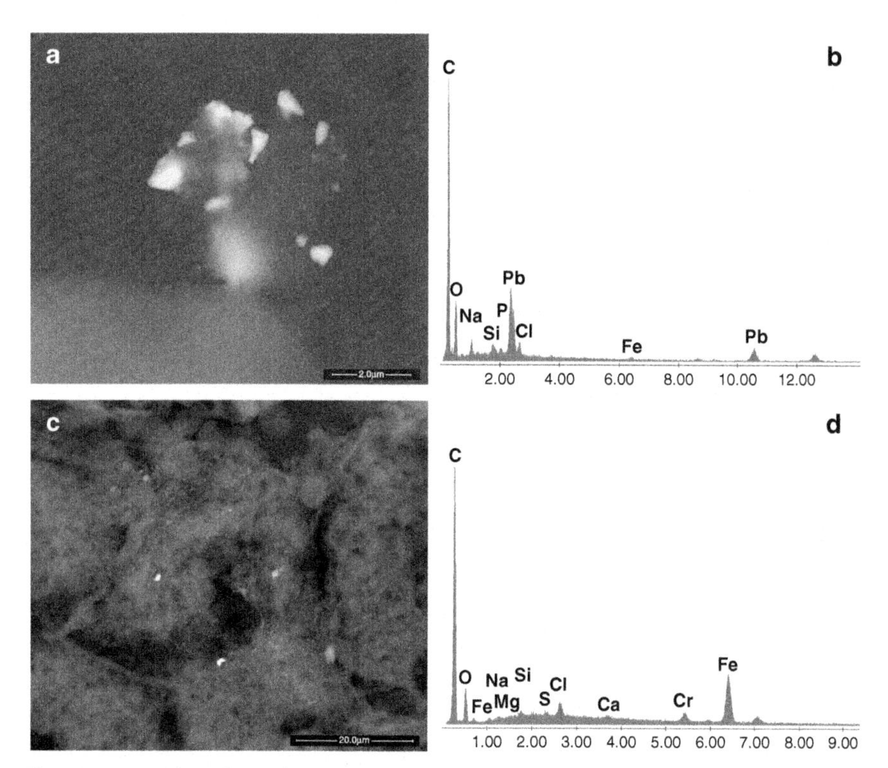

Figure 5.6 Images show the presence of particulate matter in the liver (a) and in the kidneys (c) of the baby. Its chemical composition varies from lead compounds (b) to iron compounds (stainless steel) (d).

micro- and nanosized particles. Especially in the liver and in the kidneys we saw a homogenous and important distribution of particulate foreign bodies as evidence that the fetus had been exposed to exogenous pollution through the mother's blood. We identified heavy metals such as lead in the liver and stainless-steel particles of smaller size in the kidneys (Figure 5.6).

But, apart from the unmistakable particles that the EDS analysis showed to be iron-based, everything we found as far as inorganic particulate matter is concerned was typical of an industrialized environmental pollution. Since the only communication between a fetus and the world outside the womb is what is offered by the mother, we could not help but deduce that that kind of pollutant is rather freely carried from mother to fetus.

All parts of the body contained metallic micro- and nanosized debris. Particularly revealing was what we detected in the heart: perfectly round particles typical of a formation involving high temperature, i.e., something absolutely impossible in any living organism and indisputable evidence of an external

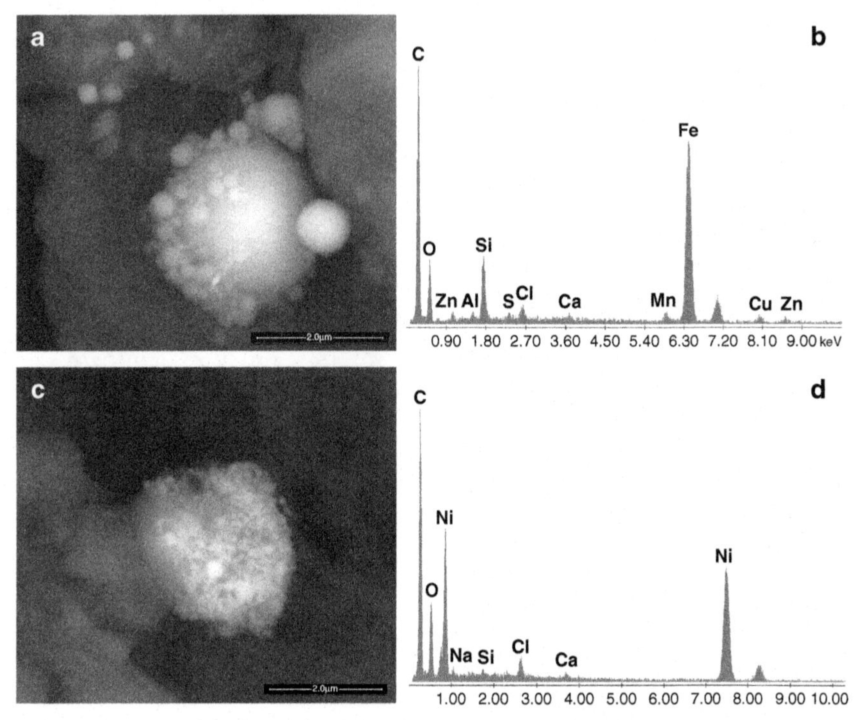

Figure 5.7 Images of debris (a, c) found inside the baby's heart. Agglomerates of spherical micro- and nanoparticles of iron-silicon-manganese-copper-zinc-aluminium-calcium (b) and nickel-calcium-silicon (d) were identified.

source (Figure 5.7). The clustering of the spherules can be an indication of a formation in a small and confined volume, perhaps the internal part of a smokestack.

Later, analysing the air filters taken from air-monitoring stations used six months earlier (mid-January) at Mantua, we found particles with the same characteristics as those photographed inside the child's tissues and, in particular, in the heart. It must not be forgotten that spherical particles produced at high temperature are generally hollow and particularly fragile, and what is most frequently found is not whole spheres but their fragments. So, what we saw in the heart tissue was, if not exceptional, certainly remarkable.

Though numerous genetic and environmental risk factors have been defined, the etiology of AML is actually unknown or, at least, uncertain and still a matter of debate, but the presence at birth and what we documented suggests the possibility of intrauterine exposure to toxic substances absorbed through the mother: not only chemical compounds but also particulate matter.

Mantua is an Italian town situated at the center of the Po Valley in a territory where winds are scarce and the high concentration of factories in a comparatively large area at its outskirts makes its environment particularly polluted. The largest analytic study on childhood AML concluded that occupational exposure of either parent of children suffering from that illness to pesticides is particularly high, and so are paternal exposure to petroleum products and postnatal exposure to pesticides [17].

A Californian study claims that children living in areas with high levels of hazardous air pollutants (HAPs) are at increased risk of developing leukemia. Although pollution is recognized as a major factor in the etiology of the illness, to our knowledge very little exists in literature about children that did not contract AML in the course of their life but were born with it, and nothing, more in particular, can be found about the presence of inorganic, obviously exogenous, particles in their organs [23, 24].

The above-mentioned results suggest that metallic particles in contact with an embryo can be really dangerous for its normal development. Can leukemia be the result of a specific biointeraction of those foreign bodies (lead, stainless steel, nickel, etc.) with specific cells such as the red cells and the blood components? Surely the presence of inorganic tiny foreign bodies in all the organs favored some patho-mechanisms and the consequent adverse biological reactions.

5.9 MALFORMED CHILDREN

In a previous work we showed that malformed children born in polluted areas carry, embedded in their tissues, the environmental particulate pollution that their mothers inhaled and/or ingested and that submicronic particles, because of their very small size, can reach the embryo through the placental circulation [25].

Although the cases examined came from different areas (Mantua, Modena, Gela [26], Malta) and may look somewhat disparate in some instances, all the necroscopy samples presented particulate matter inside (see Table 5.3). The chemical compositions of the particles were different, but, in general, the common factor was a metallic nature. The correlation between particulate air pollution and fetal health was already observed and speculated but no one verified directly inside the fetuses the real presence of the environmental contamination or traced it to industrial activities [27–32].

The following is a selection of the pollution we identified inside the internal organs of the malformed fetuses. Figure 5.8 shows 4-0.8 sized particles, respectively, of strontium-sulfur-iron-aluminium, antimony and lead.

Table 5.3 List of cases of miscarriages and malformed babies analyzed

Case	Sex	Week	Pathology
1	?	20°	abortion; androgen insensitivity syndrome (AIS) – undermasculinized genitalia
2	♂	21°	miscarriage due to chorioamnionitis
3	♂	19°	abortion; trisomy 16
4	♀	21°	abortion; agenesis of the corpus callosum
5	♀	20°	abortion; trisomy18
6	♂	22°	abortion spina bifida
7	♂	40°	dead of asphyxia due to umbilical hemorrhage
8	♀	29°	dead of placental insufficiency
9	/		miscarriage
10	♀	20°	abortion; vertebral schisis
11	♂	16°-18°	miscarriage; normal fetus
12	♂	22°	abortion; tetralogy of Fallot with mosaic trisomy 22
13	/		Neu-Lexova syndrome
14	/		Neu-Lexova syndrome
15	/		Neu-Lexova syndrome
16	/		spina bifida
17	/		spina bifida
18	/		spina bifida
19	/		spina bifida
20	/		spina bifida
21	/		spina bifida
22	/	41°	multiple malformations
23	/	23°	heart malformation
24	/	14°	normal
25	/	18°	normal
26	/		abortion; normal
27	/	14°	fetal malformation

The cases we gathered in Malta were of Neu-Lexova syndrome, a rare disease with a fatal prognosis, since all subjects are stillborn or die immediately after birth. The pathology is a genetic disorder inherited as an autosomal recessive trait characterized by severe intrauterine growth retardation, low birth weight and length. Abnormalities of the craniofacial region that may include marked microcephaly, sloping of the forehead, ocular hypertelorism and other malformations are also distinctive features. Further characteristics are the abnormal accumulations of fluid in tissues as a generalized edema, and permanent flexion with the immobilization of multiple joints, besides other limb malformations. Abnormalities of the brain, skin, genitals, kidneys and heart are also evident. All the biological samples we could examine presented

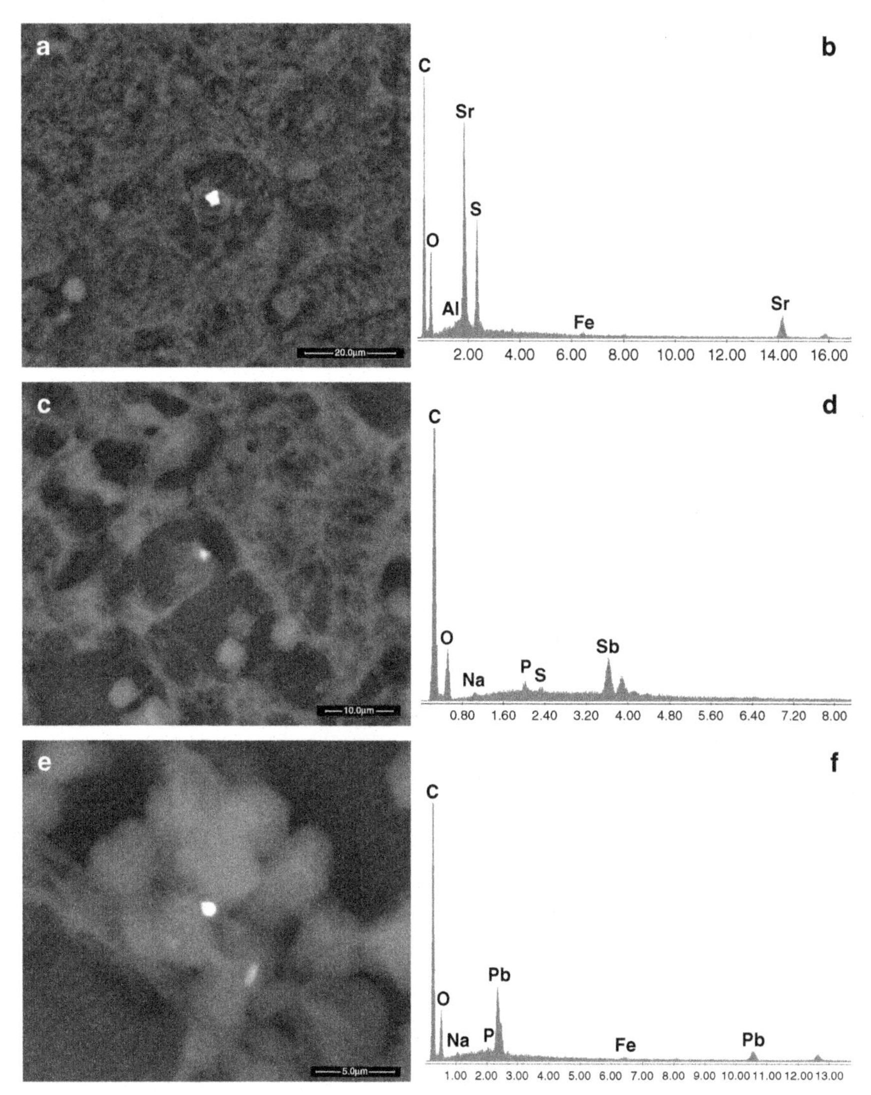

Figure 5.8 ESEM images of debris identified in the internal organs of malformed babies from Malta (a, c) and from Mantua (e). The particles are composed of strontium-sulfur-iron-aluminium (b), antimony (d) and lead (f).

particulate matter with different chemical composition, but all shared the presence of antimony particles. We speculated with the local anatomopathologists on the possible meaning of such an uncommon presence.

We thought it possible, as a hypothesis open to discussion, to attribute the cause to fireworks, as they are made to explode there with great frequency throughout the course of the year and everybody attends the fireworks, inhaling the residues.

The cases we got from Priolo (Sicily, Italy) (cases 16–20) were affected by neural-tube defects and they shared the presence of iron-sulfur particles. In the samples we analyzed we found particles composed of a variety of metals but the distribution of iron, silicon, aluminium and magnesium ($P < 0.05$, for all) was higher in neural-tube-defect fetuses than in the reference fetuses, i.e., the aborted ones. As an additional information, those Sicilian children had been conceived in an area contaminated by a large oil refinery.

As with the child from Mantua mentioned in section 5.8, the children from Priolo also had never been exposed to direct environmental pollution, but what we found in their internal organs can be referred to as indirect exposure due to a translocation from the mother's blood circulation. The interaction of developing embryos with pollutants could have prompted a disruption or, in any case, an alteration of their metabolism, but could have also represented a physical obstacle to normal growth. A submicronic particle could cause a submicronic lesion that grows as the fetus grows. As demonstrated, during pregnancy the fetus captures the inorganic and biopersistent pollutants, thus helping its mother clean her blood.

Evidence of a free entrance of debris through the fetal circulation and their entrapment inside the fetus or in the environment where he lives poses the question of whether we should continue to assume that all these malformations must be considered genetic, when it is obvious that, at least in part, they are due to incidental, contingent physical-chemical interactions during the developmental stage. In this scenario, cases like the one described as follows are understandable.

We examined the case of a baby born with a yolk-sac tumor, a rare malignant teratoma of the cells that line the yolk sac of the embryo. Physiologically, these cells become ovaries or testes, but the cancer can also occur in areas such as the brain or the chest or, as is the case we had a chance to observe, in the sacrum. Most often found in children before the ages of 1-2 years, their cause is one of the many classified as cryptogenic.

The object of our observation was a small collection of biopsic samples of the pelvis taken at surgery from an almost-one-year-old child (Figure 5.9).

All samples contained numerous submicronic particles whose chemical compositions included copper, iron and stainless steel (iron-chromium-nickel alloy). After approximately one year, the cancer reappeared in the coccyx and the subject was operated on again.

Unfortunately, we could not check the samples from the second surgery.

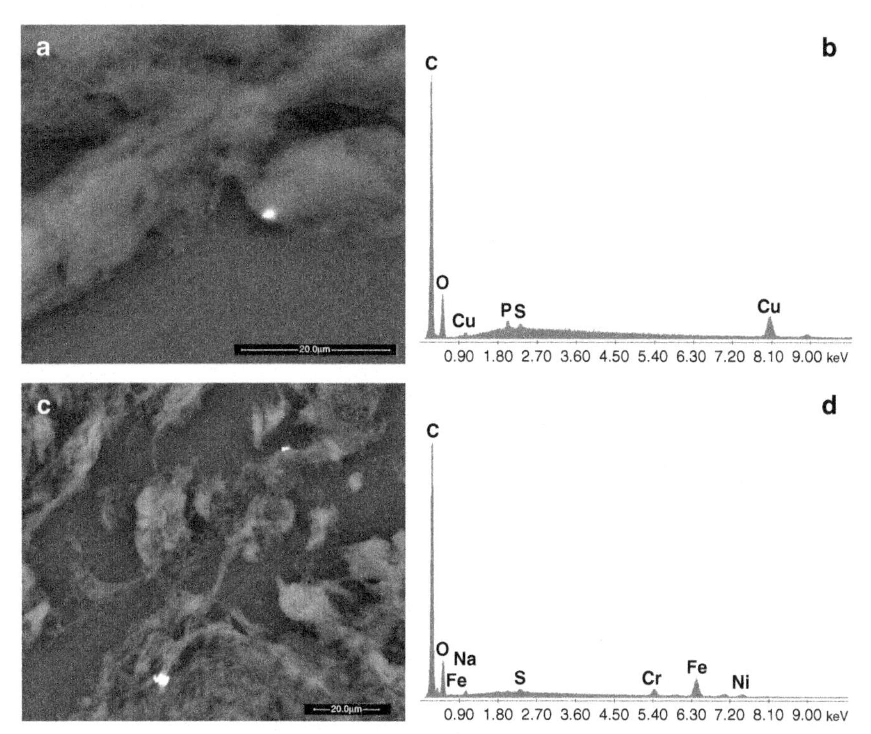

Figure 5.9 Images (a, c) show some of the debris found in the pelvis sample. Particles of copper (b) and stainless steel (d) are clearly identified.

As well as in all other cases of very young subjects, it is evident that the probability that they underwent a direct exposure to obviously exogenous particulate matter is substantial enough to have caused a cancer and, on top of it, a cancer with an unusually rapid indeed development is close to nil. Additionally, at least in this case, the origin was doubtless embryonic in nature. So, once again, because of the absence of any other reasonable explanation, we were facing a case of particles transmitted to the fetus by the mother through the fetal-placental circulation.

In a way, paraphrasing from the Bible, the phrase, "the sins of the mothers are visited upon the children" can prove true.

5.10 THE CHILD WITH BONE CANCER

An 8-year-old girl started to feel pain in her left arm and her father thought it was nothing more than a muscle sprain. But the pain did not go away and an X-ray finally showed that the left humeral diaphysis was occupied by an aggressive malignant neoplasm: an osteosarcoma.

Table 5.4 List of debris we found inside a biopsic sample of an arm bone

Particle size in μm	Debris chemistry
Standard of bone tissue	Ca,P,C,O,Cl,Na
12	Fe,C,O,Cl
0.8 μm	C,O,Ca,Pb,Si,Al,Cl,P,Cr,Ti
cluster	C,Ca,O,Si,Cl,Al,P
nanoparticles	C,Al,O,Pb,Bi,Cl,Cu,Na
nanoparticles	Fe,C,O,Cl
4	C,Ti,O,Si,Al,Cl,Ca,Fe,K,Mg
Debris with nanoparticles	C,O,Ti,Cl,Al,Ca

Inside the small fragment of bone taken during surgery and before the chemotherapy was started we found many foreign bodies as listed in Table 5.4.

What we identified were foreign bodies: metallic particles almost as usual, but, in this event, with very unusual compositions that demanded an explanation (Figure 5.10). Why were particles made of aluminium, lead, bismuth, copper or lead, chromium, calcium, titanium, silicon, aluminium or of titanium alloyed with other elements (aluminium, silicon, iron, etc.) found? Where did they come from?

We tried to make an anamnesis as accurately as possible, talking with the family, but what emerged from the longish discussions was that apparently there had been no clear exposure of the subject to any particular pollution source. Yet, those particles were of an indubitable exogenous origin.

The girl lived with her family near the sea in what did not look like a particularly polluted area, the only significant sources being a military airport in the vicinity of the house (3 km) and an international airport at 7 km.

The chemical composition we detected in the particles was strange indeed and could have been unrelated to the pollution let out by the airplane engines. Looking for a different explanation, we asked if the girl had been vaccinated (see Chapter 5), since, by a brief study of ours carried out then on 19 vaccines (now 26), we had identified lead and lead-bismuth particles in three samples. As a matter of fact, a vaccination had been issued about eight months before the diagnosis, but, because of the negligence of the doctor who was in charge of the vaccination plan, it was not possible to identify the brand name of the vaccine and, of course, its lot number.

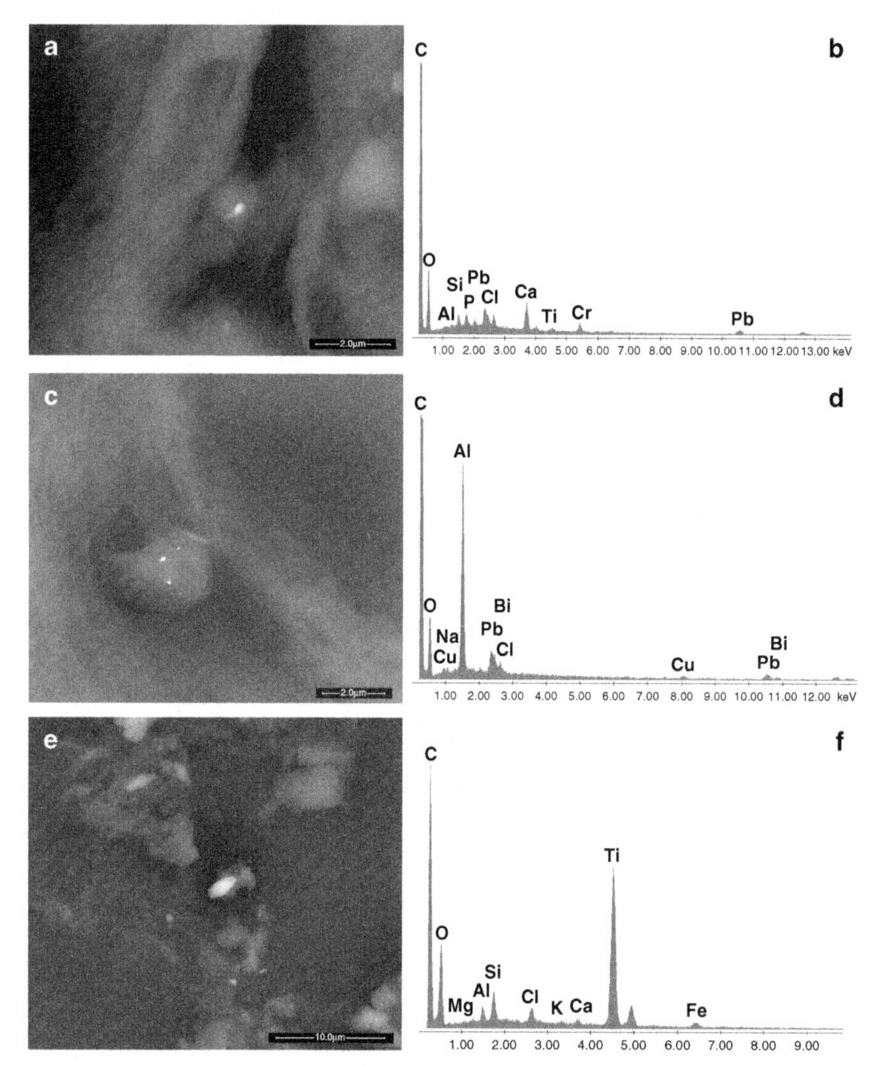

Figure 5.10 SEM images (a, c, e) of some micro- and nanosized particles we identified in the bone biopsy. Particles of alloy of lead (b), aluminium (d) and titanium (f) were identified.

Though unproven and not supported by any documented evidence, such a hypothesis seems worth consideration. The localization of the disease (close to the injection point and without any other systemic symptoms expressed) and the actual presence of particles in some vaccines as we can document ought not to be disregarded or rejected a priori.

5.11 THE CASE OF THE PATIENT KILLED BY REPEATED ENEMAS

An 83-year-old male patient was admitted as an emergency case to the hospital with colicky abdominal pain, vomiting and constipation. As a relief to constipation, the patient was given repeated hypertonic phosphate enemas, which resulted in a colonic perforation [33].

Phosphate intoxication and acute renal failure secondary to the massive absorption of phosphorus salified with sodium were further consequences. Hypotonic fluids were administered in the hope of solving the problem of anuria, but, after a 16-hour lethargy, the patient died [34–39].

Among other results, which, though important, are not relevant to our subject, post-mortem examination revealed hemorrhagic pulmonary edema, vascular congestion and perivascular edema of the brain, the liver and the spleen. The kidneys showed a focal tubulonecrosis.

What we saw with our usual electron-microscopy observations were large quantities of calcium-phosphorus crystals scattered in all samples we were given (Figure 5.11) (i.e., liver, heart, lung, kidney, brain and cerebellum). They cannot be considered as normal calcification, nor are they similar to the calcium phosphate of the bone, which is, in fact, a hydroxyapatite characterized by a calcium/phosphorus ratio of 1.67. In the crystals we found the ratio is inverted and, as can be seen in the EDS spectrum (Figure 5.11(b)), the height of the phosphorus peak is higher than the calcium peak.

The crystals that obstructed the renal tubules, thus justifying the renal failure, were too big to have entered into such small lumen. So, the only possible explanation is that they formed directly there as a simple supersaturation phenomenon of calcium and phosphorus ions.

Those ions, administered in large quantity and in hypertonic solution, altered their physiological concentration in the blood and inevitably interacted with the physiologically-present calcium prompting a precipitation. It must be remembered that calcium and phosphorus have a strong chemical affinity. Precipitation depleted the blood of calcium bringing about hypocalcaemia, which may reasonably account for the dehydration of the patient described by the clinicians: the hyperosmotic solution created by the chemical phenomenon attracted fluids to the peritoneum subtracting them from the other tissues. The formation and growth of the crystals resulted in the physical destruction of the structure, morphology and functionality of the liver, the kidney, the brain, etc.

That is an example of what happens when simple chemical and physical reactions prevail on the normal biological reactions and are not taken into due account.

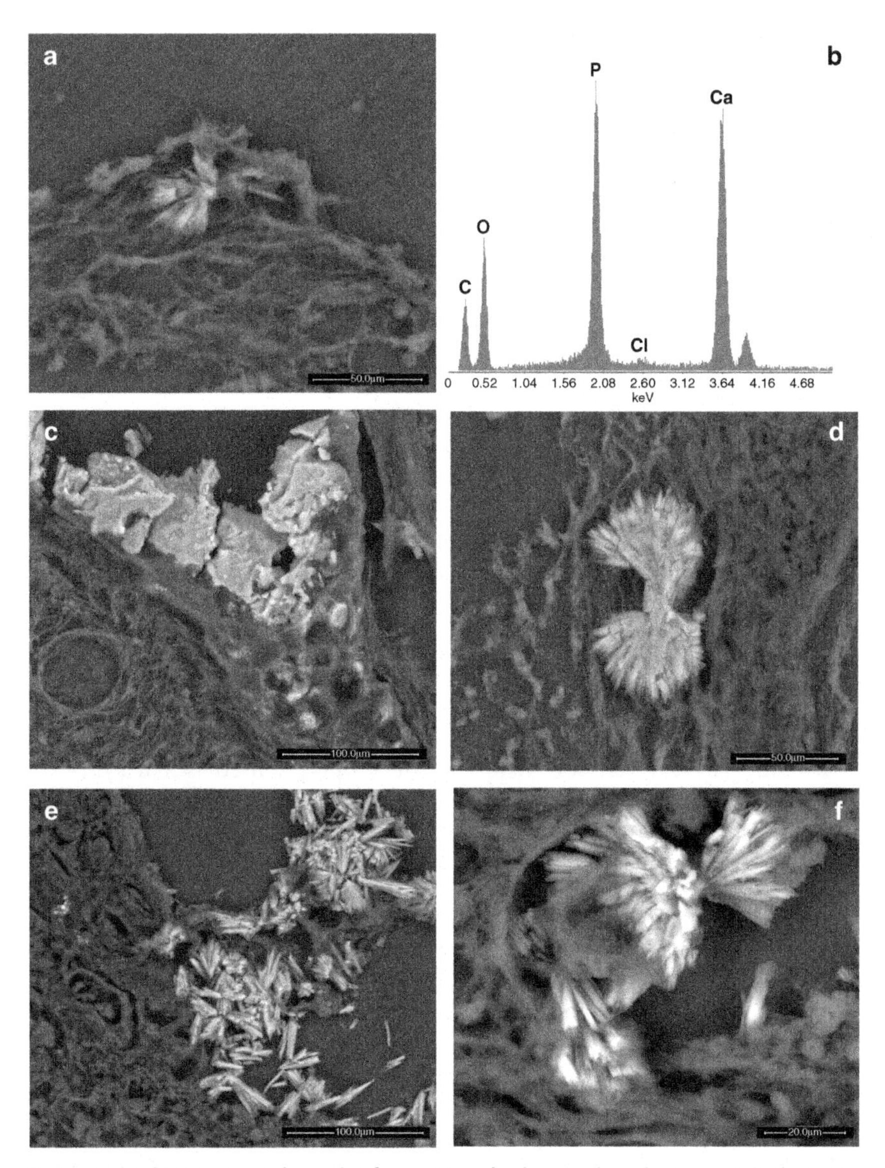

Figure 5.11 SEM images show the formation of calcium-phosphate crystals (b) in the saturated tissues: heart (a), lung (c), brain and cerebellum (d), liver (e) and kidney (f).

Particularly interesting are the calcium-phosphorus spherules we saw in the hepatic tissue (Figure 5.12). Their shape is evidently different from that of the multitude of crystals detected in the other organs, and their very likely origin is the reaction to inflammation we almost invariably see. As a matter of fact, the patient suffered from a form of hepatitis and the

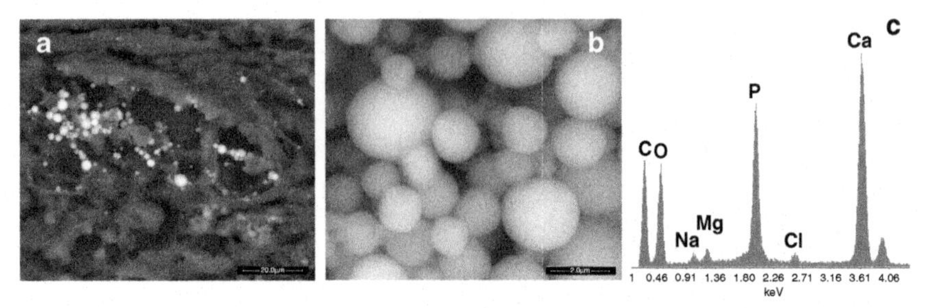

Figure 5.12 SEM images at low (a) and high (b) magnification of calcification in the liver. The calcification occurs in spherules (0.5-4 micron) composed of calcium-phosphorus-sodium-magnesium-chlorine (c).

Ca-P spherules were probably there much before the formation of the other crystals.

A discussion about those spherules will be found in the last chapter of this book.

5.12 THE BOY WHO PLAYED FIVE-A-SIDE FOOTBALL

We received a telephone call from a mother who requested some analyses for her 16-year-old son who had developed a mediastinal synovial fibrous monophasic sarcoma. Though the case was a very difficult one for a few reasons, one being the great extensiveness of the tumor, the boy was operated on successfully in a pediatric specialized hospital located in North Italy. The primary biopsy of the thoracic wall neoformation that had served for the diagnosis was the specimen we analyzed. We chose not to work on the surgical sample since the operation could be performed only after some cycles of chemotherapy that reduced the mass to be explanted but probably introduced foreign substances that for us are confounding factors.

The analyses of the tissue also revealed in this case the presence of dust. Because of the young age of the subject, it was particularly important to pinpoint the origin of the particles, so as to put him in a place to avoid further contaminations, probably, though not absolutely certain, which occurred mainly through inhalation, given the location of the pathology. Inside the tissue we found much particulate matter with different chemical compositions. The most represented particles (in many cases nanosized) were predominantly composed of calcium (for instance, Ca with C, O, Mg,

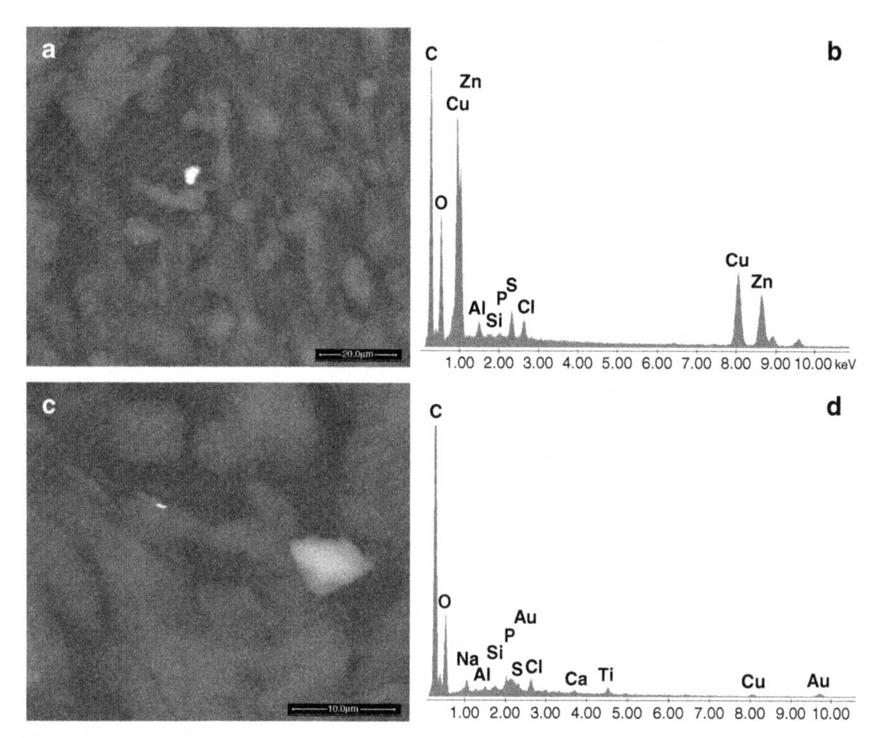

Figure 5.13 SEM images show micron (a) and submicronic ((c), left side) debris of copper-zinc-aluminium-silicon (b) and copper-gold-aluminium-silicon-calcium-titanium (d) embedded in the biopsy of the thoracic sarcoma.

Si, Al, Cl, Na, P, S, K, Ti,Fe, Mn) and of iron (mostly stainless steel), but we identified also debris of particles of zinc-copper, of silicates containing in some cases copper-gold-titanium and also tungsten-iron or tungsten alone (Figures 5.13 and 5.14). The quantity of toxic dust was sufficient to correlate its presence with the disease, but its actual origin was to be identified, not an easy task since the boy lived in Rome, a big city, but in an area with trees and away from industries. As to genetics, no one in his family had suffered from any particular disease, let alone cancer.

An accurate work of anamnesis was performed with the aid of the subject's mother. We learned that the boy had played five-a-side football from the age of six and practiced two or three times a week in a field where he played the Sunday game. The mother told us that the field was constantly covered with dust in order to prevent slipping during the game. A visit to the field revealed the actual presence of granules that we collected and

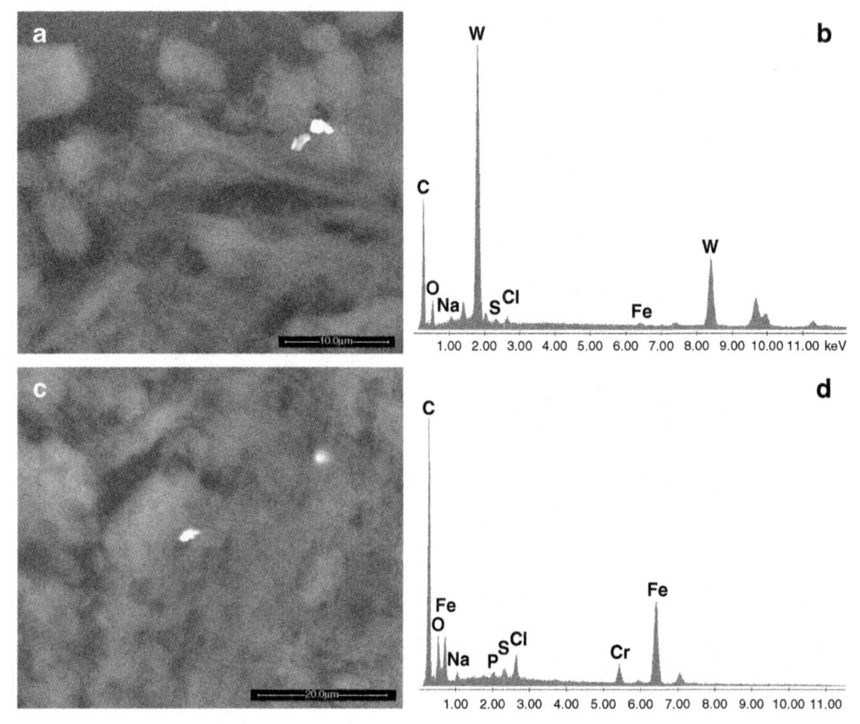

Figure 5.14 Images (a, c) of tungsten (b) and stainless-steel (d) debris embedded in the thoracic sarcoma tissue.

analyzed, even though we were informed that that kind of dust was a recent introduction and that in the past the field was covered with another type of material. The lady told us also that the previous coach had died of cancer and that there were rumors about diseases suffered by other young players, but we could not check in detail this information.

The analysis of the chemical composition of the particles identified inside the sarcoma posed some questions. The calcium-based particles were strange enough because of the quantity of elements present. They could come from the "normal" environment of just about any city, but their great variety and the presence of some elements such as titanium and manganese could be the result of a random combustion. A similar hypothesis can be put forward for the debris of copper-zinc with aluminium and silicon and of gold-copper-silicon-aluminium-titanium. Other kinds of investigations, not scientific ones, would be necessary to be sure about the origin of the dust. What we suspect is that the granules spread in the previous years on the field surface were not natural sand as claimed but ashes of an incinerator.

It is obvious, for example, that stainless steel, gold alloys and tungsten are not elements of any natural sand. So, though without final evidence, we think it possible to associate an exposure to the pollution of the field with the sarcoma the boy developed.

5.13 THE BOY WHO WENT INTO A SUDDEN COMA

A request for help for a 9-year old boy who went into a coma suddenly and without any apparent reason was sent to us. The symptoms experienced by the patient started as nasal-mucosa inflammation and itchy eyes followed by headache and, finally, coma. A frontal biopsy of the brain was performed and there was enough material for us to analyze. The normal histology did not offer any explanation, since the only evidence was signs of inflammation. Our usual observations gave more information: The biopsic area presented wide areas full of crystals of zinc-phosphorus and chlorine (Figure 5.15) or phosphorus-zinc-calcium-chlorine (Figure 5.16(a,b)). Those crystals looked as if they were of endogenous origin, but we could not understand what the source of zinc could have been and how the saturation could have occurred to prompt a precipitation. It seemed that the crystals had grown inside the brain, modifying its morphology and compromising its activity. Germs of the crystals are visible in Figure 5.16(a) as pale white points, but among these crystals there are also some debris (0.8 micron) of tungsten (or tungsten-carbide), whose presence is difficult to explain. We thought that it might be debris from the saw used to drill the skull bone, but we have no proof of that.

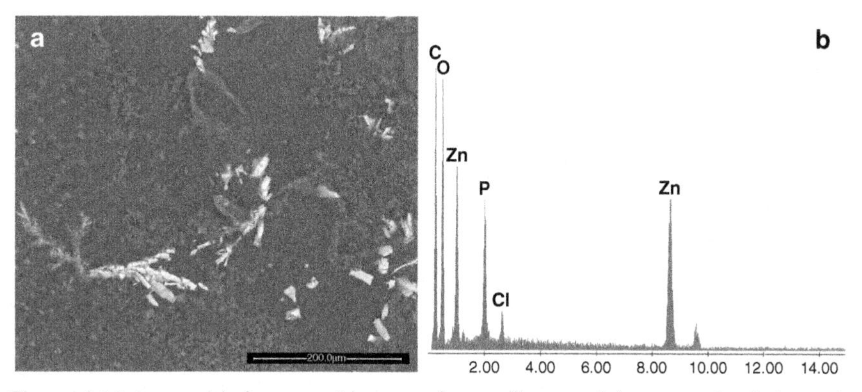

Figure 5.15 Image (a) shows a wide area of crystalline precipitates made of zinc, calcium and phosphorus (b).

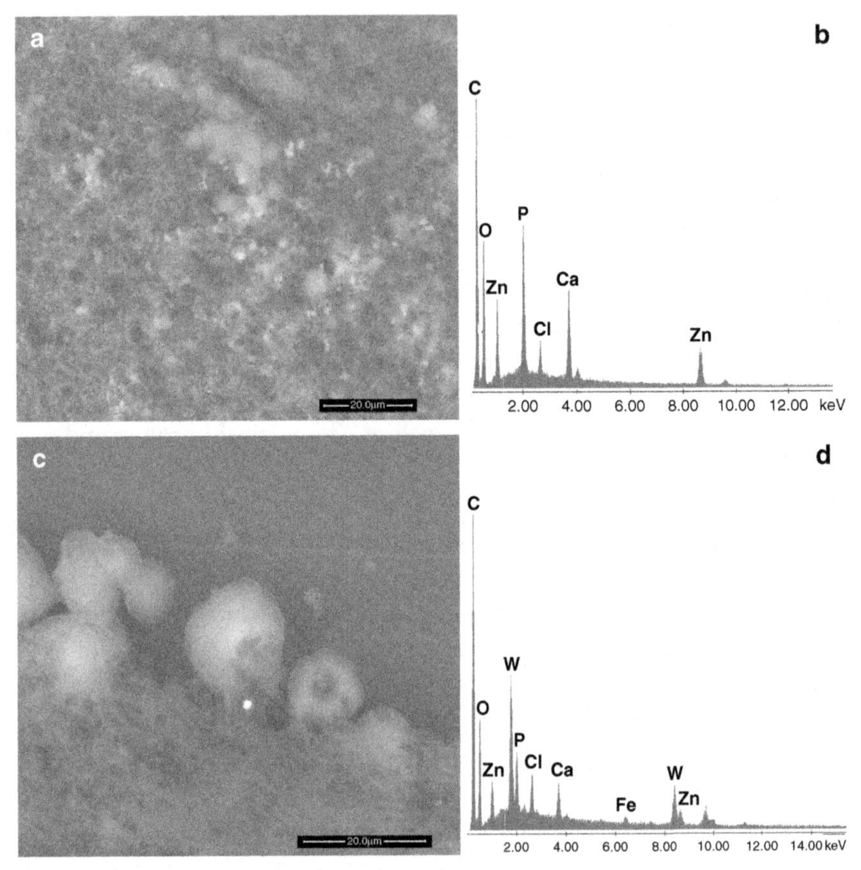

Figure 5.16 Image (a) shows an area of the brain full of crystallization germs (b). Image (c) shows a particle of tungsten-zinc-chlorine-calcium-iron (d).

What looks absolutely unusual is what probably remains of cells changed into shells of phosphorus-zinc-calcium and chlorine (Figure 5.17). It seems that a saturation concentration of those ions was reached inside the cells, so that crystals started to precipitate, occupying first the cells and then the whole area. This might be a further case, in a way similar to the one described in section 5.11, where, for reasons to be explained, the rules of simple chemical reactions prevail on the biological ones.

Since the young patient's house was very far from our laboratory, we made only an anamnesis and no direct investigation of the places where the patient lived and all the hypotheses we put forward lead us nowhere. So, this remains an unsolved case.

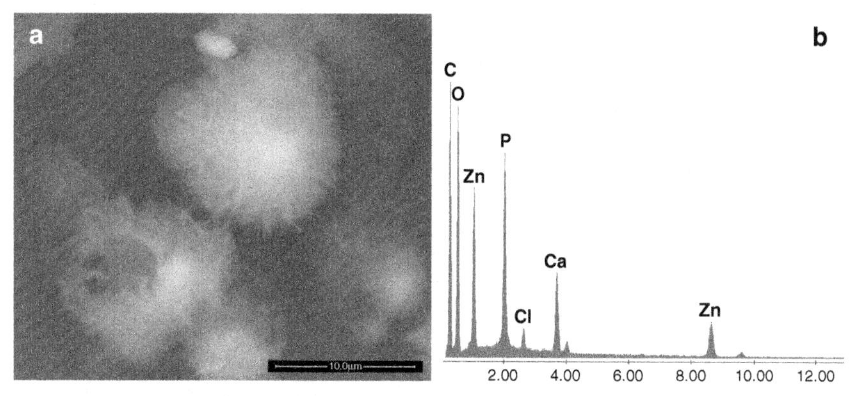

Figure 5.17 SEM image (a) shows transformed cells into zinc-calcium phosphates precipitates (b) that look like empty shells. The core of the sample is full of those precipitates, some of them even inside cells.

REFERENCES

[1] Environment Agency. Air Pollution in Europe 1990-2004. EEA Report 2/2007;1–80.
[2] Gatti A, Montanari S. Nanopathology: the health impact of nanoparticles. Singapore: PanStanford Pub; 2008. Cap.3: 72-79.
[3] http://www.eapm.eu.
[4] http://euapm.eu/what-does-personalised-medicine-promise/.
[5] Denk H, Scheuer PJ, Baptista A, et al. Guidelines for the diagnosis and interpretation of hepatic granulomas. Histopathology 1994;25:209–18.
[6] Anderson DS, Nicholls J, Rowland R, LaBrooy JT. Hepatic granulomas: a 15-year experience in the Royal Adelaide Hospital. Med J Austr 1998;18(2):71–4. 148.
[7] Brincker H. Granulomatous lesions of unknown significance: in biopsies from lymphnodes and other tissues: the GLUS-Syndrome. Sarcoidosis 1990;7(1):28–30.
[8] Friedland JS, Weatherall DJ, Leedingham JG. A chronic granulomatous syndrome of unknown origine. Medicine (Baltimore) 1990;69(6):325–31.
[9] Jacobs JJ, Gilbert JL, Urban RM. Current concepts reviews. Corrosion of metal orthopaedic implants. J Bone and Joint Surgery 1998;80A:268–82.
[10] Albores-Saavedra J, Vitch F, Delgado R, Wiley E, Hagler H. Sinus histiocytosis of pelvic lymph nodes after hip replacement. A histogenic proliferation induced by cobalt-chromium and titanium. Am J Surg Path 1994;18:83–90.
[11] Bos I, Johaninsson R, Lohrs U, Lindner B, Seydel U. Comparative investigations of regional lymphnodes and pseudocapsules after implantation of joint endoprosthesis. Pathol Res and Pract 1990;186:707–16.
[12] Urban RM, Jacobs JJ, Tomlinson MJ, Gavrilovic J, Black J, Peoc'h M. Dissemination of wear particles to the liver, spleen and abdominal lymphnodes of patients with hip or knee replacement. J Bone and Joint Surgery 2000;82-A(2):457–77.
[13] Ballestri M, Baraldi A, Gatti AM, Furci L, Bagni A, Loria P, Rapanà R, Carulli N, Albertazzi A. Liver and kidney foreign bodies granulomatosis in a patient with malocclusion, bruxism, and worn dental prostheses. Gastroenterology 2001;121: 1234–8.
[14] Darling AS. Iridium Platinum Alloys: a critical review of their constitution and properties. Platinum Metals Rev 1960;4(1):18–26.

[15] Timmons JW Jr, Burgert EO Jr, Soule EH, Gilchrist GS, Kelalis PP. Embryonal rhabdomyosarcoma of the bladder and prostate in childhood. J Urol 1975;113(5):694–7.
[16] Kapels KM, Nishio J, Zhou M, Qualman SJ, Bridge JA. Embryonal rhabdomyosarcoma with a der (16)t(1;16) translocation. Cancer Genet Cytogenet 2007;174(1):68–73.
[17] Pastore G, Peris-Bonet R, Carli M, Martínez-García C, Sánchez de Toledo J, Steliarova-Foucher E. Childhood soft tissue sarcomas incidence and survival in European children (1978-1997): Report from the Automated Childhood Cancer Information System project. European Journal of Cancer 2006;42:2136–49.
[18] Incidence des cancers à proximitè des usines d'incinèration d' ordures ménageéres. Institute de Veille Sanitarire. http://www.invs.sante.fr, 2006.
[19] Zambon P, Ricci P, Bovo E, Casula A, Gattolin M, Fiore AR, Chiosi F, Guzzinati S. Sarcoma risk and dioxin emissions from incinerators and industrial plants: a population-based case-control study (Italy). Environmental Health 2007;16(6):19.
[20] Comba P, Ascoli V, Belli S, Benedetti M, Gatti L, Ricci P, Tieghi A. Risk of soft tissue sarcomas and residence in the neighbourghood of an incinerator of industrial wastes. Occup Environ Med 2003;60:680–3.
[21] Viel JF, Arveux P, Baverel J, Cahn JY. Soft-tissue sarcoma and Non Hodgkin's Lymphoma clusters around a municipal solid waste incinerator with high dioxin emission levels. Am J Epidemiol 2000;152(1):13–9.
[22] Biggeri A, Catelan D. Mortality for Non Hodgkin Lymphoma and soft-tissue sarcoma in the surrounding area of an urban waste incinerator. Campi Bisenzio(Tuscany, Italy) 1981-2001. Epidemiol Prev 2005;29(3-4):156–9.
[23] Neglia JP. Epidemiology of childhood acute myelogenous leukemia. J Pediatr Hematol Oncol 1995;17(2):94–100.
[24] Reynolds P, Behren JV, Gunier RB, Goldberg DE, Hertz A, Smith DF. Childhood cancer incidence rates and hazardous air pollutants in California: an exploratory analysis. Environ Health Perspect 2003;111:663–8.
[25] Gatti A, Bosco P, Rivasi F, Bianca S, Ettore G, Gaetti L, Montanari S, Bartoloni G, Gazzolo D. Heavy metals nanoparticles in fetal kidney and liver tissues. Frontiers in Bioscience (Elite edition, E3) 2011; 1 (January):221-6.
[26] Bianchi F, Bianca S, Dardanoni Ga, Linzalone N, Pierini A. Congenital malformations in newborns residing in the Municipality of Gela (Sicily, Italy). Epidemiol Prev 2006;30(1):19–26.
[27] Vrijheid M, Loane M and Dolk H. Chemical environmental and occupational exposures, in: EUROCAT Special Report. The environmental causes of congenital anomalies: a review of the literature www.eurocat.ulster.ac.uk/pubdata.
[28] Glinianaia SV, Rankin J, Bell R, Pless-Mulloli T, Howel D. Particulate air pollution and fetal health: a systematic review of the epidemiologic evidence. Epidemiology 2004;15:36–45.
[29] Dolk H, Vrijheid M. The impact of environmental pollution on congenital anomalies. Br Med Bull 2003;68:25–45.
[30] Elliott P, Briggs D, Morris S, de Hoogh C, et al. Risk of adverse birth outcomes in populations living near landfill sites. BMJ 2001;323:363–8.
[31] Geschwind SA, Stolwijk JAJ, Bracken M, et al. Risk of congenital malformations associated with proximity to hazardous waste sites. Am J Epidemiol 1992;135:1197–207.
[32] Dummer TJ, Dickinson HO, Parker L. Adverse pregnancy outcomes around incinerators and crematoriums in Cumbria, north west England, 1956-93. J Epidemiol Community Health 2003;57:456–61.
[33] Bell AM. Colonic perforation with a phosphate enema. J Roy Soc Med 1990;83:54–5.
[34] Driman D, Preiksaitis HG. Colorectal inflammation and increased cell proliferation associated with oral sodium phosphate bowel preparation solution. Hum Pathol 1998;29:972–8.

[35] Beloosesky Y, Grinblat J, Weiss A, Grosman B, Gafter U, Chagnac A. Electrolyte disorders following oral sodium phosphate administration for bowel cleansing in elderly patients. Arch Intern Med 2003;163:803–8.
[36] Pitcher DE, Ford RS, Nelson MT, Dickinson WE. Fatal hypocalcemic, hyperphosphatemic, metabolic acidosis following sequential sodium phosphate-based enema administration. Gastrointest Endosc 1997;46(3):266–8.
[37] Marraffa JM, Hui A, Stork CM. Severe hyperphosphatemia and hypocalcemia following the rectal administration of a phosphate-containing Fleet® pediatric enema. Ped Em Care 2004;20(7):453–6.
[38] Orias M, Mahnensmith RL, Perazella MA. Extreme hyperphosphatemia and acute renal failure after a phosphorus-containing bowel regimen. Am J Nephrol 1999;19:60–3.
[39] Desmeules S, Bergeron MJ, Isenring P. Acute phosphate nephropathy and renal failure. New Eng J Med 2003;349(10):1006–7.

CHAPTER 6

Environmental Cases and Nanoecotoxicology

Contents

6.1 The case of a power plant 99
6.2 Contamination around urban incinerators 109
6.3 The case of the incinerator of Terni 116
6.4 Contamination by engineered nanoparticles 122
References 127

The sources of environmental pollution are growing more numerous and complex. Those involving particulate matter generated by combustive processes can be found in the environment characterized by size, morphology and chemical composition of the particles they emit, whose impact on vegetable, animal and human life, of which the environment is an essential component, can be linked to the results obtained from research.

6.1 THE CASE OF A POWER PLANT

Between the northern Italian regions of Veneto and Emilia Romagna, at the delta mouth of the river Po, there is a wetland formed by marshes, channels, lagoons and the Adriatic Sea. In the early 1990s, by political decision, the territory became a regional park. In the 1970s, before the park was established, on the island of Polesine Camerini, at the east side of the territory, an electric power station managed by the Italian company ENEL had been built in which heavy oils were burned at a maximum rate of 621 tons/hour [1].

For a long time the people living in the park and in its vicinity had complained about the fallout of oily drops that, according to them, spoiled fruit, vegetables, hanged-out washing and even corroded the body of cars. The power station, being the only industry in a relatively large radius, was often pointed out as responsible and, on the insistence of a group of people, the local criminal court started an investigation.

As mentioned above, the plant was fueled with heavy oils, a highly impure derivative of petroleum, containing, besides carbon, hydrogen and

Case Studies in Nanotoxicology and Particle Toxicology
http://dx.doi.org/10.1016/B978-0-12-801215-4.00006-6

oxygen, other elements such as lead, sulfur, barium, strontium, antimony and iron, in most cases present in different mutual combinations. After having been burned, part of those elements remains in the ashes, but part of them, randomly combined, contributes to form micron-sized and submicronic particles that are released into the environment through the stack along with vapors and gases, inevitably polluting the surrounding territory. Duly identified and classified, those particles looked for in the environment can be excellent markers, enabling to track their source.

In 2005 the cognizant public prosecutor asked us to analyze 54 different samples with the aim of seeing if they contained micro- and nanoparticles whose origin could be traced to the plant, which in that period was inactive. Those samples were the heavy oils used as fuel, ash from the combustion of those oils, filters used by the local environment protection agency to filter the air, lichens, lettuce leaves, a mattress cover and a t-shirt that, both, had been hung out to dry outdoors during one of the not uncommon shutdowns due to malfunctions of the power plant, when unburned oil drops were dispersed in the environment and fell down.

Analyzed with our usual technique, the samples of heavy oils showed a content of particles ranging from about 200 nanometers to about 40 microns in size. The chemicals found were lead, iron-sulfur, barium-sulfur-strontium, antimony, iron compounds, tin compounds, bismuth compounds, vanadium compounds and silicates (of aluminium, titanium and iron). In many instances the particles made of iron-sulfur, of barium-sulfur-strontium and of lead were roundish or perfectly spherical. The samples of oil declared to be free of sulfur showed a lower concentration of particles and no barium-sulfur-strontium, vanadium and antimony debris.

Table 6.1 lists some of the debris we identified in the ashes and their chemical compositions. The same elements were identified in the oil drops.

Table 6.1 List of some of the particles contained in the oil ashes listed by size, morphology and chemical composition

Analysis	Debris size in μm	Chemical content
1	40	C, S, O, Mg, V, Si, Fe, Na, Ni, Al
2	16 μm cluster (1-5 μm debris)	Fe, O, Ni, Al, C, S, Mg, Cr, V, Si, Zn
3	40	Fe, S, O, C, V, Mg, Ca, Cr, Ni, Na, Si
4	Ca-S rods	S, Ca, O, C, V, Fe, Mg
5	2-10	S, Ba, C, O, V, Mg, Sr, Fe, Ca, Ni, Na, Pb
6	4-20 rods	V, S, O, C, Ni, Fe, Mg, Al, Si, Ca, Na
7	40	C, S, O, Mg, V, Si, Fe, Na, Ni, Al

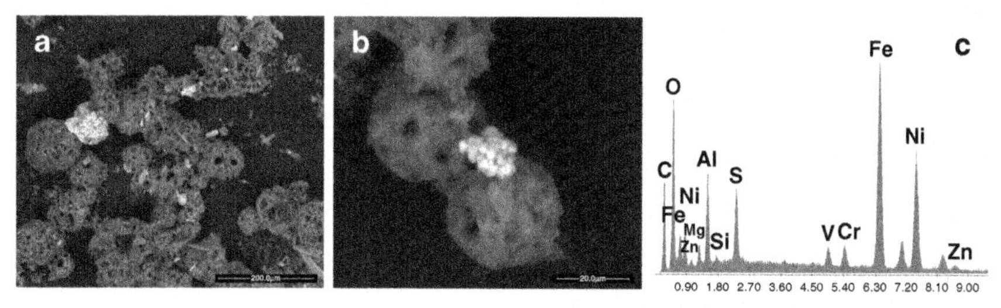

Figure 6.1 Micro-images of spherical particles (a, b) found in the ashes of the burned heavy oils, with the EDS spectrum (c) of one of the inorganic debris visible in the sample. The debris contains iron, oxygen, nickel, aluminium, carbon, sulfur, magnesium, vanadium, chromium, silicon and zinc. Note the spherical shape of all particles.

The elements detected in the ashes (iron, lead, vanadium, barium, aluminium and sulfur) were the same as those in the oils. The differences in composition of the particles from those seen in the oils were due to the heat of the combustion that broke the original particles into their elemental components, which, then, combined again to give origin to particles made of the same elements, but in different compositions. Chromium and nickel, which we found alloyed with iron, were not present in the oils and their being in the ash is ascribable to the substantial corrosion phenomena affecting the plant, phenomena largely due to the sulfur of the fuel.

Apart from carbon, which was obviously expected, what we found in the six samples of ash we checked was a great quantity of inorganic debris, often contained in carbon matrices and often in a globular form or in clusters of smaller, irregularly-shaped particles that could be fragments of larger particles. Figure 6.1 shows an example of the debris in the ashes in which organic (grey) and inorganic (white) metallic particles are visible. The debris appears spherical, porous and rather large. Some rods of barium sulfate can be seen, as well as other aggregated spherules of heavy metals.

Figure 6.2 shows that other debris are mixtures of carbonaceous and metallic particles fused together. Also, in this case the combustion melts together the components of the oil or part of them and new materials are formed.

We considered only the high-density (white) particles composed of heavy metals, ruling out the carbonaceous debris (the spherical porous particles), since they are poor markers for the traceability method we use. In fact, carbonaceous exogenous particles inside a biological matrix essentially composed of carbon are hard for us to identify.

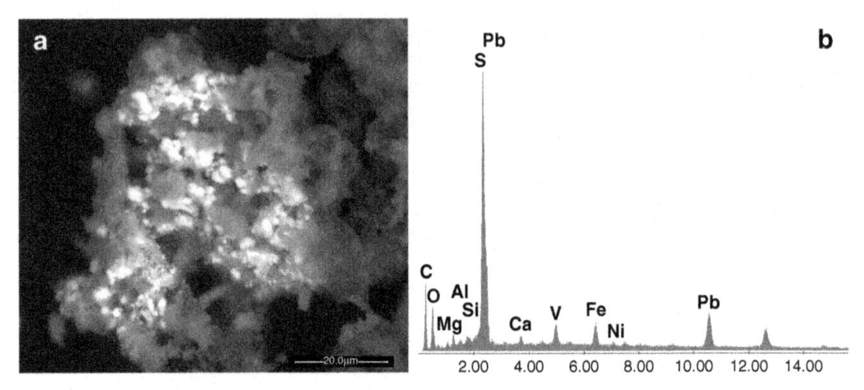

Figure 6.2 ESEM image of a 60-micron-sized particle (a) of a carbonaceous matrix with 0.8-5-micron debris of lead-sulfur-iron-vanadium-magnesium-aluminium-silicon-calcium and nickel (b) entrapped.

The study of the lichens was quite easy, because brown spots of oily carbon-based substances containing inorganic white debris (Figure 6.3) were visible without any difficulty, in many circumstances even by the naked eye. Those debris were of different compositions, in all cases not compatible with the environment of a natural park: vanadium-calcium-magnesium-zinc-aluminium-silicon-phosphorus-sulfur-potassium-lanthanum-iron, bismuth-lead-sulfur and iron-sulfur-magnesium-silicon-potassium. The presence of elements other than those we detected in the oil can be ascribed to obvious further interactions that the pollutants issued by the oil combustion have inevitably with the natural and anthropic environmental pollution.

The analyses of nine samples of lettuce leaves proved somewhat more difficult. Nevertheless, it yielded significant results. The elements we had found in the oil were present in the leaves in different combinations. That, along with the shape of the particles detected, confirmed that the origin of those debris was the power plant (Figure 6.4). It may be of some interest to report that we identified many different compounds containing lead, iron-chromium-potassium and barium-sulfur-potassium-calcium-strontium, etc., on the leaves, while others were perfectly clean. That is because some of the lettuce specimens had been collected in a greenhouse where no particulate pollutant fallout could have occurred, and for that reason, they were free from particulate matter's contamination.

The t-shirt we received had been washed and hung to dry outdoors when it was involved in one of the fallouts of oily droplets. At the center of the oily stains (containing sulfur) we saw numerous micro- and nanoparticles, many of them roundish or spherical, composed of heavy

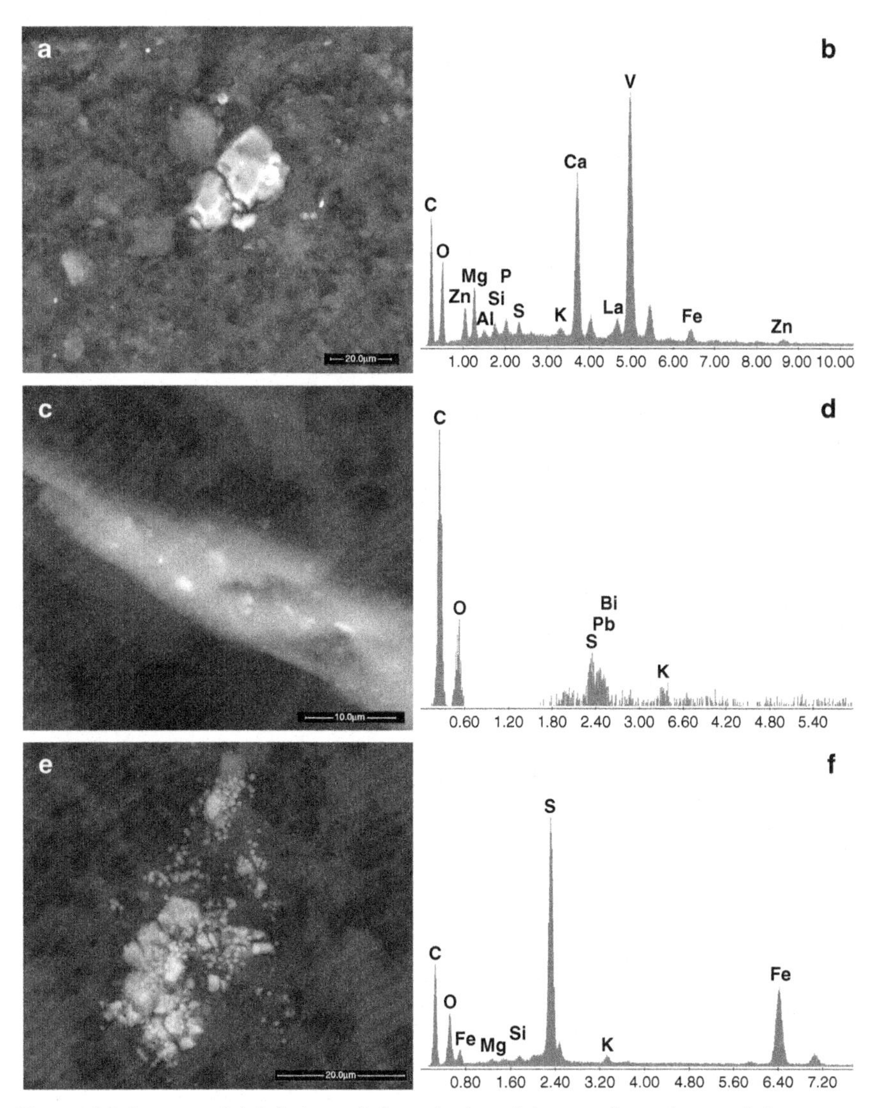

Figure 6.3 Images of debris (a, c, e) deposited on lichens collected around the power plant. In some cases (c, e) the particles are spherical.

metals (iron, nickel, strontium, vanadium, lead and bismuth). The reason why they were contained in oil is probably attributable to an occasional, not particularly rare, incomplete combustion of the fuel used by the power station (Figure 6.5).

Also the mattress cover, hit by the fallout at a considerable distance of time from the t-shirt, showed oily spots containing particles. In that case, the

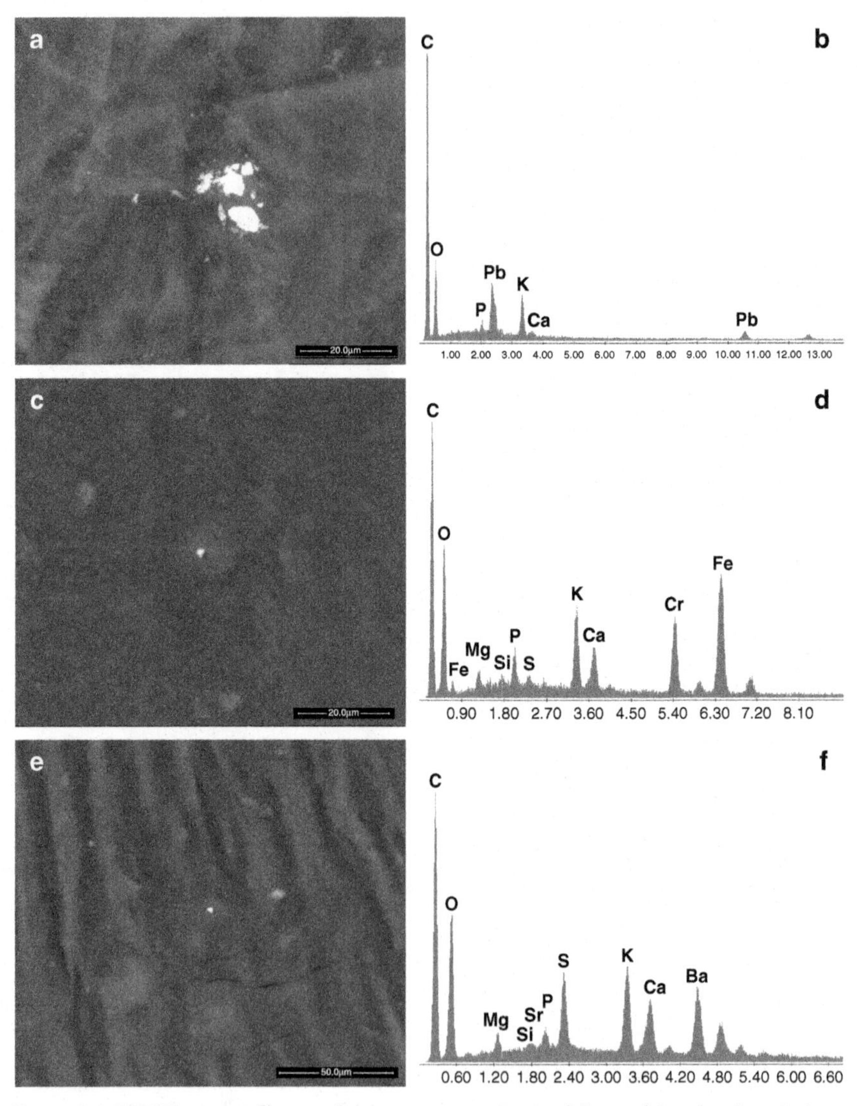

Figure 6.4 ESEM images of lettuce leaves grown under the fallout of the plant's emissions. Images (a, c, e) show lead (b), stainless steel (d) and barium-sulfur (f) debris.

particles, again roundish and spherical, were mainly composed of iron and sulfur but contained also vanadium.

The filters of the air pump used by the Environmental Protection Agency we were given had captured many spherical particles with a content of sulfur-iron, lead, tin-sulfur-barium and sulfur-calcium: elements we had also detected in the heavy oils.

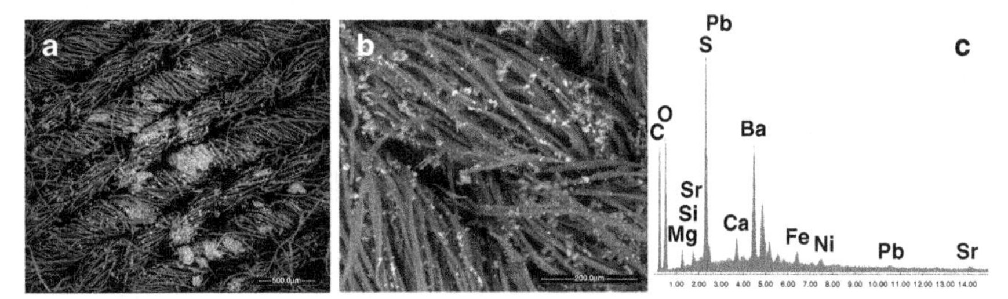

Figure 6.5 ESEM images (a, b) of the t-shirt sample exposed to the environmental pollution of the power plant and stained by oily drops containing particles of lead-sulfur-barium-iron-nickel-strontium-silicon-magnesium (c).

Since the particles detected are the result of a random combustion and a sort of chaotic chemical post-recombination, they show a complex elemental composition and each one can be composed of many elements, in our experience up to 12-14 or, though rarely, even more. That is a peculiarity that must be taken into due consideration when a polluting source must be identified and, therefore, the characterization of particulate matter dispersed in the environment is necessary. After combustion, the various elements present in the fuel are found as accidental alloys in micro- and nanoparticles whose composition may vary. Metals that have a melting point higher than the temperature at which the carbon compounds present in the process break tend to separate from them and form metallic particles. Generally speaking, all those elements can be found in the residues of combustion. Some of them, particularly the heavier among those elements, enrich the ashes more, while the lighter are more prone to be liberated into the atmosphere and combine to form condensable particles.

Combinations of those elements are what we detected in the lichens and in the other samples checked. Having found particles composed of up to 11 such elements, and, in addition to that, along with the unmistakable roundish or spherical shape, are witnesses to their combustive origin. No alloys like those exist as either natural or man-made products and – the limit being our experience – none of them was ever found in urban or other varieties of pollution.

Excluding very rare, larger samples, most of the particles we found were in the inhalable range (4-10 microns) and in the respirable range (below 4 microns). Often clusters could be seen formed by submicronic and nano-sized particles.

The location of the plant, inside a park where car traffic was very scarce, and human settlements were quite small, and other polluting sources were virtually absent, together with its relatively-to-the-context huge size, made

Table 6.2 List of pathological cases analyzed and the relative diseases of subjects living in the area around the power plant

N. cases	Type of diseases
1	Adenocarcinoma, prostatitis and pulmonary neoplasia
9	Goitre, thyroiditis, adenoma and carcinoma
1	Kidney neoformation
2	Liver cancer and liver neoformation
3	Lung carcinoma
2	Brain neuroblastoma, schwannoma
1	Ovarian cancer
7	Neoplasia and adenocarcinoma of sigma, stomach, colon and ulcerous colitis
1	Dermatitis with keratosis
2	Hodgkin's and non-Hodgkin's lymphoma
1	Breast neoplasia
1	Pancreas neoplasia
1	Bone cancer

our research easier. The shape and size of the particles in conjunction with the few elements constantly present in obviously different combinations were excellent markers and convinced us that the power plant should be held responsible for the pollution the citizens complained about. The court at first instance found ENEL responsible for the pollution.

This judgment was upheld on appeal by the Court of Venice. ENEL then applied to the Court of Rome, Italy's highest Court and the final avenue of appeal, but its application was dismissed.

If, as sentenced, that power plant had caused an irreversible pollution to a territory, it was only reasonable to think that its inhabitants could have suffered harmful consequences as, in fact, many of them maintained. In fact, the public prosecutors actually gave us biopsic and post-mortem pathological samples taken from local patients [2]. Table 6.2 shows the list of the pathological samples investigated with the frequency of each disease.

Eventually, we analyzed 35 specimens from people affected by and, in some circumstances, that died of diseases that could tentatively be attributed to the pollution generated by the plant. All the cases we dealt with had been declared cryptogenic as to their origin.

All samples had been taken years ago to be analyzed according to methods that had nothing to do with those peculiar to nanopathology, so some samples were not suitable for our electron-microscopy analysis, since they were stained or did not contain the interface between the healthy and pathologic area, the one most indicated for the detection of particles. The cases were selected by the public prosecutor without any preliminary anamnestic investigation to verify the actual possibilities of exposure to the plant contamination the patient could have undergone.

With the limitations mentioned above, 32 samples could be analyzed and two of them, a dermatitis with keratosis and a pancreas cancer, showed no particles, which is only natural, since, as is obvious, particles cannot be blamed for any disease. But their absence can also be due to a (from our point of view) incorrect sampling not performed in the right anatomical place or to a cancer developed from cells that had been penetrated by particles and had reproduced according to the mutation they had undergone. When the latter is the case, the cancerous tissue cannot contain any particles.

In nine cases the particulate matter detected was mainly composed of calcium phosphate, a classical biological response to an inflammatory reaction. It cannot be excluded that those calcifications had formed around metal particles, something that we could verify at least in four patients. In the first of them, a case of neuroblastoma, iron particles were identified; in another case (thyroiditis), we spotted tin and copper debris; in the third case (thyroiditis), there were tungsten particles; and in the fourth case (breast carcinoma), we saw copper and zinc debris. Besides iron, the other metals could not be considered typical of the pollution coming from the power plant, even if they confirmed that an exposure to a pollution was still present.

All other samples contained particles that proved rather homogeneous as to composition, shape and size, the like of which had been found very often in what had been analyzed for the trial.

Figures 6.6–6.8 show some pathological tissues with embedded particles of heavy metals.

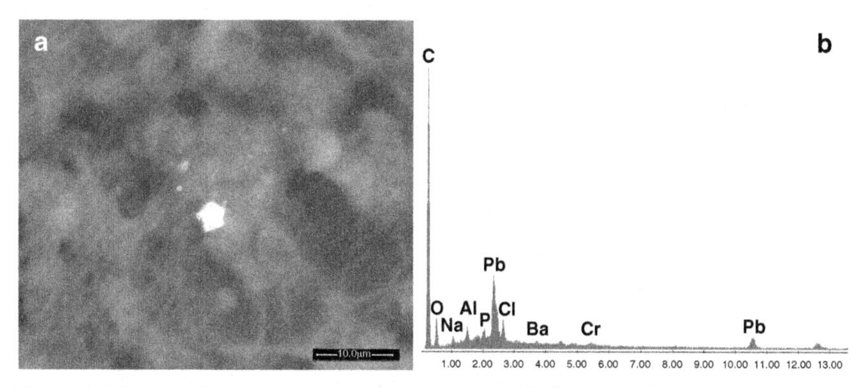

Figure 6.6 Image of micro- and nanoparticles embedded in a lymph node in a patient affected by Hodgkin's lymphoma (a). The debris are composed of lead, chlorine, aluminium, phosphorus, barium and chromium (b).

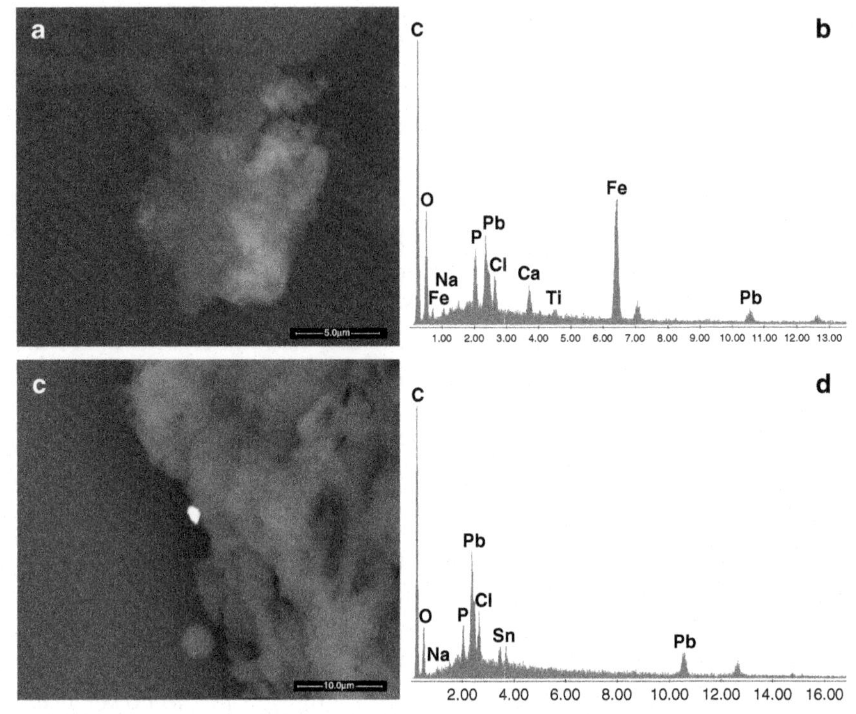

Figure 6.7 Images of debris identified in the gastric mucosa biopsy (a) and the prostate biopsy (c), both tissues being inflamed, in the same patient. In the first image (a) a cluster of nanoparticles of iron, lead, phosphorus, chlorine, calcium, titanium and sodium (b) is visible on the left part of the tissue. Image (c) shows a micrometric particle of lead, chlorine, phosphorus and tin (d).

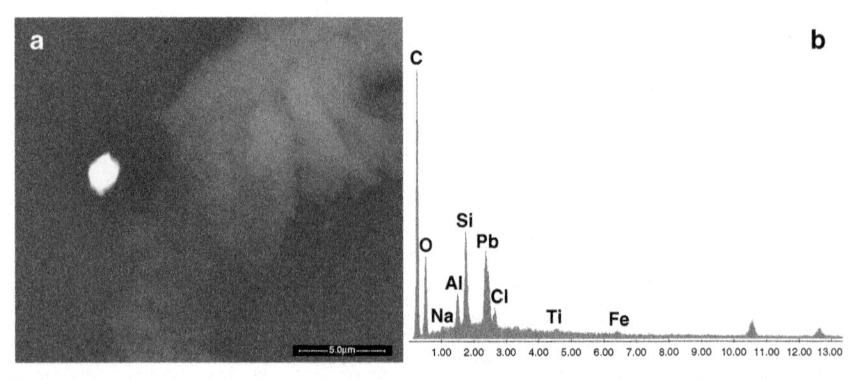

Figure 6.8 Image of a two microsized debris (a) of silicon, lead, aluminium, titanium and iron (b) identified in a lymph node of a patient affected by gastric adenocarcinoma.

It is obvious that the method described can be used to check the pollution coming from other combustive sources such as, for example, incinerators, in whose case traceability can be harder, though absolutely possible, to assess, since the materials burned are far from homogeneous, unless those plants are dedicated to a single type of waste, something that applies only to some cases of industrial incineration and is not particularly common. Lack of homogeneity is one of the characteristics to be taken into account when municipal waste incineration is at stake. In any case, tracing the source of particulate pollution requires the availability of a sufficient quantity of samples, as well as patience and experience. Even harder is work with human subjects.

Today people move very frequently and, as a consequence, they are exposed to numerous, often very different, types of pollution. To make things more complicated, food must be considered as well. Our diet consists only partially of local products, those products that could be polluted by local sources. Much of our food and many ingredients come from very distant places and often bring along pollutants that have nothing to do with the environment of residence, thus introducing confounders. As a matter of fact, at least for those reasons, the results of epidemiology surveys carried out around a particular plant to assess its possible responsibility in causing a specific pathology or a group of pathologies, though in a way indicative of a situation, should be considered with the utmost prudence and never be taken alone. In addition to that, we have also come across questionable or thoroughly misleading results of epidemiology studies.

Just as an example: industrial plants issuing ultrafine particles were considered harmless because the incidence of pulmonary pathologies affecting people living in their vicinity did not exceed what is considered to be expected in a "normal" environment taken as reference. As a matter of fact, ultrafine particulate matter does not linger in the alveoli and in a matter of tens of seconds reaches the bloodstream, thus not being able to affect the lungs.

On the other hand, pathological conditions typical of that kind of pollution are not taken into consideration: male sterility, miscarriages, fetal malformations and neuro-endocrine pathologies among them.

6.2 CONTAMINATION AROUND URBAN INCINERATORS

Man is the only "energivorous" tenant of this planet and now, with the rapid increase of population – at this time in excess of 7 billion, while when Columbus landed in America the Earth hosted less than half a billion people, 1.65 billion in 1850, risen to 2.5 billion in 1950 – and with the at least as rapid increase of demand of an "easier life," energy sources present in (in and not only on) the planet start visibly to dwindle away. The vast majority of energy is still extracted from the Earth, i.e., from an isolated system containing a finite quantity of energy resources that sooner or later are obviously bound to run out. This is true for many reasons, side by side with the demand for energy that the production of waste has demanded, and more visible in the so-called third world, which, in its eagerness to imitate the richer countries, is copying their worst faults, neglecting their virtues, whatever they are.

A huge amount of waste – especially the easily burnable fractions made up of plastic materials that, being obtained from oil are, in a way, fossils – can be used to produce energy. Burning that material, transformed into rubbish, is the easiest way to both have it disappear (but the principle of conservation of mass states something quite different) and to obtain at least part of the energy it contains, a squeezable energy actually much smaller than the one that had been used to produce it.

While some more illuminated countries are trying, in fact successfully, to reduce the production of urban waste and to forbid the marketing of products that prove to be "unfriendly" toward the environment, others are not so successful and in some instances do not even try to be.

In those countries, incinerating waste has been seen as the solution to the problem, and a solution with the bonus of producing energy, at that. Next to municipal waste incinerators, discarded items from hospitals and industrial scraps are burned in dedicated plants, and cement works, with their high operating temperatures, are being currently used to dispose of virtually any kind of waste, legal limitations being minimal and their compliance not properly strict, with the residual ash mixed with their final product.

It is inherent in the process that incineration involves high-temperature treatments and it is obvious that, according to the principle of conservation of mass, mass remains unchanged independently of the techniques used and the chemical and physical conditions adopted. Actually, the outcome of incineration is roughly twice the mass of the waste burned because of the additions needed for natural and technical reasons (atmospheric oxygen, water, sodium carbonate and bicarbonate, ammonia, activated carbon, methane, etc., according to the system employed).

If organic compounds can be modifiable with comparative ease by chemical and physical reactions and ideally transformed into more or less harmless substances – though literature is not particularly reassuring on this subject: at least not always – analogous modifications aiming at obtaining harmless scoriae from inorganic materials is far less possible.

Anticipating what the outcome of an incinerator will be is something that can be done only with a non-negligible margin of error, since, according to the kind of waste treated (municipal, hospital, industry waste) and, in each class, their inhomogeneous and, in some cases, ever-changing composition, the pollutants generated and, as a consequence, dispersed in the environment, are the most varied. As far as particulate matter is concerned, filtration of the fumes as is currently done is not particularly efficient, since it acts only on primary filterable particles above a certain size and is useless for primary condensable and secondary particles. For that reason, assessing the risk to human health that the technique implies not only should not be neglected but must be done with the utmost care [3–11].

Smaller plants burning biomasses from which electric and thermal energy is obtained are also being built. Strictly speaking, biomass is biological material belonging to or produced by a living organism. So, forest and agriculture residues, yard clippings and wood chips can be rightfully called biomass provided they are consumed in a time that is longer than or, at most, equal to the one they need to grow again to the same size. And animal, in particular poultry, droppings are also biomass. But in some, in fact very frequent, circumstances, biomasses are not available in a sufficient quantity. So, urban waste and other items such as, for instance, used tires, are lawfully accepted as a sort of surrogate. For that reason, in many instances, it is hard to see a real difference between a biomass and regular waste incinerators if we set aside size which, in the case of biomass plants, is generally much smaller.

Even "uncontaminated" vegetables contain elements that are not carbon, hydrogen and oxygen alone, but are substances coming from the soil and, as a consequence, depending on the soil chemistry and on what falls on the plants, be it something administered on purpose such as, for example, pesticides, or something that drops on them by accident like the many pollutants present in the atmosphere. It is true that their concentration is relatively low and, therefore, the inorganic primary particle pollution they cause is not particularly important. Nevertheless, it is there and accumulates.

Just as an example of what can be found, what follows is a summary of our analyses carried out on filters of the air pump positioned onboard our aero-model (Figure 6.9) that flew over the incinerator of Modena (Italy). In this case the location of the plant is in the outskirts of the city in a place where there are no other industrial emissions. Other results reported briefly are those obtained by analyzing the surface of cherry leaves picked at a short distance from the incinerator, a plant that also burns special toxic wastes.

Figure 6.9 Photo of the aero-model with remote control we constructed when we took part in the European project DIPNA to detect particulate matter in air. The arrow indicates the inflow of the air in the internal pump where a filter entraps the debris.

It is necessary to remember that this pollution is strictly related to what was burned in the days immediately before the analysis. In other periods with other waste to be treated, pollution could be different.

Figure 6.10 reports the morphology and chemical compositions of some of the dust identified on a filter after a 15-min flight around the incinerator stack on a day when no smoke seemed to be emitted (usually most smoke emissions occur during the night) (Table 6.3).

Table 6.4 shows the list of debris identified on leaves grown at a distance of 200 m from the incinerator. The compositions identified are rather complex. We report just some of the many kinds of debris issued by that stack. Some show the presence in combination of rather unusual elements (e.g., hafnium) possibly coming from the combustion of toxic waste processed in that plant. Figure 6.11 shows a selection of particles identified on some of the leaves analyzed.

It is obvious that the pollutants detected, many of them of submicronic size, can be inhaled, breathed or ingested by people who live in the area affected by the emission fallout.

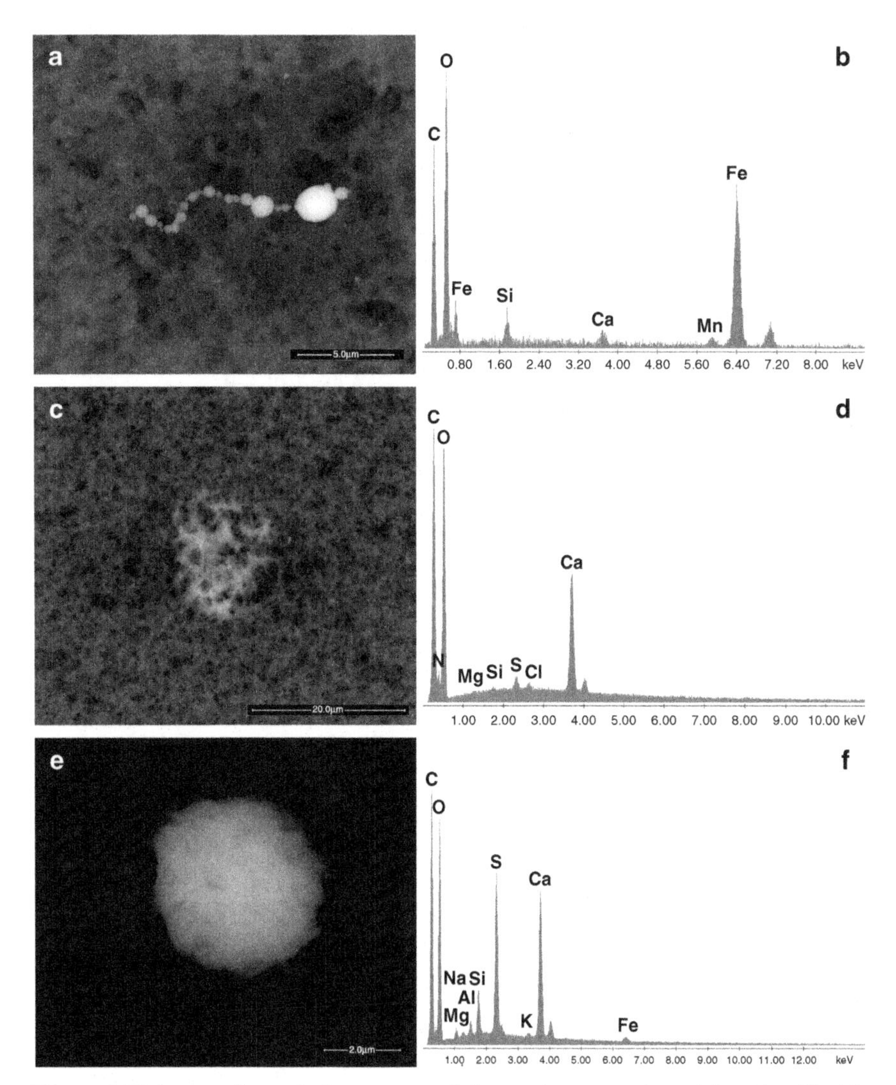

Figure 6.10 Images (a, c, e) show some of the environmental particles identified in the filter of the air pump mounted onboard the aero-model. (a) shows a sequence of oxygen-carbon-iron-silicon-calcium-manganese micro- and nanoparticles (b); (c) is a porous spherule of carbon-oxygen-calcium-sulfur-chlorine-magnesium-silicon (d); (e) is a 5-micron carbon-oxygen-sulfur-calcium-silicon-sodium-magnesium-aluminium-potassium (f) spherule.

Table 6.3 List of elemental composition of debris identified on the air-pump filter after a 15-min flight around the waste-incinerator stack on a day without apparent smoke

Analysis	Debris in μm	Elemental composition
1	aggregate	C, O, Ca, N, S, Cl, Si, Mg
2	5	C, O, P, K, S, Cl, Mg, Si
3	nano	C, O, Si, S, N
4	aggregate 3.5	O, C, Si, Al, Fe, Mg, K, N, S, Cl, Ti
5	spherule 4	C, O, Ca, N, Cl, K, Mg, Si
6	1	O, C, Fe, N
7	0.7	C, O, Fe, Si, Cu, S, N
8	0.5	C, O, Fe, S, Si, Cu
9	spherule 5	C, O, S, Ca, Si, Al, Na, Mg, K, Fe
10	5	Fe, O, C
11	0.5	O, C, Si, Al, Mg, K, Ca, Fe, N
12	0.5	C, O, Si, Zr, N, Al, Fe
13	0.3	C, O, Pb, N

It is not surprising that the pollution caused by that incinerator could have consequences on the people living in its vicinity.

A couple of elderly farmers who lived close to the urban waste incinerator of Modena (Italy) suffered from a collection of apparently unrelated pathological symptoms. Besides that, it is interesting to report that they had to stop raising rabbits because the animals did not reproduce anymore.

Table 6.4 List of elemental compositions of debris identified on leaves around the waste-incinerator plant

Analysis	Debris in μm on leaves	Elemental composition
1	4	C, O, K, Si, Ca, Mg, Al, P, S
2	2	C, O, Fe, K, Ca, Cr, Si, Mg, Al, P, S, Ti
3	3	C, Fe, O, Si, K, Ca
4	0.2-10 μm steel spherules	Fe, O, Cr, C, K, Ca
5	cluster of steel	Fe, O, Cr, C, K, Ca
6	6	C, Zr, Si, O, Hf, K, Ca, Fe
7	11	C, O, S, Fe, K, Cu, Si, Zr, Mg, Ca, Ba, Zn
8	cluster of nanoparticles	C, Fe, O, Si, K, Ca
9	aggregate	Si, C, O, Al, K, Fe, Mg, Ca, Ti, S
10	2.5 μm spherule	O, Fe, C, K, Ca, Si, P
11	aggregate of spherical particles - 0.5	S, Fe, O, C, Si, K, Ca
12	aggregate	Pb, S, C, O, Ba, Si, K, Ca, Al, Fe, Cu, Na

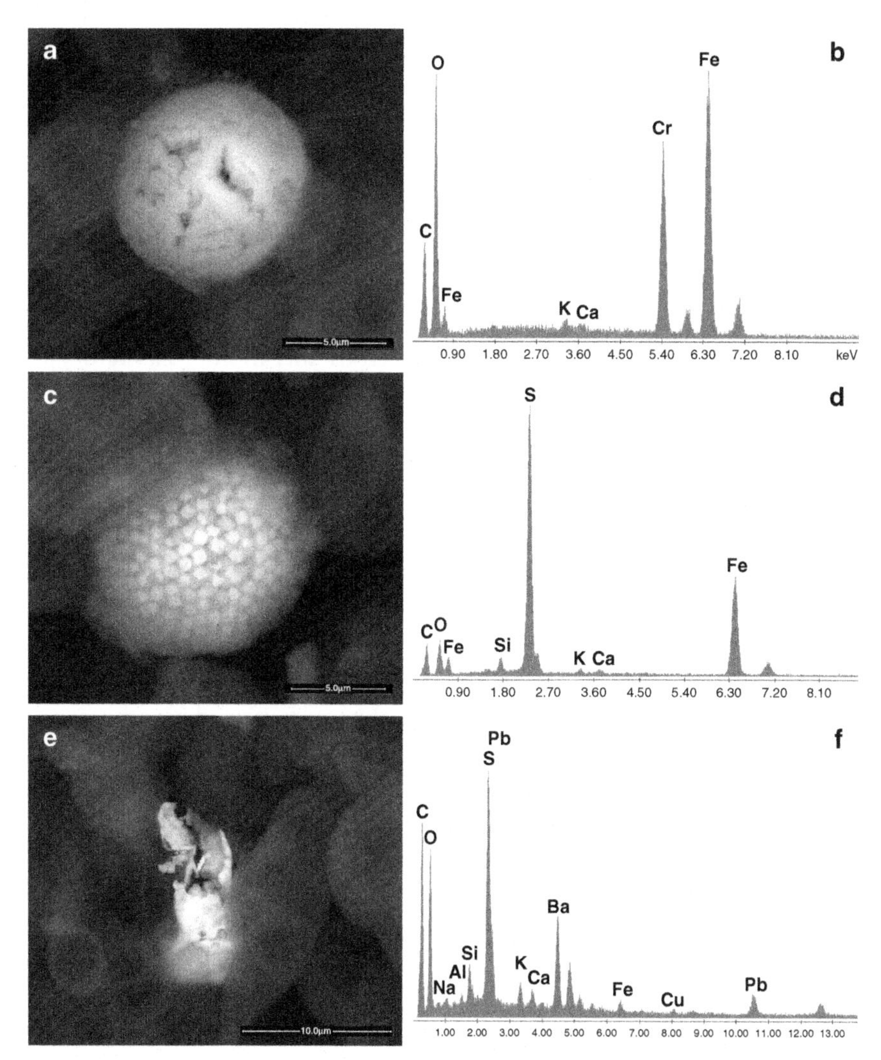

Figure 6.11 Images (a, c, e) show other debris found on a leaf grown at a distance of 200 m from the incinerator. (a) shows a stainless-steel spherule (b) with crystalline grains; (c) shows a sulfur-iron particle (d) composed of spherical grains; (e) shows debris typical of a random combustion (f).

The most seriously ill of the two was the husband who, among other pathologies, suffered from a basal-cell nodular carcinoma of the nose. The histological specimen of the cancer showed the presence of many debris with a size ranging from 1 to 8 microns, mostly of metals such as iron-copper-sulfur, barium-sulfur, iron-cerium, bismuth-chlorine. Besides cerium, two more rare-earth elements, lanthanum and praseodymium, were identified. So, the hypothesis that he had inhaled and ingested the pollution generated by the combustion of waste could be logical, and just as logical is the possibility that the particles were the cause or, in any case, a very import contributory cause, of the cancer.

6.3 THE CASE OF THE INCINERATOR OF TERNI

Terni is a small industrial town in central Italy. Although it has just over 100,000 inhabitants, there is more than one source of industrial pollution there. Also in this case, the local court asked us and another laboratory to check the pollution produced by one of the sources, the municipal waste incinerator, and to evaluate its impact on the environment, the workers and the population. Our task was, of course, to check particulate pollution. We were also requested by the local court to evaluate a great number of analytical documents about the plant, its procedures and its emissions. The result was that the incinerator was impounded by the court because of its non-observance of the legal emission limits of NOx, SOx and particulate matter. Then, we were asked to analyze a considerable number of samples collected directly at the incinerator site, and, after that, we were asked to analyze pathological tissues taken from local people affected by cancer.

So, we analyzed the dust taken in the shed where untreated waste was stored before being moved to the furnace room, the bottom ashes and leaves of plants growing around the incinerator.

The ashes collected in the furnace were composed of many, whole spherical particles, often nanosized, being they much less fragile than larger spheres, in single and aggregated form. Those aggregates were roughly spherical and highly porous. Iron, silicon and lead were the most common elements, but they were alloyed in various combinations with titanium, barium, chromium, nickel, zirconium, manganese, aluminium, zinc, tin, vanadium, strontium, antimony, chlorine, etc. We also found particles composed just of barium and sulfur and others of silver and chlorine. We also identified a particle containing thorium, a radioactive element. Virtually identical was the dust collected on the surface of leaves of vegetables grown in the vicinity of the incinerator. It is important to make it clear that what we observed in that particular instance was not much more than a sort of snapshot of the momentary situation of the incinerator, as the composition of the dust and ash produced by the plant depends on what is occasionally being fed into the furnace, and what is being fed is far from homogeneous and constant. Figure 6.12 shows some of the particles we identified in the ashes.

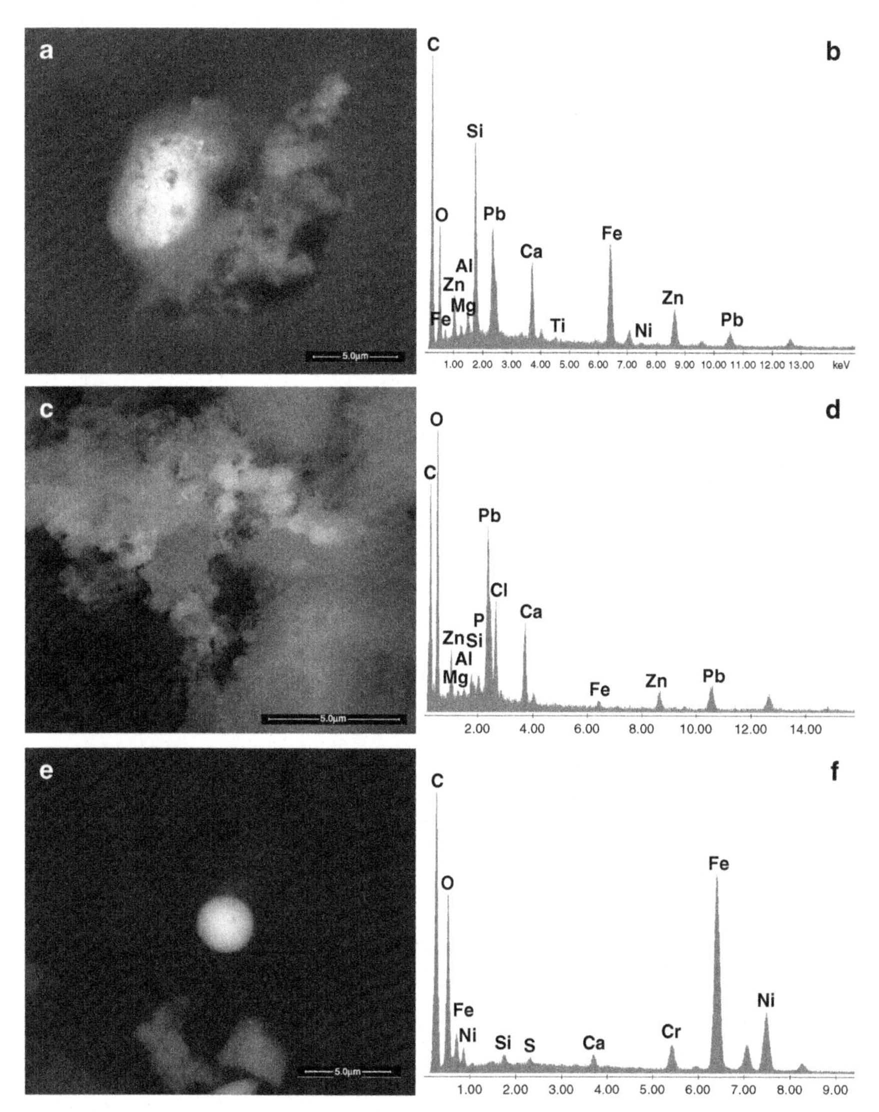

Figure 6.12 SEM images of some particles identified in the ashes of the waste incinerator. (a) shows dust composed of carbon-silicon-lead-iron-oxygen-calcium-zinc-aluminium-magnesium-potassium-titanium-nickel (b); (c) shows an aggregate of nanoparticles composed of oxygen-carbon-lead-chlorine-calcium-zinc-silicon-aluminium-magnesium-phosphorus-iron (d); (e) shows a 4-micron spherule of stainless steel (f).

With the data collected in this survey, we began to analyze the human pathological tissues we were given. That is always the most critical and delicate part of the work, since these analyses represent the exposure the patient underwent and can show a causative link between emissions, their toxicity and pathology. For that reason, whatever the case, the results are invariably challenged by a bevy of lawyers. In that instance, we analyzed 15 cases and for every one we had often more than one tissue available. In fact, it is frequent that a patient is affected by a primary pathology accompanied by other, multiple, systemic symptoms. This multifaceted aspect of some pathologies is almost as a rule neglected by medical doctors who concentrate their attention on the major pathology and disregard other symptoms that they believe to be of minor importance or that do not belong to their specialty. That means that the full picture is often lost. In some cases, dust can be the origin of totally different pathologies. Once we came across a subject who was affected by intestinal cancer, and in his biopsic samples we found certain particles. The same patient, though, died of a dissecting aortic aneurysm and in the specimens of his aorta we found concentrated the same particles. A hypothesis that, in our opinion, should be considered is a mechanical origin of the aneurism: those particles were pushed off from the left side of the heart with a certain, non-negligible, speed and, being rather hard, could have scratched the aortic arch. If this was the case, the same pathogen could have generated two thoroughly different pathologies. Certainly not always, but in some circumstances, in our opinion not that uncommonly, thinking of particulate matter as the origin of a disease or a syndrome can offer an objective explanation to all the symptoms and help to find a solution. Starting from the "entrance door" of that particular pollution (nose, mouth, vagina [see Chapter 4, the dirty sperm], drug injection, etc.), following the possible ways of dispersion in the organism and considering the possibility of a local accumulation with its ensuing symptoms, a once "mysterious" pathology may become much more understandable and, as a consequence, easier or, anyway, more possible to treat successfully. The 15 cases analyzed are listed in Table 6.5.

Figure 6.13 shows spherules of rare-earth elements of combustive origin found in a bronchioloalveolar carcinoma that could be probably related to the incineration of some special waste. Figures 6.14(a,c) presents some of the numerous particles identified in a surgical sample affected by bladder carcinoma. The images show 3-5-micron-sized aggregates of lead and stainless-steel debris. Their presence in the bladder could be witness to their first passage through the kidneys. Figure 6.15 shows submicronic and

Table 6.5 List of pathological cases investigated

Case	Sex	Organ	Pathology
1	M	Lung	Bronchioloalveolar adenocarcinoma
2	F	Breast	Ductal breast Carcinoma
3	M	Intestine	Intestinal Adenocarcinoma
4	M	Kidney	Adenocarcinoma
5	M	Anus	Adenocarcinoma
6	M	Sigma	Adenocarcinoma
7	M	Parotid	Acinic cell carcinoma
8	M	Bladder	Papillary carcinoma
9	M	Bladder	Papillifer carcinoma
10	M	Bone marrow	Myelodysplastic syndrome (chronic myelomonocytic leukemia)
11	M	Brain	Subependimoma – (IV ventricle neoplasia)
12	M	Skin	Bowen's Disease
13	M	Skin	Bowen's Disease
14	M	Skin	Squamous-cell carcinoma
15	M	Skin	Skin fibrosarcoma

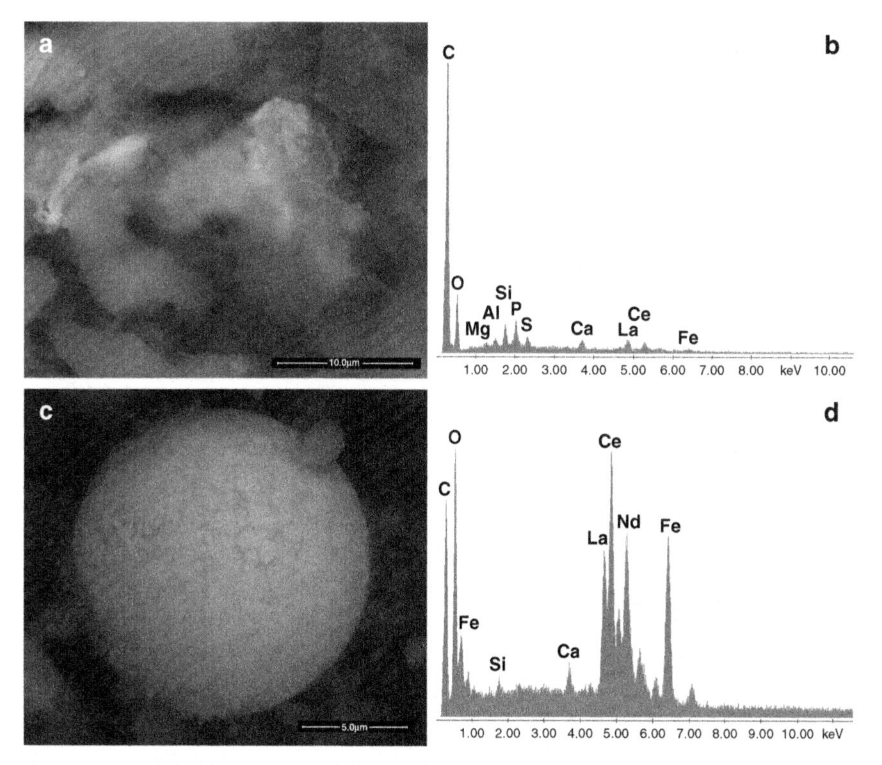

Figure 6.13 ESEM images (a, c) of debris identified in the bronchioloalveolar carcinoma. Particulate matter in a spherical shape composed of cerium-lanthanum-neodymium-iron-calcium-silicon-aluminium was also identified (b, d).

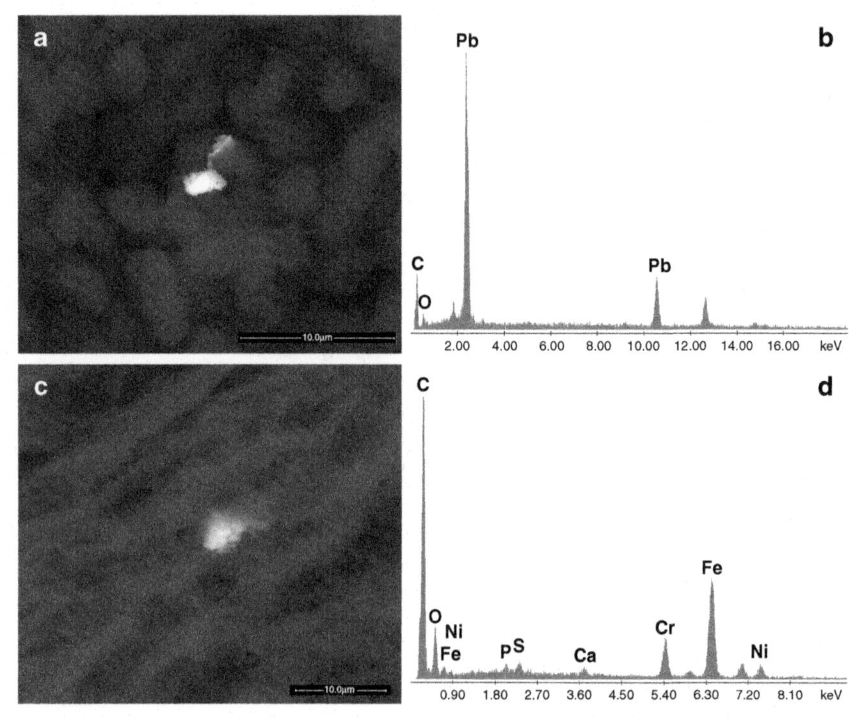

Figure 6.14 Images (a, c) show aggregates of submicronic particles with different compositions identified in a tissue affected by bladder adenocarcinoma: lead (b) and stain-steel (d).

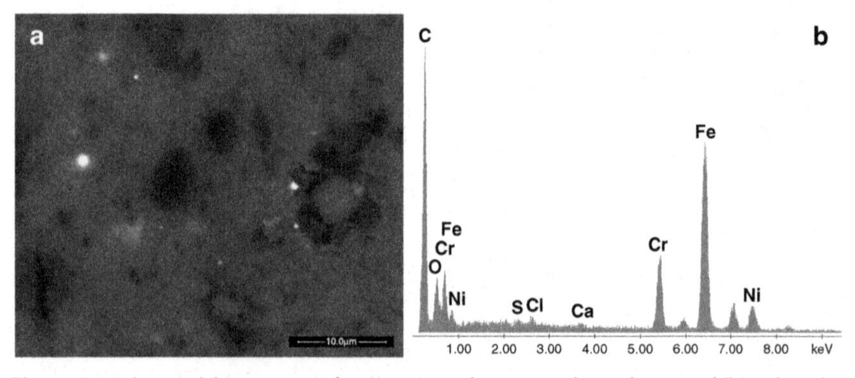

Figure 6.15 Image (a) presents submicronic and nanosized stainless-steel (b) spherules in the anal tract tissue affected by adenocarcinoma.

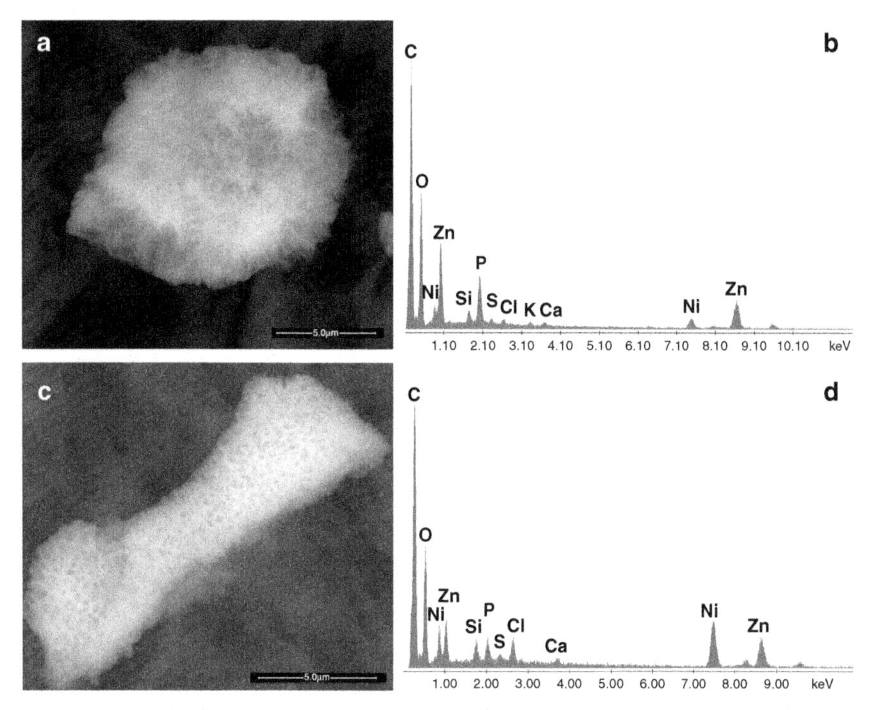

Figure 6.16 ESEM images (a, c) show unusual organic/inorganic particles identified in the subependimoma. They are composed of carbon-oxygen-nickel-zinc-phosphorus-sulfur-chlorine (b, d).

nanosized stainless-steel spherules in a carcinoma tissue of the anal tract. This location could support the hypothesis of an origin by ingestion, maybe through eating vegetables and fruit grown under the pollution fallout. We cannot exclude that the ingestion of animals fed with polluted grass could induce similar effects.

Figure 6.16 shows an unusual organic/inorganic presence in the brain subependimoma. These entities are composed of nickel-zinc-phosphorus-sulfur-chlorine. The compositions of the two debris are similar, but the ratio among the elements is different; this aspect suggests a combustive origin. A comparison could be made with the case of the child who fell into a coma (see Chapter 5).

Also in that case, similar inorganic/organic crystallization was visible, inducing us to think of an accumulation of zinc, phosphorus and, in this case, nickel ions, that, exceeding the saturation concentration, precipitated as insoluble particles.

6.4 CONTAMINATION BY ENGINEERED NANOPARTICLES

As part of a national research project on the theme of nanoecotoxicology, a two-room greenhouse was set up (Figure 6.17). Rice, basil, tillandsia, micro-tom, i.e., a dwarf cultivar of tomato (*Solanum lycopersicum L.*), and moss where grown in both rooms, each of which also hosted insects (*Ceratitis capitata*) and earthworms. One of the rooms was intentionally polluted with cobalt, nickel, silver, titanium-oxide and cerium-oxide engineered nanoparticles dispersed in the confined atmosphere after a combustion process, sprayed in a very fine aqueous suspension and added to the water necessary to keep the soil moist. Because of that pollution we wore special clothes (Figure 6.18). The other room was kept clean [12].

As expected, we had to deal with the usual behavior of nanoparticles that show a strong tendency to aggregate, thus becoming microparticles, which are useless for nanotoxicity tests. Part of the experiments concerned aquatic tests with marine-animal models such as sea urchins (*Paracentrotus lividus*), brine shrimp (*Artemia salina*), zebra fish (*Dania rerio*) and barnacles (*Balanus amphitrite*).

Figure 6.17 Two-room greenhouse built to assess the toxicity of engineered nanoparticles dispersed in the air and soil. The small wood box on the right simulates a small incinerator burning nanomaterials whose fumes are driven to one of the rooms.

Figure 6.18 Protection mask with air supplemented to the operators working inside the polluted room of the greenhouse.

The results of this three-year project can be summarized in nine points:

1. Once dispersed in the soil, titanium-oxide as well as cerium-oxide nanoparticles induce an abnormal root over-proliferation in tomato, rice and basil plants, probably due to a coalescence of the nanoparticles that occlude the root pores. Figure 6.19 shows the reduced growth of rice plants exposed to titanium-oxide nanoparticles compared to the other plants [14].
2. In a test in which rice plants were infected by *Xanthomonas oryzae pv. oryzae* (xoo) before being exposed to nanoparticles, that infection causes a yellowing of the leaf apex. The extension of the lesion depends on the virulence of the pathogen. The plants grown in the presence of cobalt nanoparticles added to the water and dispersed in the soil presented reduced lesion lengths related to the non-exposed rice plant, suggesting an inhibitory role (or toxic effect) of cobalt nanoparticles on xoo growth and rice infection. In this case, we see a positive effect. In other experiments carried out with other nanoparticles, we noted a lesser growth of the treated plants than of the reference ones [13].
3. The tests carried out on the micro-tom plants showed an uptake of nanoparticles through the roots and their translocation to aerial organs including leaves. Titanium-oxide nanoparticles induced an evident effect on the root morphogenesis by stimulating hair formation, an effect

Figure 6.19 Rice plants after addition of nanoparticles in the supplement water. The first box (foreground) contains the untreated plants, the second one those treated with cobalt nanoparticles, the third one those treated with cerium oxide and the last one, with reduced growth, with titanium oxide.

iron-oxide nanoparticles could not induce. It is not clear if this effect resulted from mechanical perturbation (thigmomorphogenesis) and/or lack of oxygen or mineral uptake (suffocation) due to external (i.e., on the epidermis) TiO_2-nanoparticles deposition, or from genuine uptake of particles. By means of transcriptome analyses, a genetic compression was verified due to nanoparticle exposure [14].

4. Cobalt nanoparticles, due to their high bioavailability, seem to have the most stressful effect on soil microbial biomass as shown by the increase of available carbon and a decrease of available nitrogen pool. This suggests a trend toward the sterility of the soil. The dispersion of those particles in the soil have entered into an earthworm's intestine, without showing any acute toxic effect [15,16]. Chronic exposure to the above-mentioned nanoparticles induced a strange and unexpected formation of calcium-phosphate spherules in earthworms similar to those identified in human cancerous tissues. The finding, being exceptional, is under evaluation with new tests.
5. Spanish moss (*Tillandsia usneoides*) proved to be an excellent bioindicator for nanoparticles. In fact, our electron-microscopy tests demonstrated the capture and retention of nanoparticles in the plant's tissue surface [16].
6. The engineered metal-oxides nanoparticles affected the C/N microbial ratio and increased the metabolic quotient (qCO_2), probably due to

microbial stress and changes in the composition of microbial communities inhabiting the soil.

7. The tests with aquatic models showed that the effects of exposure to silver nanoparticles induced a trend to the impairment of immune responses, possibly related to the degree of inhibition of the cholinesterases activities [17–22].
8. Exposure of sea urchin's larvae and gametes to silicon-oxide, iron-oxide and cerium-oxide nanoparticles caused a delay in larval development and a skeletal degeneration: malformations in arm length and morphology. Similar malformations were identified in natural conditions in foraminifers that we collected in the Italian Adriatic Sea at a distance of about 200 m from the beach. That area is affected by the fallout of the emissions coming from an incineration plant. Figure 6.20 shows normal

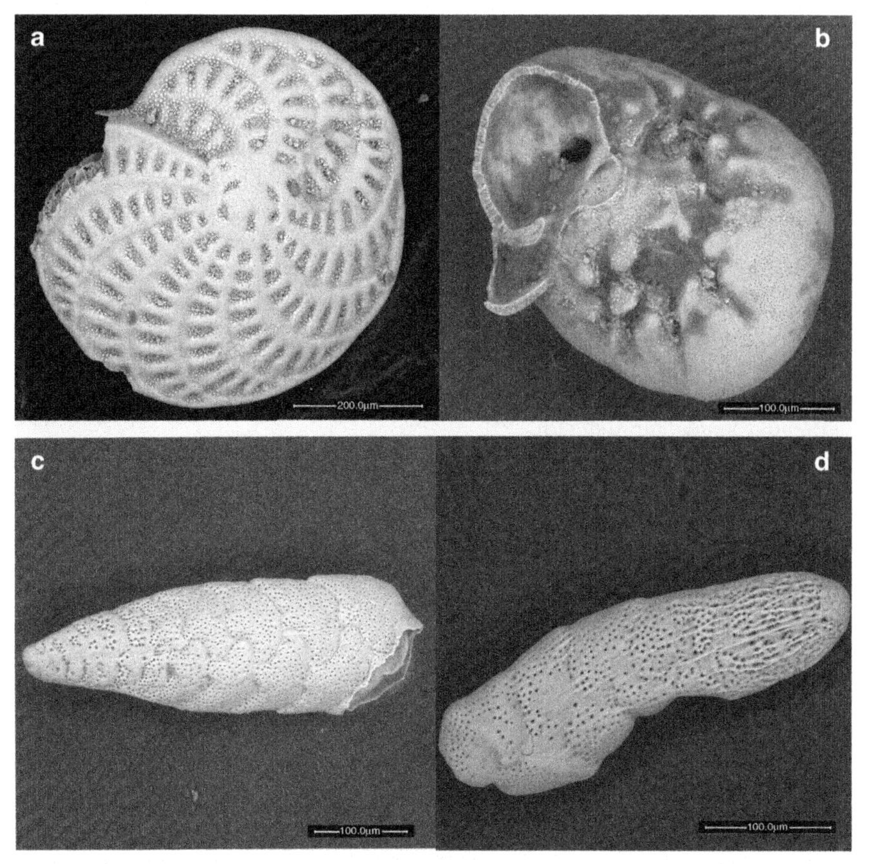

Figure 6.20 Healthy (a, c) and malformed (b, d) foraminifers found in the Adriatic Sea, respectively in a clean area and in an area polluted by the fallout of an incineration plant.

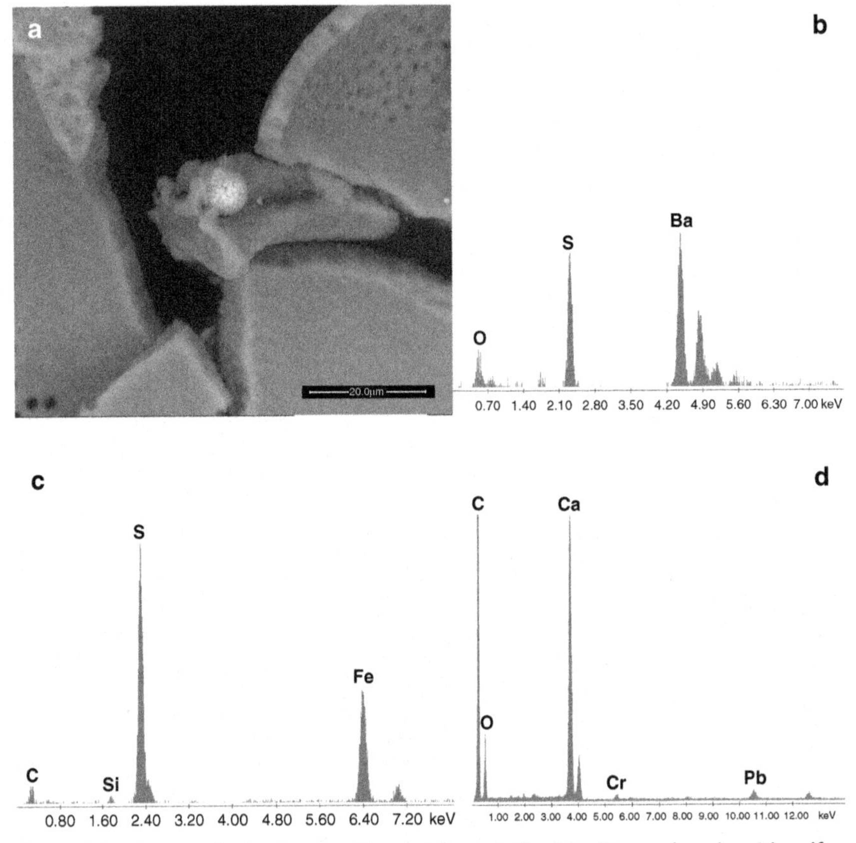

Figure 6.21 Image of a broken, malformed foraminifer (a) with a spherule with sulfur-iron grains (b). (c, d) are the EDS spectra of other debris found inside other malformed foraminifers; they are composed of carbon-oxygen-calcium-lead-chromium (c) and carbon-oxygen-barium-sulfur (d).

vs. malformed foraminifers. Figure 6.21 shows an iron-sulfur spherule found inside a foraminifer's body. Figure 6.21(c) and Figure 6.21(d) show the chemistry of other pollutants we identified inside malformed, sectioned foraminifers.

9. It was demonstrated that nanoparticles can enter the food chain, and become bioavailable to marine organisms, thus affecting their functions, including larval development. The study showed that a dispersion of nanosized particles, be they incidentally-produced or engineered, can affect the food chain and represent a potential biological risk for marine organisms and, potentially, for the whole ecosystem.

The tests verified that engineered nanoparticles can be entrapped in the soil and in plant bodies as well as in insects and aquatic animals. The nanointeraction can cause damages that at a developmental stage can induce malformations.

What we saw in our tests shows evident similarities to what happens to superior animals and humans in terms of illnesses and malformations when incidental particles are at stake (sea urchins, foraminifers, babies; see Chapter 4).

REFERENCES

[1] http://www.nanodiagnostics.it/images/EnelDefinitiva.pdf.

[2] Knox EG. Oil combustion and childhood cancers. J of Epidemiology and Community Health 2005;59:755–60.

[3] Bakoglu M, Karademir A, Ayberk S. An evaluation of the occupational health risks to workers in a hazardous waste incinerator. J Occup Health 2004;46(2):156–64.

[4] Bocio A, Nadal M, Garcia F, Domingo JL. Monitoring metals in the population living in the vicinity of a hazardous waste incinerator: concentrations in autopsy tissues. Biol Trace Elem Res 2005;106(1):41–50.

[5] Cormier SA, Lomnicki S, Backes W, Dellinger B. Origin and Health Impacts of Toxic By-Products and Fine Particles from Combustion and Thermal Treatment of Hazardous Wastes and Materials. Environmental Health Perspectives 2006;114(6):810–7.

[6] Cordier S, Chevrier C, Robert-Gnansia E, Lorente C, Brula P, Hours M. Risk of congenital anomalies in the vicinity of municipal solid waste incinerators. Occup Environ Med 2004;61(1):8–15.

[7] Elliot P, Shaddick G, Kleinschmidt I, Jolley D, Walls P, Beresford J, Grundy C. Cancer incidence near municipal solid waste incinerators in Great Britain. Br J Cancer 1996;73(5):702–10.

[8] Gochfeld M. Incineration: health and environmental consequences. Mt Sinai J Med 1995;62(5):365–74.

[9] Hours M, Anzivino-Viricel L, Maitre A, Perdrix A, Perrodin Y, Charbotel B, Bergeret A. Morbidity among municipal waste incinerator workers: a cross-sectional study. Int Arch Occup Environ Health 2003;76(6):467–72.

[10] Gustavsson P. Mortality among workers at a municipal waste incinerator. Am J Ind Med 1989;15(3):245–53.

[11] Leem JH, Hong YC, Lee KH, Kwon HJ, Chang YS, Jang JY. Health survey on workers and residents near the municipal waste and industrial waste incinerators in Korea. Ind Health 2003;41(3):181–8.

[12] Gatti A, Massamba I, Capitani F, Commodo M, Minutolo P. Investigations on the impact of nanoparticles on environmental sustainability and ecotoxicity. EQA Environmental quality 2012;8:1–8.

[13] Degrassi G, Bertani I, Devescovi G, Fabrizi A, Gatti A, Venturi V. Response of Plant-bacteria interaction models to nanoparticles. EQA Environmental quality 2012;8:39–50.

[14] Giordani T, Fabrizi A, Guidi L, Natali L, Giunti G, Ravasi F, Cavallini A. Alberto Pardossi Response of tomato plants exposed to treatment with nanoparticles. EQA – Environmental quality / Qualité de l'Environnement / Qualità ambientale 2012;8:27–38.

[15] Vittori L, Carbone S, Gatti A, Fabrizi A, Vianello G. Toxicological effects of engineered nanoparticles on earth worms (lombricus rubellus) in short exposure. EQA Environmental quality 2012;8:51–60.

[16] Vittori L, Carbone S, Gatti A, Vianello G, Nannipieri P. Toxicity of metal oxide (CeO_2, Fe_3O_4, SnO_2) engineered nanoparticles on soil microbial biomass and their distribution in soil. Soil Biology and Biochemistry 2013;60:87–94.
[17] Gambardella C, Ferrando S, Gallus L, Gatti A, Ramoino P, Usai C, Vittori L, Carbone S, Falugi C. Nanosilver pesticide-like toxic effect assessed by using Lemna minor. Int, J, of Environmental Analytical Chemistry - Manuscript ID GEAC-2012-0504 2012.
[18] Gambardella C, Aluigi MG, Ramoino P, Diaspro A, Bianchini P, Gatti A, Rottigni M, Tagliafierro G, Falugi C. Developmental abnormalities and cholinesterase activity alteration in sea urchin embryos and larvae obtained from sperms exposed to engineered nanoparticles. Aquatic Toxicology 2013;130-131:77–85.
[19] Falugi C, Aluigi MG, Faimali M, Ferrando S, Gambardella C, Gatti A, Ramoino P. Dose dependent effects of silver nanoparticles on the reproduction and development of different biological models. EQA Environmental quality 2012;8:61–5.
[20] Myrzakhanova M, Gambardella C, Falugi C, Gatti A, Tagliafierro G, Ramoino P, Bianchini P, Diaspro A. Effects of nanosilver exposure on cholinesterase activities, CD41 and CDF/LIF-like expression in zebra fish (Danio rerio) larvae. Hindawi Publishing Corporation. BioMed Research International 2013;1–12. ID 205183.
[21] Gambardella C, Falugi C, Gatti A, Feimali M, Gallus L. Toxicity and transfer of metal oxide nanoparticles from microalgae to sea urchin larvae. Chemistry and Ecology 2014;16:1–9.
[22] Olasagasti M, Gatti A, Capitani F, Barranco A, Pardo MA, Escuredo K, Rainieri S. Toxic effects of colloidal nanosilver in zebrafish embryos. Journal of Applied Toxicology 2014;34(5):562–75.

CHAPTER 7

War Cases and Terrorist Attacks

Contents

7.1 Introduction 129
7.2 The war environmental dust 131
7.3 The Italian case: diseases among soldiers after the Balkan war 135
7.4 The case of Soldier 1 139
7.5 The case of Soldier 2 140
7.6 A case of a soldier with contaminated semen and artificial insemination 142
7.7 A case of a soldier with aspergillosis complications 144
7.8 The case of a civilian who worked in Sarajevo during the siege and war 145
7.9 The cases of two reporters who worked in the Gulf and Balkan war theaters 147
7.10 Quirra and the Quirra Syndrome 149
7.11 The cases of rescue workers during the terrorist attacks on the Twin Towers 152
7.12 Have Hiroshima and Nagasaki been misinterpreted? 155
7.13 A brief conclusion 159
References 160

7.1 INTRODUCTION

Immediately after returning home or, in some cases, when still present in the war fields of the first Gulf War fought between 1990 and 1991, some soldiers started to show collections of "unusual" symptoms. In most cases, those symptoms showed aspects of psychological disorders, some were neurological pathologies, but some were something else: cancer. Not uncommonly, the three of them were present together. It was not long before new problems became manifest: just as an example, malformations in children conceived a short time after the soldier-parent's return back home were more numerous than expected.

At the beginning, the existence of the problem was officially denied, and neurological and psychological symptoms were diagnosed as due to stress (post-traumatic syndrome). Chronic fatigue, headache, muscle and joint pain, diarrhea, frequent fevers, skin rashes, shortness of breath, chest pain, sleep disturbance, irritability, depression, fainting and loss of memory were and still are symptoms that most ill veterans show, and that collection, not necessarily including all varieties, was called SSIDC, i.e., symptoms and

Case Studies in Nanotoxicology and Particle Toxicology
http://dx.doi.org/10.1016/B978-0-12-801215-4.00007-8

signs of ill-defined conditions, by the British. More or less rightfully, more or less reasonably, all those symptoms may be explained by the stress the patients underwent. Of the other pathologies, though, those for which stress is much harder to blame, nothing was said.

Given the awkward behavior of officials and in the absence of data, the opinions of journalists, militants, politicians and scientists in need of funds, i.e., something that had little or, more often, nothing to do with facts, contributed, which, of course, did not help to clarify the issue. So, blame was laid in turn on the vaccines administered to soldiers, the drugs they took (some of them decidedly off-label to say the least), the substances which their tents were sprayed with, the radioactivity of depleted uranium some weapons are made of, just to blame but a few of the presumed killers. Needless to say, none of the allegations was supported by substantial evidence.

To rule out without hesitation some hypotheses, it should have been enough to observe that many civilians living in the same war theaters contracted the same diseases as the soldiers, though none of them had been vaccinated, had taken the drugs the soldiers had taken or had slept in sprayed tents.

Some theories enjoyed only a fleeting popularity, while the thesis that was most successful and enduring was that bringing up the radiations produced by depleted uranium.

In contrast to what the scientific method prescribes, i.e., that a theory is considered correct only if it answers all questions, this was not the case with radioactivity. Among other questions that received answers of dubious validity, at least three remained unanswered: the first was that the symptoms shown by the soldiers were not those typical of radioactivity, the second was that none of the workers employed in the factories where depleted uranium was used to manufacture weapons showed symptoms similar to those of the soldiers, then grouped under the name of Gulf Syndrome and the third was that no radioactivity was ever detected in the tissues examined. Nevertheless, the uranium radioactivity hypothesis, often taken for granted even by some medical doctors, no matter if scientifically untenable, continued to retain much of its popularity.

Regardless of sociological value and just sticking to facts, soldiers and civilians shared the same pathologies, not rarely cancers and, among cancers, mainly, though absolutely not only, leukemia and lymphomas. So, it was reasonable to think that, if the pathologies were the same, so must be the pathogens.

7.2 THE WAR ENVIRONMENTAL DUST

In the early 2000s we started to analyze samples of cancerous tissues taken mostly from Italian veterans and, occasionally, from French, Canadian and American veterans, finding in all cases the undue presence of inorganic micro- and nanoparticles. In none of them nor in the more than 200 similar cases studied at the time of writing this book did we find uranium or traces of radioactivity. It is important to understand that that does not necessarily mean that uranium has not somehow entered the organism of soldiers and civilians alike: It just means that that uranium was not there as particles. Its detection in the urine of some soldiers [1–4] simply demonstrates a well-known phenomenon: the kidneys, where uranium, though not in particulate form, accumulates, can surely eliminate it.

Stating first that almost all cases that are sent to us are preselected, in all pathological specimens examined, hard or soft tissues, blood or seminal fluid, we found inorganic debris, in some instances showing very unusual compositions, i.e., impossible to find in any handbook about materials. To explain those combinations, we considered that the explosion of the new high-potential bombs (e.g., depleted uranium or tungsten) produces very high temperatures [5], for depleted-uranium bombs about 3,036-3,063°C, values that are above the melting point and even the sublimation temperature of most chemical elements. Once the projectile has reached the target, the two of them are reduced to a sort of gaseous matter that quickly solidifies as hollow micro- and nanometric droplets that are scattered all around, whose composition depends only on chance, according to the elements that come occasionally in touch with each other: the same phenomenon as happens with all high-temperature generations of particles. Since the formation temperature, at least that at the core of the explosion, is particularly high, the particles born there are very small, while the size of the other particles increases as the distance from the impact point grows higher.

The fact that they are respirable and their potential danger to human health had already been pointed out in a document drawn up by the US Army in 1978 [6], but made public much later. Those scientists observed under an electron microscope the dust generated by depleted-uranium penetrators as a consequence of hard impacts and reported the very small size of the particles generated: below 0.1 micron ("an unexpected phenomenon"); their hollow, thin-walled, spherical shape; their fragility with the consequence of breaking easily into smaller fragments; and their composition, in that case uranium, iron, silicon, aluminium and tungsten [7].

The reason we never detected uranium in the composition of the particles found in soldiers can be explained by the fact that that element has a very high specific gravity, and a small volume, a volume incomparably smaller than that of the whole bulk involved in the particle generation's phenomenon, is enough to make an efficient penetrator. So, the quantity of uranium actually introduced into the environment as compared to all other materials is negligible. In addition to that, the heaviness of that metal makes it (in the unlikely case it can remain alone) and its particulate combinations fall to the ground within a short distance from the explosion spot, and it is in that limited surrounding area that the US Army scientists found them. So, the chances for soldiers and civilians to inhale or ingest it are close to nil. In any case, an old study on depleted-uranium explosions by Gilchrist et al. [8] showed that soon after the test about half of the airborne-depleted uranium observed was in the respirable size range, of which 43% was dissolved in simulated lung fluid within seven days. After that time, what was left was essentially insoluble. That quick solubility can explain the elimination through urine and its absence in embedded foreign bodies. One of the possibilities is that some uranium particles, few as they are likely to be, are actually present in tissues that did not turn or have not turned yet to pathological and, for that reason, were not given to us.

The behavior of the particles produced by high-temperature explosions is the same as that of any other particle the same size, remaining suspended in the atmosphere for a long time, being carried for long distances and eventually falling to the ground with all the polluting consequences that ensue. And one of the consequences is that people living away from the war theater, possibly even in territories not involved in the conflict, may come also across those pollutants. Given the durability of most of those particles, modern wars produce effects on health extending long after their conclusion and far beyond the territories intentionally involved.

It is only obvious that civilians are the victims of those situations as much as the military. One of the differences is that they, being permanent residents in the polluted territories, will suffer the consequences of the war's side effects much longer than the soldiers, most of whom move away after the end of their term of service. Another difference is that most pathological cases affecting civilians (e.g., fetal malformations) are not counted, thus escaping any attempt of epidemiological research and often also of medical care, even if it must be admitted that that given to soldiers does not seem to be particularly successful.

We had the chance to get samples from the Iraqi war field. Just before the elimination by explosion of a massive amount of weapons carried out inside a burn pit in the Italian base of Tallil in the vicinity of Nasiriya (Iraq), we placed some passive sensors at 100, 200 and 300 m from the core of the explosion and collected the dust generated. We identified micro- and nanoparticles with the "usual unusual" chemical compositions that could be explained only with the random combination of the elements present in the weapons.

Table 7.1 presents a selection of the chemical compositions of the dust collected 100 m away from the explosion site.

In the far-from-impossible event that particles like those enter the organism, leaving aside the foreign-body reactions, a toxicologist would have

Table 7.1 List of chemical compositions of the most represented particles identified on passive sensors located at 100-200-300 m close to a burn pit after a planned elimination of weapons in Nasiriya (Iraq)

N.	Particle size in μm	Chemical composition
1	30–1	C, Si, Ca, O, Al, Fe, Mg, S, Na, K
2	10	Sr, C, S, O, Ca, Ba
3	0.8	C, Au, O, Si, Ag, Cu, Zn, Al, Na, Ca, Fe
4	1	C, S, Fe, O, Si, Na, Ca
5	30	Fe, C, O, Ti, Si, Al, Cl, S, Mg, Na, Ca
6	0.1	Fe, O, C, Cl, Si, S, Ca, P, Al, Na, Mg
7	20	Si, Zr, C, O, Al, Fe, Ca
8	0.5	C, Pb, O, Si, Cr, Ca
9	1	C, O, P, Si, Ca, Cl, S, Ce, Nd, La, Al, Mg, Na, K, Fe
10	1.5	C, Cr, O, Si, S, Ca, Cl, Al, Mg, Na
11	2	C, Ag, Hg, O, Si, Ca, Cl, Al, Mg, Na, Fe
12	1	Ti, O, Si, C, Al, Ca, P, Mg, S, Ce, Fe, K
13	0.5, spherules	S, Si, C, Fe, O, Ca, Al, Mg, K, Cl, Ti
14	4	C, Cu, Zn, O, Si, Ca, S, Cl, Al, Mg, Ti, Fe
15	7	Sr, C, S, O, Ca, Mg
16	5	Cd, C, Ca, Si, S, O, Al, Cl, Mg, Fe, Na, Cu
17	1.5	Zr, Y, Gd, Dy, Er
18	4	Th, U
19	0.3, spherules	C, Pb, O, Zn, Sn, Ca, Si, Mg, Fe
20	0.8–1	C, Bi, Pb, Al, Si, Ca, O, Fe, Zn, Mg

some difficulties in predicting the toxicological effects of such collections of elements, considering them one by one and in combination.

Table 7.1 shows some of the chemical compositions of the dust generated by the elimination of weapons in the burn pit and dispersed in the environment (Figure 7.1A). These are unusual compositions that vary from iron to zirconium, from cadmium to lead, from strontium to gold, from silver to mercury. Some particles also contain rare-earth elements (analyses 9, 17). Radioactive elements such as uranium, thorium, gadolinium, erbium, etc. (analysis 17,18) were also identified (Figure 7.1B). This pollution can invade also towns, headquarters, field hospitals and those who live in the area. This aspect or side effect is never properly considered by those who start a war.

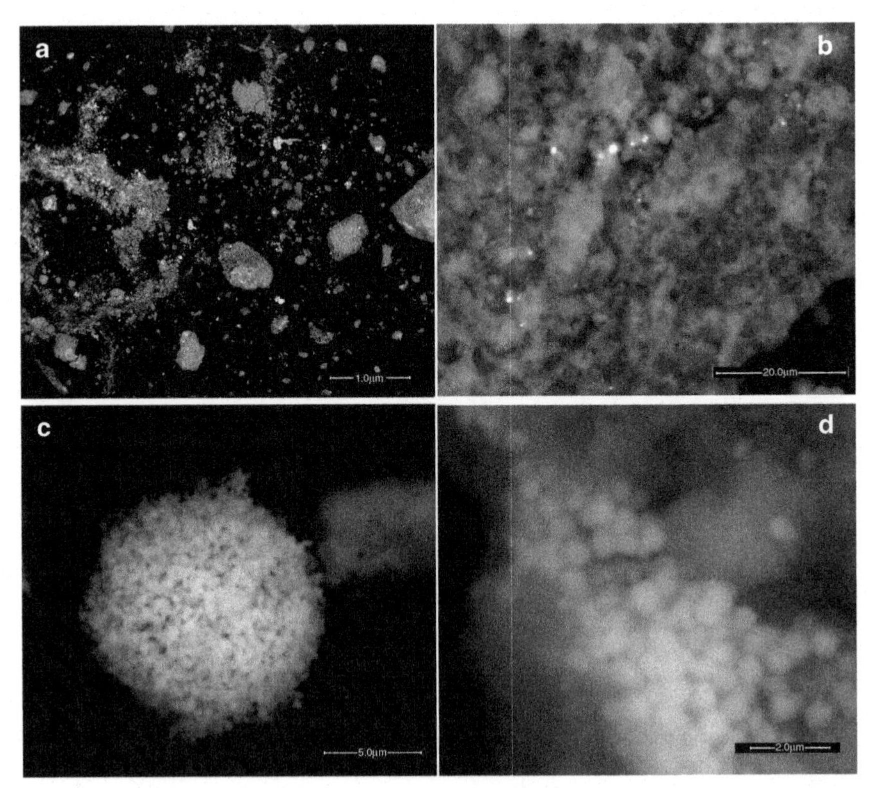

Figure 7.1A ESEM images showing the dust deposited on a passive sensor after the explosion at a 100-m distance from the burn pit at different magnifications. Micro- (a, b) and nanosized particles (c, d) are present in different compositions and morphologies.

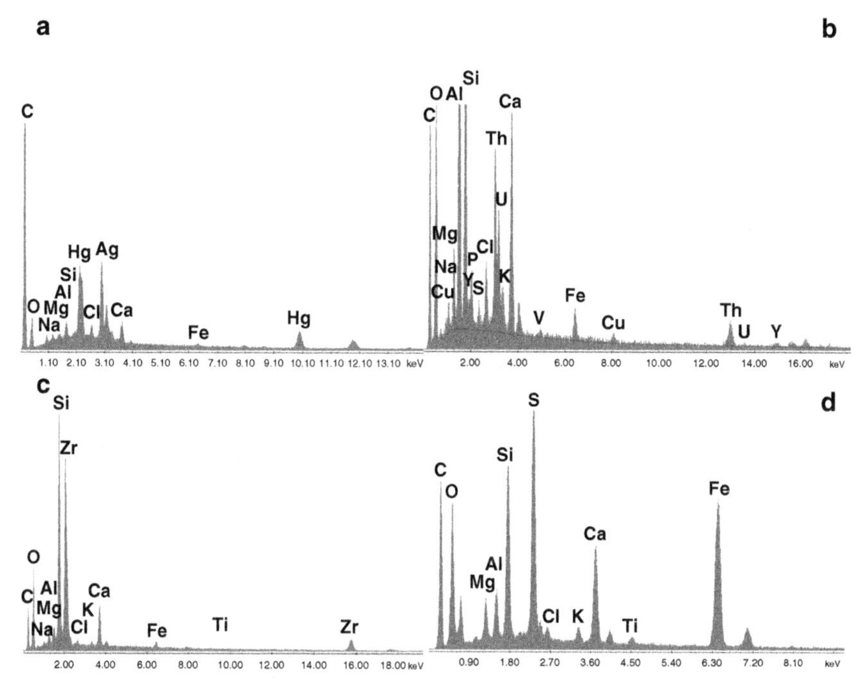

Figure 7.1B EDS spectra of some debris identified among the dust collected after the explosion. Debris composed of carbon-silver-mercury-silicon-aluminium-magnesium-sodium-chlorine-calcium-iron, (a); silicon-aluminium-calcium-thorium-uranium-magnesium-phosphorus-chlorine-yttrium-sodium-iron-sulfur-potassium-vanadium-copper (b); silicon-zirconium-calcium-iron-magnesium-aluminium-magnesium-sodium-chlorine-calcium-potassium-iron (c) and sulfur-silicon-iron-calcium-aluminium-magnesium-chlorine-potassium-titanium (d).

7.3 THE ITALIAN CASE: DISEASES AMONG SOLDIERS AFTER THE BALKAN WAR

At the time of writing we have more than 60 cases of soldiers still under statistic investigation and we keep receiving requests for analyses. For that reason, the cases we are going to consider in this chapter are just 141. Some of those do not concern soldiers who served in war theaters (Balkans, Somalia, Afghanistan...) but who did in military shooting ranges.

The list of Table 7.2 shows the types of diseases and the frequency of the illnesses identified.

The analyses showed the systematic presence of micro- and nanosized foreign bodies, mostly of metallic nature [9–14], that, because of their elemental composition, cannot be inhaled or ingested in a normal urban environment.

Table 7.2 List of types of pathologies examined with their frequency related to examined cases

Pathologies	N. cases	%
Lymphoma, leukemia, blood diseases	46	32.62
Cancer/tyroid diseases	22	15.6
Cancer/digestive system diseases	14	9.93
Cancer/testicle	10	7.09
Cancer/lung/pleura	8	5.67
No diagnosis	9	6.38
Cancer/kidney	5	3.55
Cancer/skin	5	3.55
Cancer/liver	4	2.84
Cancer/brain	4	2.84
Cancer/bladder	3	2.13
Cancer/bone	2	1.42
Cancer/nose-pharinge	2	1.42
Cancer/pancreas	2	1.42
Cancer/prostate	1	0.71
Cancer/muscle	1	0.71
Gulf War Syndrome	1	0.71
Cancer/eye	1	0.71
Cancer/adrenal gland	1	0.71
Total	**141**	**100**

In this section, we present a summary of research we carried out with other participants as part of an Italian government commission study.

A medical doctor who was in charge of a center where most of the Italian soldiers who returned from the Balkans were referred, suspected that monoclonal gammopathy, a blood disorder of undetermined significance some soldiers suffered from, could turn later into a myeloma, made a mixture of eight soldiers' blood serum and selected six other groups: one healthy group exposed to aromatic ammines in their working place (B), though in a protected way, two groups affected by infectious diseases (C, D) and three healthy control groups (E, F, G). Table 7.3 presents the list of the groups selected for comparison with the legend.

A 20-microl drop from the eight-patient blood-serum mixture centrifuged sample was smeared on a cellulose filter and analyzed under the FEG-ESEM microscope and the EDS system.

The analysis of the serum carried out on the centrifuged blood gave the results shown in Table 7.4. The preliminary investigations were carried on at two different unified magnifications. In fact, in some cases, microscopic

Table 7.3 List of groups of patients analyzed

A	Soldiers with monoclonal gammopathies
B	Oil-chemical workers
C	Patients affected by HBV and HCV
D	Patients with CMV e EBV
E	Reference people within 30 year age
F	Reference people within 65 years
G	Reference healthy people of countryside of every age

Legend:
A: Soldiers with monoclonal gammopathy at their return from peacekeeping missions.
B: Workers of chemical industries controlled for possible exposure to aromatic ammines. They are protected with masks during their work.
C: Patients infected with hepatitis virus B and virus A.
D: Patients infected by cytomegalovirus (CMV) and Epstein–Barr (EBV) virus.
E: Healthy people living in a medium-size city (<200.000).
F: Healthy people living in Padua.
G: Healthy people of all ages living in downtown or in countryside.

debris were easily identified at low magnifications, while, increasing the magnification, submicronic particles also became visible.

The analyses of the blood serum showed some interesting findings:

1. The chemical-industry workers (B) were all young; they were exposed to aromatic ammines but worked protected by a nose-mouth mask. That

Table 7.4 Results of debris identified in the different blood samples at two different magnifications

Group	Area analyzed	N. particles	Magnification
A	$9{,}8 \times 10^6\ \mu m^2$	57	50×
A	$1{,}6 \times 10^6\ \mu m^2$	41	200×
B	$9{,}8 \times 10^6\ \mu m^2$	0	50×
C	$9{,}8 \times 10^6\ \mu m^2$	52	50×
C	$9{,}8 \times 10^6\ \mu m^2$	156	50×
C	$1{,}6 \times 10^6\ \mu m^2$	22 micron size + 107 submicronic	200×
D	$9{,}8 \times 10^6\ \mu m^2$	106	50×
D	$1{,}6 \times 10^6\ \mu m^2$	78	200×
E	$9{,}8 \times 10^6\ \mu m^2$	164	50×
E	$1{,}6 \times 10^6\ \mu m^2$	100	200×
F	$9{,}8 \times 10^6\ \mu m^2$	38	50×
F	$1{,}6 \times 10^6\ \mu m^2$	49	200×
G	$9{,}8 \times 10^6\ \mu m^2$	26	50×
G	$1{,}6 \times 10^6\ \mu m^2$	9 large + 7 submicronic	200×

is probably we why did not find any particulate pollutants in their blood. It must be remembered that the system we use is unable to discriminate carbonaceous pollution in a biological medium.

2. The soldiers with monoclonal gammopathy and with a high value of beta-2-microglobuline had some, mostly metallic, debris in their blood samples. They were all young and well trained, and, before the mission they took part in, none of them had altered blood values. The particulate contamination of their blood proved to be higher than that observed in the group of healthy people living in the countryside (G) and of older people (F), but the quantity of debris deposited in the internal organs and in the bone marrow of all groups is unknown.
3. The tests on the serum of patient infected by cytomegalovirus and Epstein–Barr (106 large micron-sized debris and 78 submicronic ones) and of patients affected by the viruses of hepatitis (A and B) (156 large micron-sized debris and 107 submicronic ones) were particularly interesting. We identified a great number of micro-, submicronic and nanoparticles, mostly of ceramic nature.
4. People living in an industrialized city can accumulate dust in the blood (164 large micron-sized debris and 100 submicronic ones) and their healthy state can be compromised any time.
5. Young people living in small cities and old people living in the countryside present a very modest pollutant presence in their serum.

The following formula gives an idea of the content of foreign bodies presented in the blood serum analyzed:

Oil workers<healthy people<healthy people (65)<soldiers<healthy people-city<hepatitis- and cytomegalovirus-infected patients

The study verified that the soldiers had contaminated blood but was unable to assess the quantity of debris already embedded in other organs, lymph nodes, etc.

The most interesting result came from the infected blood. We supposed that, in those cases, the disease was only due to a virus. This finding drives us to consider as an alternative hypothesis that the disease was triggered by a large quantity of particulate matter in the blood whose massive presence can impair the immunosystem that, in such a condition, is unable to mount and depress the virulence of the attacking pathogen agents. That can represent another really innovative field of research that, if demonstrated, can represent a revolution in toxicology.

7.4 THE CASE OF SOLDIER 1

A test pilot of military helicopters returned from a number of missions (Iraq, Balkans, Somalia, etc.) and started to suffer from intermittent fever episodes, chronic fatigue, night sweats and swollen neck-and-groin lymph nodes. He was admitted to a military hospital where he spent three months without getting a diagnosis. Under our suggestion, he moved to another hospital where samples of his lymph nodes were taken and checked, and a diagnosis of lymphadenopathy was issued.

We were given two samples of inguinal lymph nodes, which, on electron-microscopy observation, showed a great number and variety of particles ranging from 15 microns down to 0.1 (Figure 7.2). Many of them were spherules, a clear indication of their high-temperature origin, and their chemistry included a fair number of elements. We did not analyze only the subject's lymph nodes but also his sweat (Table 7.5) and his seminal

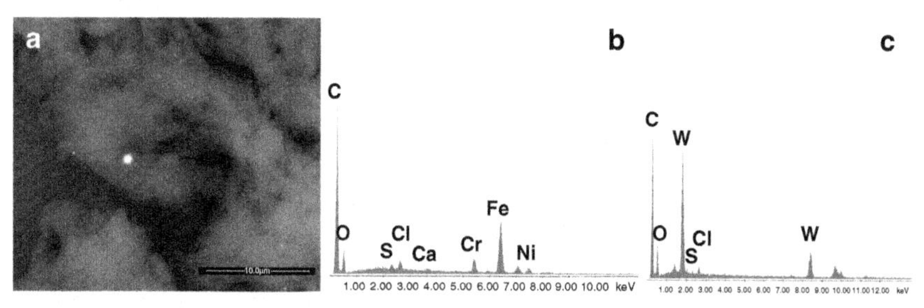

Figure 7.2 Image (a) shows stainless-steel (iron-chromium-nickel) (b) nanosized spherules embedded in the biological tissue. Other particles were composed of tungsten (c).

Table 7.5 Analyses of debris identified in the soldier's sweat

N.	Particle size	Chemical composition
1	sweat	O,C,Cl
2	spherule, 5 μm	Si,O,C,Cl
3	nanoparticles 0.1–0.3 μm	O,C,Fe,Cl,Na
4	spherule, 5 μm	Si,O,C,Cl
5	nanoparticles-cluster	O,C,Fe,Cl,Na
6	0.2 μm	C,O,Au,Ca,Cl,Na,K
7	nanoparticles	O,C,Fe,Cl,Na
8	0.8 μm	C,O,Au,Cu,Ni,Cl,K
9	1 μm	C,O,Au,Cu,Ni,Cl,K
10	2.5 μm	C,O,Bi,Cl
15	debris, 1 μm	O,C,Cr,Cu,Fe,K

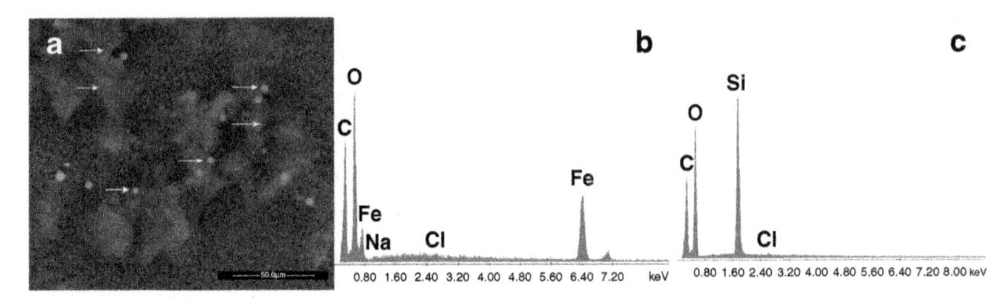

Figure 7.3 Image of iron particles (b) and silicon spherules (c) detected in the soldier's sweat drops (a).

fluid and all samples were found to be contaminated by inorganic, biopersistent, micro- and nanosized particulate matter. In the sweat (Figure 7.3) we identified iron and silicon or silica particles. The seminal fluid contained also spherules of calcium phosphate.

The small size and the spherical shape of many particles and the relatively numerous variety of chemical elements are clear evidence of the kind of exposure suffered by the subject. Among the elements found, there was tungsten, present in a number of particles, a metal with a melting temperature of 3,422°C, a temperature that must at least have been reached or, more likely, exceeded, when those particles formed.

Considering the size of the particles, it is probable that most of them, the smaller ones, were been inhaled, while the coarser were ingested, reaching in all cases the lymph nodes.

The presence of small particles in the sweat could mean that a limited and perhaps partial elimination from the body is possible [15, 16], a hypothesis we verified in New York firefighters.

7.5 THE CASE OF SOLDIER 2

We received a biopsy sample of a bladder with the diagnosis of low-grade papillary transitional neoplasm. The patient was a peacekeeping mission veteran (a ranger) who, at the time when this book was being written, had already undergone 15 surgical operations, most of them by endoscopic techniques.

In this case, the two largest particles we detected had a size of 5 microns, but the vast majority of them were not larger than a few hundred nanometers, many of which had a rather homogeneous chemical composition (tungsten, oxygen, chlorine, sulfur, calcium). This chemical and dimensional homogeneity could be evidence of a single exposure to a great quantity of dust or to repeated exposures to the same kind of pollution. As occurs in many other circumstances, stainless-steel particles (iron, chromium, nickel) were also found. Of course, they could have reached the bladder at a different time than the smaller ones.

Similar to the case described above as Soldier 1's, a great number of particles were nanosized and spherical-shaped. The presence of tungsten shows that the temperature at which the particles had formed must have been particularly high (Figure 7.4).

As can be seen, two soldiers who had served in peacekeeping missions in different countries at different times present similar pollution (stainless steel and tungsten) in their bodies but a different dispersion in their tissues. That difference, involving different organs, caused different pathologies.

After five years from the time when we made those analyses, we met the soldier again. The situation had worsened heavily and involved other

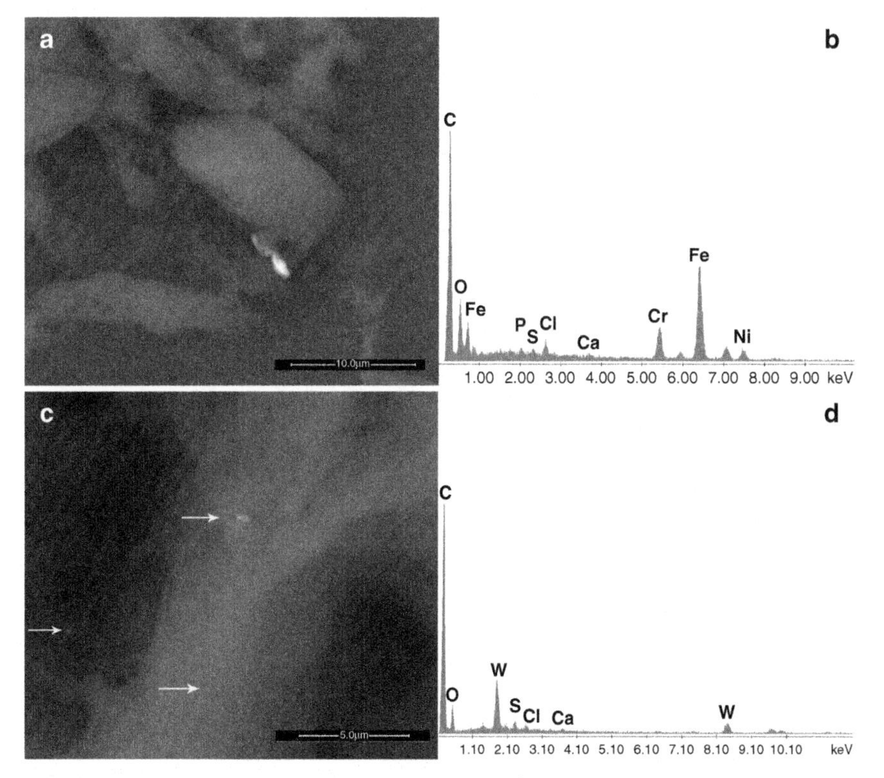

Figure 7.4 Images (a, c) show micrometric debris of stainless steel (b) and tungsten (d) found in the biopsy of a bladder internal neoformation. The bladder was full of tungsten nanoparticles.

organs. After some endoscopic operations in the bladder to eliminate some neoformations, the patient developed a cancer in the right kidney and the organ was surgically explanted. A nephrostomy catheter was put in place and connected to an external urine-collection bag. The X-ray image of the lungs had shown the presence of metastases that could not be treated with chemotherapy in order not to compromise the functionality of the left kidney. No survival prognosis was issued.

7.6 A CASE OF A SOLDIER WITH CONTAMINATED SEMEN AND ARTIFICIAL INSEMINATION

A soldier who served in the Balkans in peacekeeping missions received a diagnosis of primary thrombocythemia and oligozoospermia.

Since he wished to procreate and was dubious about possible risks for the baby's health, he asked us to check some of his biological samples. So, we analyzed his blood, his osteomedullary biopsy and his seminal fluid. The particles we found in the bone-marrow sample ranged in size between 0.7 and 8 microns and the elements they contained were rather numerous: iron, chromium, nickel, silicon, sulfur, phosphorus, copper, titanium, aluminium, calcium, magnesium, potassium, bismuth, chlorine, zirconium and antimony (see Table 7.6).

The blood sample was contaminated with particles whose size ranged from 0.8 to 5 microns. Some of them were spherical and the elements they were composed of were iron, chlorine, sulfur, silicon, phosphorus, potassium, calcium, zirconium, sodium and zinc. The spherical shape of the particle of Figure 7.5(a) demonstrates its combustive/explosive origin.

The seminal fluid contained many metallic debris ranging from 2 microns down to 0.8 microns whose compositions were particularly unusual:

Table 7.6 List of some of the particles identified in the bone-marrow biopsy

N.	Particle size	Chemical composition
1	5 μm	Fe, C, Cr, O, Ni, Si, S, P, Cu
2	0.8 μm	C, Ti, O, S, Al
5	nanocluster,	Fe, Cr, Ni
6	0.2	Fe, Cr, Ni
8	8 μm	Ca, Si, C, Ti, Al, O, Fe, K, P, S, Mg, Cu
9	2 μm	C, Bi, Cl, O
10	2 μm	C, Zr, O, Na, Ca
11	0.7 μm	C, Sb, O, S

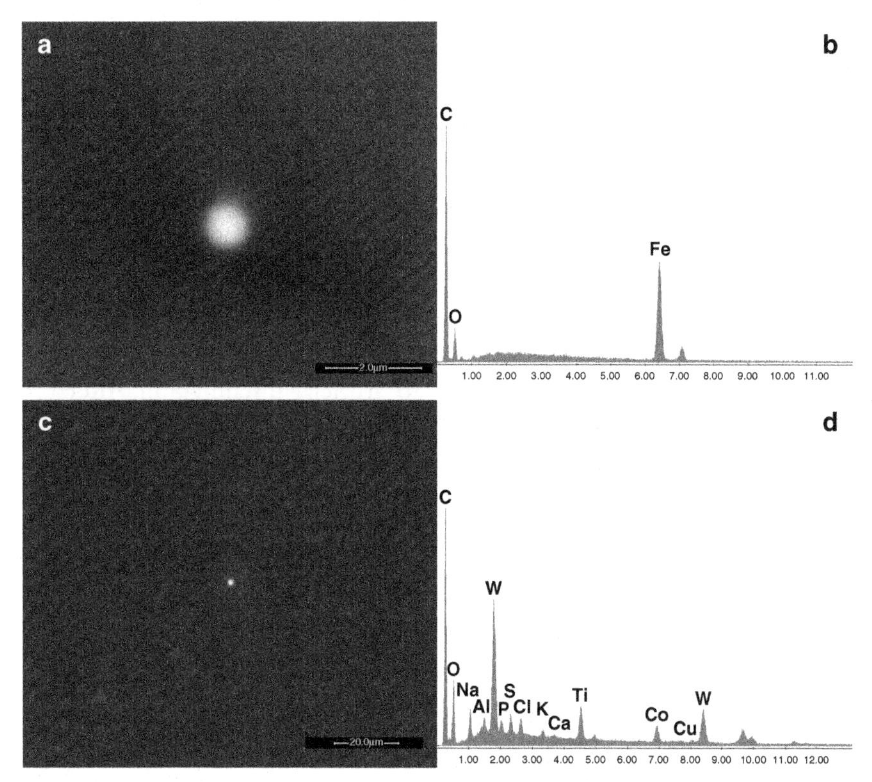

Figure 7.5 Image (a) shows an iron (b) spherule of 0.8 micron found in a soldier's blood sample. Image (c) is related to a submicronic particle found in the spermatic fluid composed of carbon-tungsten-oxygen-titanium-sodium-aluminium-phosphorus-sulfur-chlorine-potassium-calcium and cerium (d).

1 - tungsten, titanium, sodium, sulfur, aluminium, chlorine, cobalt, potassium, calcium, copper; 2 - iron, sodium, silicon, chlorine, potassium, phosphorus, sulfur, calcium, aluminium, magnesium, copper; 3 - iron, molybdenum, tungsten, chromium, vanadium, phosphorus, sodium, copper; 4 - sulfur, barium, silicon, sodium, chlorine, calcium, phosphorus, aluminium, iron; 5 - zirconium, sodium. It is evident that such compositions are the results of random combinations like those occurring in uncontrolled combustions/explosions.

Having found particles in his marrow and in his blood, we suspected that his sterility could be due to the possible presence of particles in his seminal fluid, something that could not be ruled out. As a matter of fact, his seminal fluid contained quite a number of particles ranging 0.8-7 microns. The elements we detected were rather numerous: iron, zirconium, sodium, phosphorus, silicon, sulfur, chlorine, manganese, bismuth, tungsten, cobalt, titanium, aluminium, potassium, calcium, copper, magnesium, vanadium, molybdenum, chromium and barium.

Given what we had found, it was obvious that unprotected sexual intercourse would have transferred particulate matter to the partner, thus contaminating her vaginal tissues and triggering biological local reactions (burning semen disease) as we had already observed in other circumstances.

Additionally, we could not be sure about the effects of those particles on the embryo.

We shared our results with two doctors of the hospital of Modena (Italy) and decided together to take seminal fluid from the subject, clean its plasma fraction from the particles and resort to artificial insemination. After the separation of the spermatozoa from the plasma, we could see that the contaminants remained only in the plasma, while the spermatozoa were clean. The insemination was successful and healthy twins were born after a full-term, uneventful pregnancy.

7.7 A CASE OF A SOLDIER WITH ASPERGILLOSIS COMPLICATIONS

A few months before his death, a 38-year-old, non-commissioned officer active as squad leader/instructor at the Sardinian shooting range of Capo Teulada and a veteran from Bosnia, Kosovo and Albania where he had gone after the end of the war, was certified unfit for duty. The reasons were an incapacitating poor health and chronic fatigue, health conditions that continued to worsen until he died without getting a diagnosis.

A post-mortem was performed and what was observed was a severe bone-marrow aplasia, a pulmonary angioinvasive aspergillosis and the presence of inorganic foreign particles in the lung macrophages and in the intestinal endothelia.

The hematopoietic cells of the bone marrow had been replaced by adipose tissue. Collections of macrophages were present in both lungs surrounded by solid, blackish and partially birefringent, coalescent particles, in the majority of cases located close to lymph vessels and devoid of a surrounding inflammatory reaction. Similar, though smaller, particles were detected in the intestinal endothelia. Most of the inorganic foreign bodies were silicon particles alloyed with some of the numerous elements found (chromium, chlorine, sodium, bismuth, phosphorus, aluminium, magnesium, zinc, iron, lead, zirconium, strontium, titanium, calcium and potassium), in some cases combined together. These findings, on the whole,

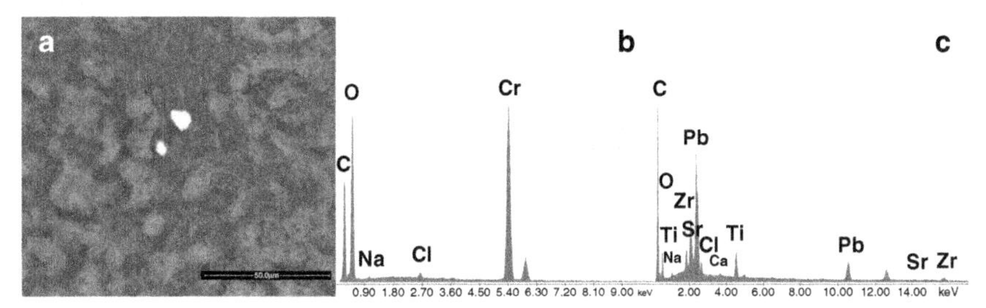

Figure 7.6 Image (a) shows two of the particles found in the liver. They are composed, respectively, of chromium (b) and lead, zirconium, titanium, strontium, chlorine, calcium and sodium (c).

point to a diagnosis of acquired immunodeficiency possibly induced by the exogenous, foreign materials. Put forward as a hypothesis without any evidence, since our analytical method cannot detect organic matter, agents used in chemical or biological warfare could have been adsorbed on the particles the patient had doubtless inhaled and ingested, thus contributing to the complexity of the pathology.

Our analyses showed the presence of foreign bodies in the liver (Figure 7.6), in the pulmonary macrophages and in the intestinal endothelia full of silicon-based debris.

Our hypothesis is that the wide dispersion of foreign bodies throughout all organs impaired the immunosystem, which could not prevent Aspergillus from entering the bloodstream via the lungs. Without the organism mounting an effective immune response, fungal cells were free to disseminate throughout the body and infect major organs such as the heart and the kidneys.

Unfortunately, a precise diagnosis was made only after the soldier's death.

7.8 THE CASE OF A CIVILIAN WHO WORKED IN SARAJEVO DURING THE SIEGE AND WAR

What follows is not the case of a soldier but of a civilian. It is only natural, in fact, that a polluted environment is equally harmful to anyone. The subject of these analyses was a person who had spent 13 years in the Balkans as an executive consultant to international agencies and local governments.

He felt a sort of excrescence on his skull and a visit to a hospital revealed a bone lesion partially covered by meninges. The diagnosis of the biopsy he underwent was of plasmacytoma, a rare form of cancer indeed.

As usual, we performed our electron-microscopy observations on the neoformation whose sample the hospital had made available, and what we found was particulate matter both spherical and irregularly shaped. The size ranged between 0.1 and 17 microns (the latter a rather large, unusual size) with the nanometric particles often clustered together.

As to their compositions, they were the most various. Iron, chromium and nickel, the typical composition of stainless steel, were present in a fair number of particles and that is something we come across often. The same thing can be said of the calcium-phosphorus particles we found there, as we find in many cancerous tissues. What was certainly rarer was a particle composed mainly of chromium, another made mainly of bismuth and compositions like iron-strontium-titanium, silver-calcium-copper, gold-sodium-copper (Figure 7.7).

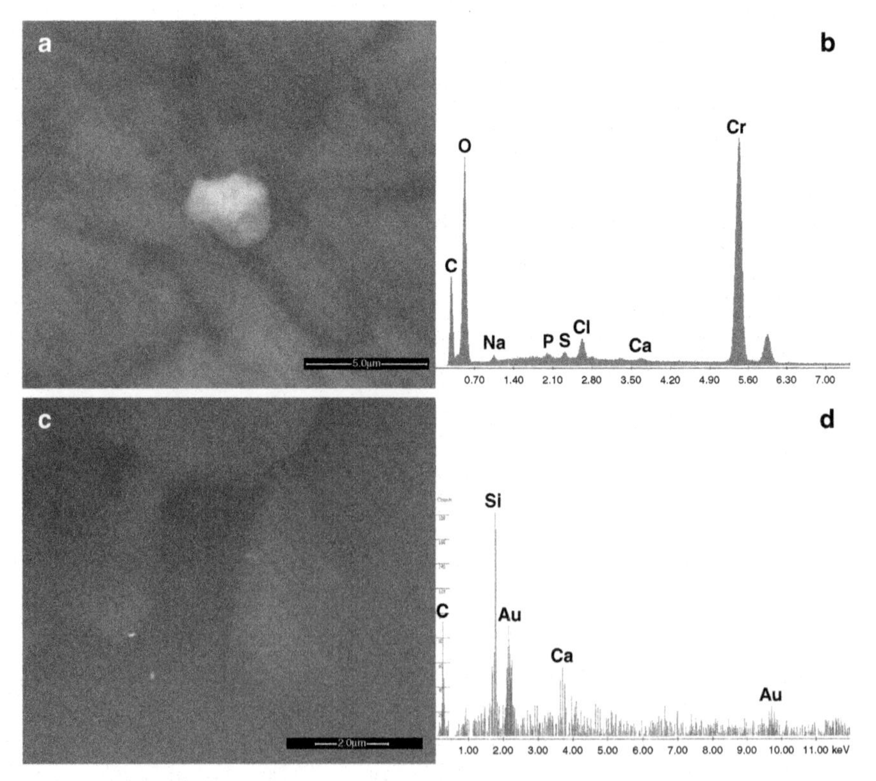

Figure 7.7 Images (a, c) show particles found in the plasmacytoma tissue. (a) shows a spherical chromium debris (b) with a sort of nanocrystalline structure; while (c) shows 200-nm-sized golden debris (silicon-gold-carbon-calcium) (d).

If it is easy to say that the fine particles we saw came either via respiration or via ingestion (almost certainly both ways), it is harder to explain how the 17-micron particle (iron-chromium-nickel) reached the skull bone. But the fact is, it was there.

Just after we completed our analyses, we were informed that two more persons who shared the same office with our patient had developed a lymphoma. Besides sharing their workplace, the three subjects shared the experience of having been present in Sarajevo during the whole siege (April 1992 –February 1996), a long period during which bombing was an almost daily event with all the dust it produces. A further possible exposure to particles may have come from a tunnel placed in the vicinity of the three subjects' workplace. After the end of the war many trucks transported "dusty materials" there about which no information is available.

7.9 THE CASES OF TWO REPORTERS WHO WORKED IN THE GULF AND BALKAN WAR THEATERS

The first case we are going to describe is about a war reporter who developed a lung carcinoma after his mission in Iraq to cover the first Gulf War. During his 6-month stay, the reporter used to take to the streets of Baghdad and its surroundings by bike, thus breathing air full of the dust hanging on the bombed city. So, what we saw under our electron microscope was by no means unexpected. The lung sample we got after a lung resection showed the massive presence of debris. Some probably belonged to the desert sand he had inhaled. Other debris were the usual, combustion-originated, spherules, many of which particularly small (0.1-0.8 microns). Having found zirconium as a component of the particles, a transition metal with a high melting point (its oxide's is even higher), was further evidence that that dust was formed by explosions of high-tech weapons.

In detail, inside the sample we found particles of silicon, aluminium, sodium, calcium, magnesium; sodium, aluminium, tin, magnesium; titanium, silicon, sodium, titanium, calcium, magnesium, potassium, sulfur, iron, chromium; silicon, sodium, calcium, chromium, iron, magnesium, potassium, aluminium, sulfur. Most of them were spherical. As usual, we do not mention carbon and oxygen, in this case, as it is always the case, present in the spectrum because they belong to the tissue. Of course, that does not mean that the two elements cannot be components of the particles analyzed.

It is interesting to observe that the two samples of this case we had a chance to analyze belonged both to the superior lobe of the left lung, and

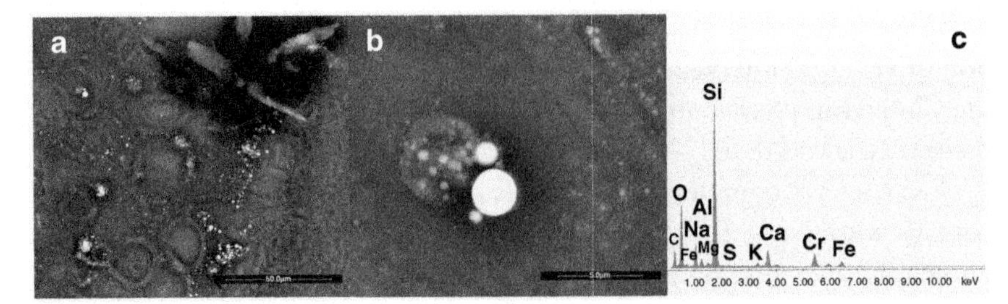

Figure 7.8 Image (a) shows the wide distribution of the debris in the lung; while (b) shows the spherical morphology of the debris at high magnification. (c) is the spectrum of the spherules that contains silicon-oxygen-sodium-calcium-chromium-iron-magnesium-aluminium-potassium.

that the carcinoma was evident only there. The other sample showed the presence of a collection of particles (Figure 7.8) responsible for the inflammatory condition that usually precedes the transformation into cancer.

The second case concerns another reporter who covered the same war theater as the former and, in addition, the Balkans. In this instance, we analyzed two samples: one of the left colon affected by adenoma, and one of the rectum with a diagnosis of adenoma and adenocarcinoma.

As to size, the colon sample contained particles between 1 and 7 microns, while the two rectum sections we examined contained particles covering a wider span (0.4–13 microns). Though the intestine is obviously the target of choice for ingested particles, like all other organs it can receive materials entered through the lungs. In this case, it is evident that particles that large could not have been inhaled but must have entered the organism through ingestion.

From the composition point of view, two particles of rather similar size and located close to each other are particularly noteworthy (Figure 7.9).

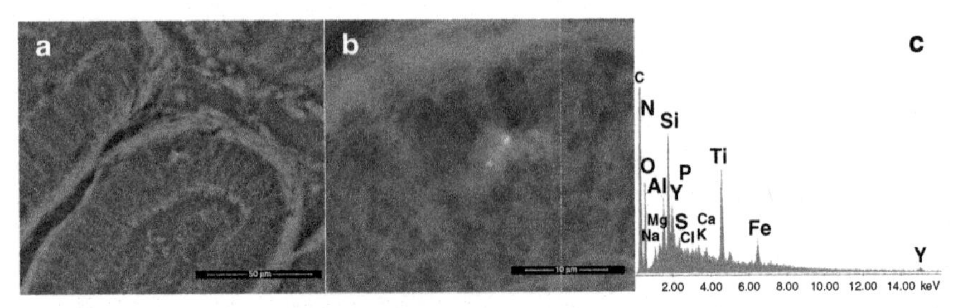

Figure 7.9 Images (a, b) show the colon sample with silicon-based foreign bodies embedded at different magnifications. Their chemical composition is: carbon-silicon-titanium-oxygen-aluminium-phosphorus-yttrium-sulfur-magnesium-sodium-chlorine (c).

The smaller one (0.7 microns) contained silicon, titanium, aluminium, phosphorus, yttrium, sulfur, magnesium, sodium, chlorine and nitrogen. The other (1 micron) was composed of iron, silicon, aluminium, sulfur, phosphorus, magnesium, potassium, calcium, chlorine and titanium.

Other metallic particles have also been detected, containing iron-chromium-nickel and iron-titanium.

The biopsy samples of the two journalists showed the presence of silicon-based, metallic particles and calcification, an usual finding in cancer cases.

7.10 QUIRRA AND THE QUIRRA SYNDROME

The southeast of the Italian island of Sardinia hosts the largest shooting range in Europe called PISQ (Poligono Sperimentale Interforze del Salto di Quirra). It is an area where, since 1956, new weapons are tested and obsolete ones destroyed, and it is also a training and experimental base used by the Italian Army, Navy and Air Force, by the NATO Alliance Forces and is also rented to arms industries. Though not much inhabited, the territory includes some villages where a majority of peasants and shepherds live. For many years and with an increasing frequency, unofficial reports about malformed animals, cancers and malformations in humans went around, but both military and civilian authorities denied their truthfulness, dismissing everything as unfounded rumors.

In the 2000s, that territory was finally brought to the attention of the media due to the above-mentioned apparently abnormal incidence of pathologies to which the name of "Quirra Syndrome" was given, for which the media blamed the use of depleted-uranium weapons. Specific studies on this subject were published in specific literature [1–3].

The military area is divided into two zones: a relatively high-altitude one, the "land range," and a seaside zone, the "sea range." Launching areas for surface-to-air and surface-to-sea missiles are located in the sea range (110 km^2) and in an area reserved to the tests on solid fuels for rocket engines, while the land range (116 km^2) is used for training with anti-tank rockets and bombing, as well as for the destruction of large amounts of weapons no longer in use.

It is evident that such activities involving explosions and the use of organic substances (fuel) produce pollutants.

We described this situation in a chapter of the book *Nanopathology* [9] in 2008, but many other events have occurred since.

In 2010, a survey was carried out by the veterinarians of the local public-health agency among the cattle and the result was the discovery of a greatly abnormal rate of barrenness, miscarriages and fetal malformations. As to humans, the scarceness of the population there makes it difficult to get statistically significant data, but it is impossible not to take note of the roughly two thirds of the shepherds living within a radius of about three kilometers from the military base suffering from forms of cancer (mainly leukemia) and the quantity of more or less seriously malformed children.

Environmental assessments and measurements were performed in the area after the Italian judiciary asked to find an official cause of the Quirra syndrome.

Abnormal cases of cancer affecting the hemolymphatic system were reported around the small village of Quirra. An unexpectedly high number of malformed children were born in the municipality of Escalaplano only in a restricted interval of time. And a higher than normal number of cases of cancer was reported among the staff who served at the base.

Besides the chemistry and the radiological characteristics of depleted uranium (whose use the Italian military authorities have always denied), a number of other possible causes to the "Quirra syndrome" were indicated:

- Toxic and teratogenic chemical substances generated by the fuel used by missiles.
- Radiological contamination by thorium used in the ballistic missiles that were tested.
- Electromagnetic pollution from military radars and electronic warfare devices [3].
- Arsenic contamination, from past mining activity in the area.

Though there is no solid evidence to support the charge, except for arsenic, which does not cause hemolymphatic tumors, the other three could reasonably be the cause or the concurrent cause.

We approached the problem in an integrated way. First, we started with the environmental pollution present in the air, on the grass and on the leaves in the upper part of the territory, where most animals pastured. In the lower part of the shooting range, we analyzed the rocks and the shelter behind the surface-to-air missile launching area. We also analyzed the pathological tissues of subjects who lived in the area and became ill, and of the animals who pastured there as well as their products.

The image of Figure 7.10 shows a malformed lamb born in 2011 in the area of Quirra. It had only one eye and it was malformed. We analyzed its

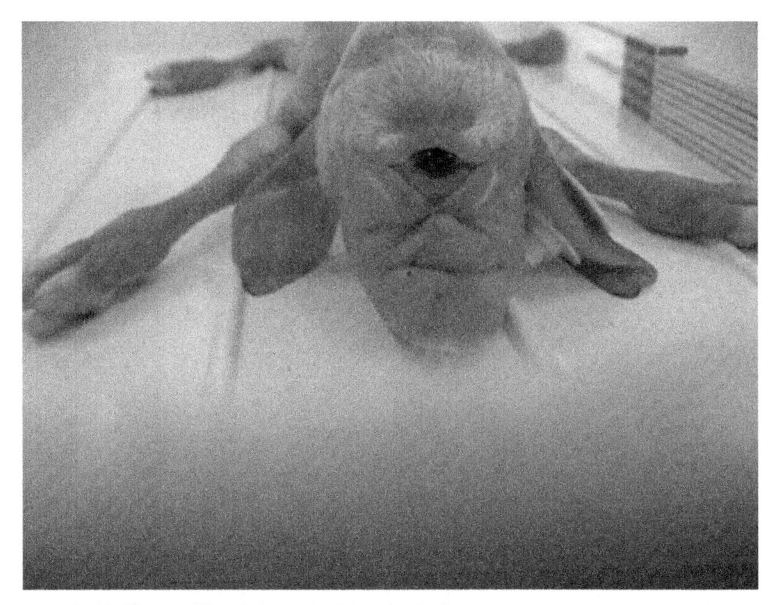

Figure 7.10 Malformed lamb born in 2011 in Quirra.

liver, its kidneys, its heart, its brain and its testicles. In all samples we found particulate foreign bodies. Table 7.7 summarizes the debris found in the brain.

Figure 7.11 shows some debris found in the lamb's body: brain and testicle. These debris are composed of particles obviously accumulated during the embryonic stage.

Similar debris were also found in the humans living in the area who developed cancer.

In our 10-year integrated monitoring experience in the shooting range, we observed that pollution is generated by the military activities. The

Table 7.7 List of particles found in the brain

.nnm	Particle size	Chemical composition
1	Cluster of debris, 100 μm	Si, Al, K, Fe
2	0.5 μm	Si, Al, Fe, K, Ti
3	0.8 μm	Ti, Fe, Si, Al, O
4	1 μm	Sb
5	2 μm	Cr, O
6	5 μm	Al, Si, Fe, K, Ti
7	10 μm	Ca, Al, Fe, Si, Ti, Mn
8	5 μm	Ti, Si, Al, Fe, K
9	1.5 μm	S, Pb, Fe, Cu, Sn

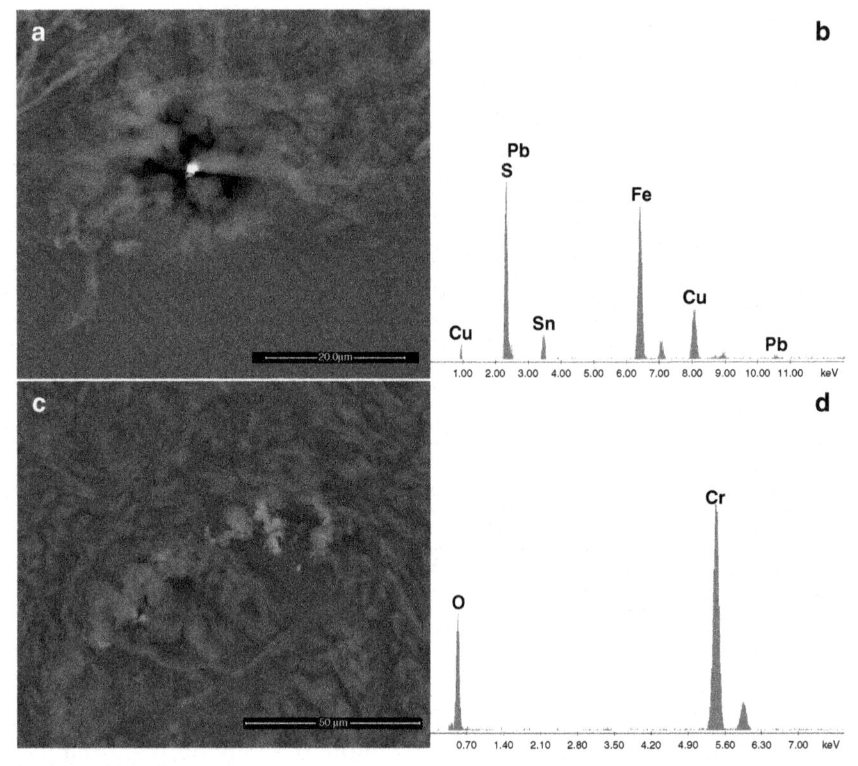

Figure 7.11 Images (a, c) indicate debris that were found, respectively, in the malformed lamb's brain and in the testicle. The foreign bodies were composed of lead, sulfur, iron, copper and tin (b) and chromium (d).

particulate pollutants produced are different from time to time (from activity to activity) especially as to chemical composition and size. In order to assess rigorously the actual risks to humans and animals it is necessary to know the types of activities carried out, something that, for many reasons, is impossible. For obvious safety reasons, we agreed with the public prosecutor that entry to the territory should be forbidden to humans and animals.

7.11 THE CASES OF RESCUE WORKERS DURING THE TERRORIST ATTACKS ON THE TWIN TOWERS

Beyond its much-debated political and human aspects, the attack on the Trade Center in New York City on September 11, 2001 can be seen from a different standpoint. It happens sometimes that tragedies provide information of great importance and are, in a sense, laboratories where otherwise, and thankfully, impossible experiments, though not by design, are carried out.

That day two airplanes crossed two huge buildings and the impact developed a temperature high enough to vaporize massive amounts of the most diverse materials. If the temperature was extremely high at the core of the phenomenon, of course its value decreased with distance, and this offered the condition for very inhomogeneous dust to be generated. In a way, this was a magnification on a huge scale of what happens with depleted-uranium bombs and with any other explosion.

According to physics, the finer and lighter particles were thrown violently off and immediately suspended in the relative vicinity of what is now known as Ground Zero, but very soon all that got dispersed either by the wind and by diffusion in a radius of at least some kilometers, though the highest concentration was for a while located at the spot of the incident. Part of that dust was sucked in by the air conditioners and it is not impossible that some of it lingered long in the pipelines. We do not know whether that was actually the case, but if the clogged air-conditioner filters were disposed of in incinerators, the dust would have taken part in another combustion and would have been reintroduced into the environment as particles different in size from the initial ones and with the same chemical elements combined in different ways. If disposed of in landfills, inevitably the dust would have been lifted by the wind and by the tractors that usually crush waste. So, in one way or another, the New Yorkers and all those who had to deal closely with the event came into contact with such an unusual, in fact never-seen-before, dust.

Those people soon started to show symptoms, the most immediate and common being coughing, but those warnings were not understood for what they meant. There were high rates of respiratory troubles, sinus problems or nasal/post-nasal irritations, shortness of breath, throat irritation and wheezing, persistent cough, asthma and headache [17–19], but the official attitude was a sort of trust that all those were brief disorders bound to heal on their own. As occurs regularly, and soldiers are an example, in that case many people involved in the disaster also showed psychological problems as well as physical health troubles and, just as regularly, conditions whose origin had nothing to do with psychology, but psychology problems were just a consequence, were treated with sedatives, tranquilizers and other psychotropic drugs that just worsened the situation the way it happens whenever a wrong or insufficient diagnosis is issued, sometimes masking at least for awhile the telling symptoms. The usual, easy, diagnosis was PTS (post-traumatic syndrome). But other pathologies were diagnosed as well in the next year and in 2008 a health registry was open in order to face situations the New York citizens demanded be addressed [20–26].

But even when the symptoms were not so grossly misunderstood, in the best of cases drug treatments just managed to mitigate temporarily some disorders, but in fact proved ineffective. So the symptoms became chronic or were just the starting point to more serious illnesses.

Besides diagnostic problems, the reason for the general pharmacological failure is obvious: probably not all, but a considerable number of pathologies was due to inhaled and ingested dust, and at that time, as is still the case to date, there are no drugs capable of making inorganic, non-biodegradable particles embedded in the tissues disappear. So, the only thing it is possible to do with what is available is palliate the symptoms.

We were asked to work with the firefighters who lent their work to the disaster site. They had not worn respirators as it would have been necessary and they all inhaled very fine dust. Not having been taken the right precautions, they also ingested it along with the food they consumed on the spot. The consequence, before evolving to far more serious illnesses, was that they manifested signs of severe, persistent and uncontrollable cough and chronic fatigue.

A special project of detoxification was supported by the Council of Lower Manhattan. Since the ill firefighters showed signs of some relief after physical exercise, saunas and the intake of vitamins they had started to take regularly, what we did was analyze their sweat, a sweat that was curiously, sometimes, dark and brownish.

In that liquid we found particles. Descriptions and images can be found in Chapter 8 of *Nanopathology* [9]. That analysis is particularly important because it witnesses to a partial capacity of the body to eliminate not only molecules but also very fine particles, in that case responsible for the unusual color.

We do not know how efficient sweat is as a means to remove particulate matter because we are not aware of its quantity in the organism. We cannot tell where the particles we detected came from, if it was some organ or the blood, and we cannot tell how long those particles had been in the bodies of those firefighters. The only indisputable thing is that they were there and that, somehow, sweating may help to get rid of them.

In 2014 Mount Sinai Hospital's World Trade Center Health Program reported that more than 2,500 rescuers and responders were diagnosed with cancer – that's a significant increase from the previous year's 1,140 cases [27–31].

One of the lessons the tragedy of 9/11 taught is that terrorism has consequences that were unimaginable until a few years ago. Most of the dust generated by the collapse of the Twin Towers is not degradable: so, once it has been formed, the only possible, though partial, remediation is its natural dilution in the atmosphere.

Having predicted a predictable situation and no one having taken the correct countermeasures is very sad [32].

7.12 HAVE HIROSHIMA AND NAGASAKI BEEN MISINTERPRETED?

In the section above we wrote that, at least in some circumstances, catastrophes can be seen as experiments otherwise inaccessible and even inconceivable. What happened in 1945 between August 6 and August 9, first in Hiroshima and then in Nagasaki, may be taken as an example: about 630,000 people taken as guinea pigs yielded most of what we know about the effects of radioactivity on humans. Although, we are not sure that everything was interpreted and understood the right way to become a now-undisputed heritage of the received medical wisdom.

In that short lapse of time, two technically different bombs, both producing an extremely high temperature (about 10 MK) and a great quantity of heat, reduced to aerosol much of what was there and can be found in any city. No one can tell how much dust the two explosions produced, not to mention the fires that followed and that lasted relatively long, nor did anyone wonder. No doubt: given the temperature and the extent of the phenomenon (the mushroom clouds (Figure 7.12) were more or less 18-km high and two whole medium-size cities were involved), probably the greatest quantity of nanoparticles ever generated in such a short time was produced, most of them non-degradable. The self-evident consequences are that an impossible-to-quantify fraction of those particles must have been inhaled and ingested by the survivors, and not only within areas limited to those of the two cities, and that much of that dust, now extremely diluted, must still be present somewhere on this planet. That obviousness has been utterly disregarded.

To boot, that ultrafine dust was also radioactive and some of it, visible as black rain, fell to the ground of what was left of Hiroshima not long after the explosion. Figure 7.13(a) shows the traces of that rain on a wall (courtesy of the Director of the Hiroshima Peace Memorial Museum). As in all other similar cases, as soon as the rain dried the dust was lifted and air-borne.

Figure 7.12 Images show the historical mushroom-clouds generated by the explosions of the atomic bombs in Hiroshima and Nagasaki [33, 34].

The Hiroshima Museum gave us a few samples of that dust: some deposited on the jacket of a young boy (Figure 7.13(b)) and some deposited on miscellaneous objects (a roof tile, also melted, a safe, a ventilation window).

Most people involved in the explosions showed symptoms that are recognized as typical to radio-activity exposure, but many symptoms reported

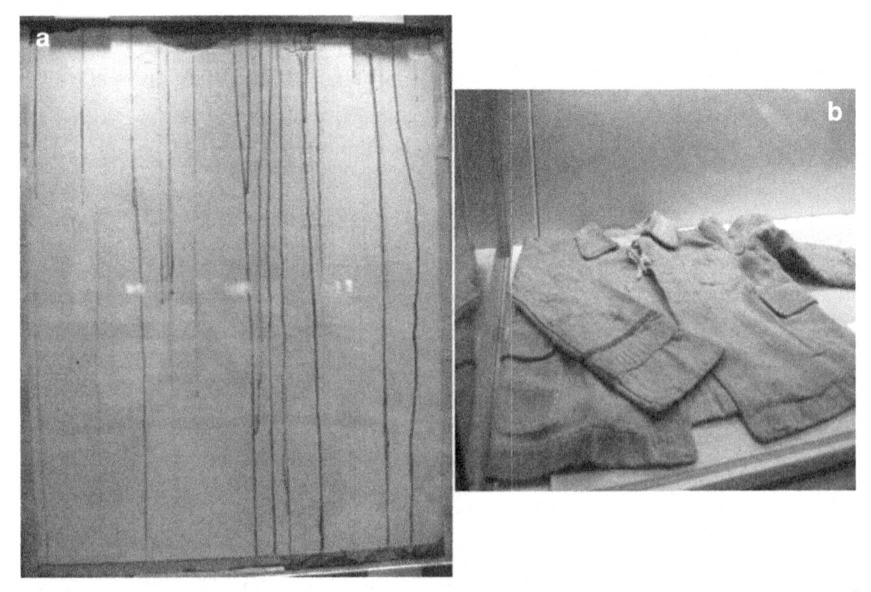

Figure 7.13 Images of objects present in the A-bomb Museum of Hiroshima: residue of a house wall where lines of the "black rain" are visible (a) and child's jacket that survived the explosion (b).

by the local population are those suffered by whom is exposed to dust or, in some cases, they are very similar.

It is easy to see that the dust must have been inhaled and ingested by those who survived the explosion and also by people who lived relatively far from there. Unfortunately, nobody took into any consideration such a truism.

The particles we analyzed were obtained by putting a carbon adhesive disk on the objects' surface. They represent examples of pollutants that the survivors could have inhaled (Figures 7.14A and 7.14B).

Though on a totally different scale, what happened in New York on 9/11 is in a way similar to what happened 56 years earlier in Hiroshima and Nagasaki: people inhaled great quantities of fine and ultrafine dust concentrated in a very short time. Then, inhalation and ingestion continued for no one knows how long. If that phenomenon was the cause of a considerable number of pathologies in New York, there is no reason to suppose that the

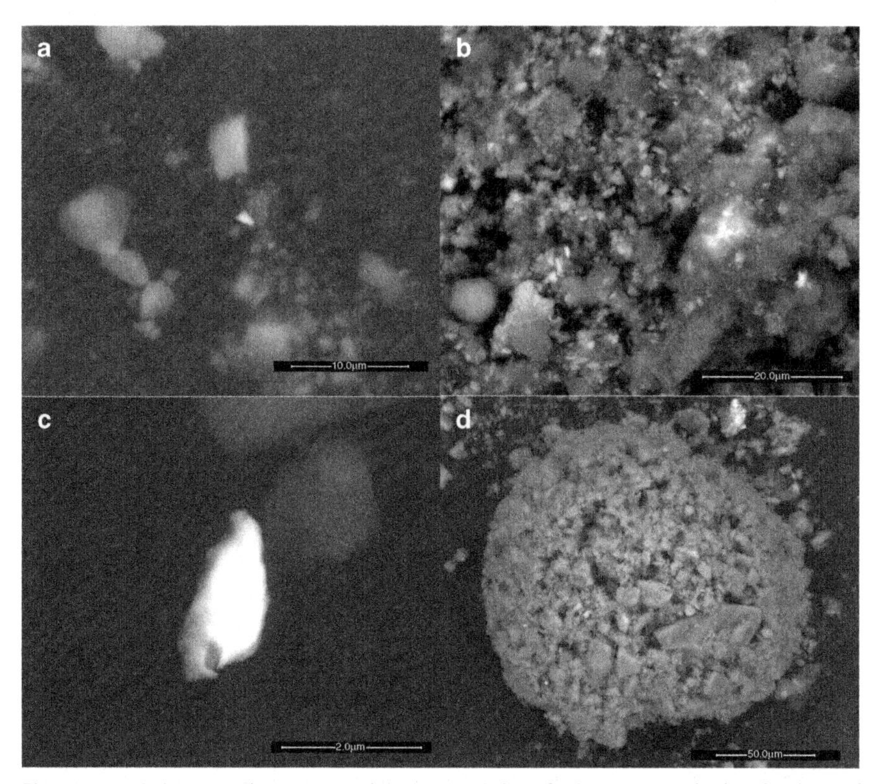

Figure 7.14A Images show particulate matter identified, respectively, by the line of black rain on the wall (a), on the jacket (b), on a safe surface (c) and on a melted roof tile (d).

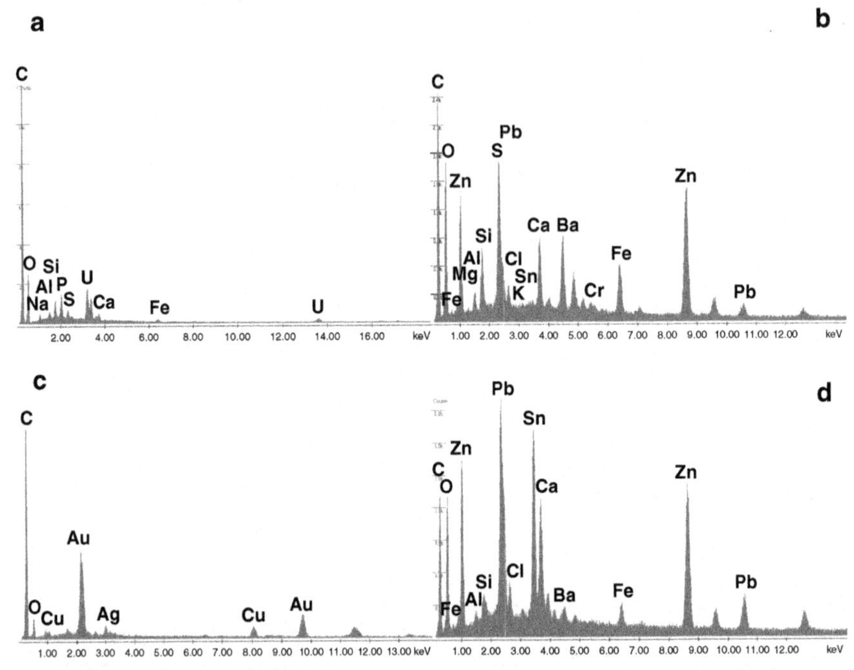

Figure 7.14B Images show EDS spectra of some debris identified in the objects mentioned in Figure 7.14A. They are composed of uranium-phosphorus-silicon-aluminium-sulphur-sodium-calcium-iron (a), lead-sulphur-zinc-barium-calcium-silicon-iron-aluminium-chlorine-magnesium-potassium-tin-chromium (b), gold-silver-copper (c) and lead-tin-zinc-calcium-chlorine-silicon-aluminium-barium-iron (d).

same pathogen was harmless in Japan. To make things worse in Hiroshima and Nagasaki, the quantity of dust was certainly far greater; because of the much-higher temperature the particulate matter's size was certainly smaller and, as a consequence, much more penetrating, most likely as far as cell nuclei; and, because of how it was generated, the dust was radioactive.

We believe that too much has been taken for granted and some things are based more on faith than scientifically-based objectivity. So, we think it is necessary to define events and data in a more complete way than has been done so far.

To that aim, we propose to investigate the issue through these three convergent approaches:

1. Analyze the objects stored at the Hiroshima Museum looking for particulate matter and characterize it by size, shape, composition and frequency.
2. It is reasonable to suppose that, because of the Coriolis effect, the dust produced at Hiroshima and Nagasaki has traveled toward the North

Pole and has fallen on the ice. An ice core taken at the depth corresponding to the period of the two explosions and just after could very likely contain their dust. That ice could be melted and the particles could be easily recovered and characterized.

3. Analyze the cancerous or, anyway, the pathological biopsy samples preserved in paraffin blocks at the Hiroshima Hospital [35] and compare the particles present there, if any, with those possibly found in the ice cores.

We cannot know what the results of such a research will be, but we are convinced that the study is worth carrying out, if only to confirm that what science has believed so far is right.

Again, seeing the Japanese tragedy as a cold laboratory, those two "experiments" showed clearly how dangerous high radiation doses can be, which supplied an evidence that could be easily extrapolated: even low doses can be very harmful if the exposure is repeated. For that well-founded reason the pieces of equipment that now work with radiation use the lowest possible quantities and people who are exposed to radioactivity for professional necessities are kept under constant control. Our hope is that 9/11, as Hiroshima and Nagasaki did, has taught something about the behavior of micro- and nanoparticles.

7.13 A BRIEF CONCLUSION

One may wonder why soldiers and civilians suffer from different pathologies if the pathogen is in any case dust?

The differences are due to exposure, which is hardly the same even when two people are exposed to the same dust in the same place at the same time. As the degree of complexity of a living being grows, the differences in biological behavior between individuals become more and more pronounced. And man is doubtless the most complex and complicated being living on this planet.

Nanopathology is a newborn discipline and we can put forward a hypothesis, but polymorphism could play an important role in the different reactions to the same stimulus represented by the same dust. But being "the same" is a very rare occurrence. Except for engineered particles, the dust we are confronted with is extremely variable by size, shape, chemistry and density and the changes in those four parameters are very quick. So, if for convenience particles may be considered "one" pathogen, we must be conscious of its inherent huge differences.

Another important aspect is represented by the health conditions of the subject. If, for instance, he/she is a smoker and his/her respiratory system is more or less compromised with the vibratile cilia unable to get rid efficiently of the particles entered in the bronchi, the probability of contracting respiratory disease will increase. If an inflammation, even a mild one, is ongoing in an organ, it is only natural that a larger-than-usual amount of blood will flow to that organ. The consequence is that more particles will reach that tissue and will be captured there and not released. Age too must be considered. Due to their small size, children are more easily attacked by particles.

Chance too may play a role. It is impossible to guess where a particular volume of blood is going to go and, with it, the particles it may contain. And if some particles manage to enter cell nuclei, a thoroughly stochastic event, the biological/pathological scenario will take a dramatic turn.

Finally, there is the much-disregarded aspect of synergism among pollutants. Often particles, besides being mutually different, are present in the environment along with a huge variety of polluting chemicals and, as a matter of fact, nobody can tell exactly what those combinations can cause in an organism.

As to war pollution, the problem is that war activities create new toxic waste that can contaminate the environment and its inhabitants leaving a long-lasting legacy that every involved country "is poorly equipped to manage" [36].

REFERENCES

[1] Durakovic A, Horan P, Dietz LA, Zimmerman I. Estimate of the time zero lung burden of depleted uranium in Persian Gulf War veterans by the 24-hour urinary excretion and exponential decay analysis. Mil Med 2003;168(8):600–5.
[2] Galletti M, D'Annibale L, Pinto V, Cremisini C. Uranium daily intake and urinary excretion: a preliminary study in Italy. Health Phys 2003;85:228–35.
[3] LM, Heller J, Kalinsky V, Ejnik J, Cordero S, Oberbroekling KJ, et al. Military deployment human exposure assessment: urine total and isotopic uranium sampling results. J Toxicol Environ Health A 2004; 67(8-10): 697-714.
[4] McDiarmid MA, Squibb K, Engelhardt SM. Biologic monitoring for urinary uranium in Gulf War I veterans. Health Phys 2004;87:51–6.
[5] Technical report of the Air Force Armament Laboratory – Armament development and test Center, Eglin Air Force Base, Florida, USA, From October 1977 to October 1978, Project n° 06CD0101.
[6] Michael AP, Cornette JC. Morphological Characteristics of Particulate Material Formed from High Velocity Impact of Deleted uranium Projectiles with Armor Targets, AFATL-TR-78-117, Eglin AFB, FL, United States Air Force Armament Laboratory, Environics Office, November 1978).

[7] http://www.gulflink.osd.mil/du_ii/du_ii_tabl1.htm].
[8] Gilchrist RL, Glissmyer JA, Mishima J. Characterization of Airborne uranium from Test Firings of XM774 Ammunition, PNL-2944. Richland, WA: Battelle Pacific Northwest Laboratory; November 1979.
[9] Gatti A, Montanari S. Nanopathology: the health impact of nanoparticles. PanStanford pub. Singapore. 2008:5;161-200.
[10] Gatti A. New constituents and particle sizes herald new health dangers from pollution. NANO Magazine 2014;29:1–3.
[11] Gatti A, Montanari S. Engineered nanoparticles, natural nanoparticles and nanosized-by-products. NANO magazine 2012;26:18–9.
[12] Gatti A. 9/11 and nanoparticles: when one doesn't like to be right. NANO magazine for small science 2012;26:20–1.
[13] Gatti A, Montanari S. Nanopathology: A Controversial Aspect of Nanomedicine. NANO Magazine 2011;22:18–21.
[14] Gatti A, Montanari S. Nanoparticles: a new form of terrorism? in Proceedings of the NATO-ASI Conference in Chisinau, Moldova 07-17 June 2010 entitled "Technological innovations in detecting and sensing of chemical biological radiological, nuclear threats and ecological terrorism. edited by A.Vaseashta, E. Braman, P. Susman, Publisher SpringerScience & Business Media BV, Dordrecht, the Netherlands. 2012: 45-53.
[15] Gatti A, Montanari S. Unintended Nanoparticles: The most dangerous yet? Military Problems and Nanotechnology Solutions. NANO magazine 2009;15:30–4.
[16] Gatti A, Montanari S. Nanopollution: the invisibile fog of future wars. The Futurist 2008;(May-June):32–4.
[17] Friedman SM, Farfel MR, Maslow CB, et al. Comorbid Persistent Lower Respiratory Symptoms and Posttraumatic Stress Disorder 5-6 years Post-9/11 in Responders Enrolled in the World Trade Center Health Registry. American Journal of Industrial Medicine 2013 Jun 21.
[18] Brackbill RM, Hadler JL, DiGrande L, et al. Asthma and Posttraumatic Stress Symptoms 5 to 6 Years Following Exposure to the World Trade Center Terrorist Attack. JAMA: The Journal of the American Medical Association 2009;302(5):502–16.
[19] Huang MJ, Li J, Liff JM, et al. Self-Reported Skin Rash or Irritation Symptoms Among World Trade Center Health Registry Participants. Journal of Occupational and Environmental Medicine 2012 Mar 22.
[20] http://www.nyc.gov/html/doh/wtc/html/registry/registry.shtml.
[21] Jordan HT, Stellman SD, Prezant D, et al. Sarcoidosis Diagnosed After September 11, 2001, Among Adults Exposed to the World Trade Center Disaster. Journal of Occupational and Environmental Medicine 2011;Aug :19.
[22] Bowler RM, Harris M, Li J, et al. Longitudinal Mental Health Impact Among Police Responders to the 9/11 Attack. American Journal of Industrial Medicine 2011;Dec: 27.
[23] Jordan HT, Miller-Archie SA, Cone JE, et al. Heart Disease among Adults Exposed to the September 11, 2001 World Trade Center Disaster: Results from the World Trade Center Health Registry. Preventive Medicine 2011;Oct: 28.
[24] Li J, Brackbill RM, Stellman SD, et al. Gastroesophageal Reflux Symptoms and Comorbid Asthma and Posttraumatic Stress Disorder Following the 9/11 Terrorist Attacks on World Trade Center in New York City. American Journal of Gastroenterology 2011;Sep: 6.
[25] Lipkind HS, Curry AE, Huynh M, et al. Birth Outcomes Among Offspring of Women Exposed to the September 11, 2001, Terrorist Attacks. Obstetrics & Gynecology 2010;116(4):917–25.
[26] Li J, Cone JE, Kahn AR, et al. Association between World Trade Center Exposure and Excess Cancer Risk. Journal of the American Medical Association 2012 Dec 19;308(23):2479–88.

[27] Thorpe LE, Friedman S. Health Consequences of the World Trade Center Disaster: a 10th Anniversary Perspective. Journal of the American Medical Association 2011;306(10):1133–4.
[28] Jordan HT, Brackbill RM, Cone JE, et al. Mortality Among Survivors of the Sept 11, 2001, World Trade Center Disaster: Results from the World Trade Center Health Registry Cohort. Lancet 2011;378(9794):879–87.
[29] Cone JE, Farfel M. World Trade Center Health Registry—A Model for a Nanomaterials Exposure Registry. Journal of Occupational and Environmental Medicine 2011;53(6 Suppl):S48–51.
[30] http://www.foxnews.com/politics/2012/09/10/federal-government-to-acknowledge-ground-zero-cancer-link-for-first-time/#ixzz267MeKCjW.
[31] http://nypost.com/2014/07/27/cancers-among-ground-zero-workers-skyrocketing.
[32] Gatti A. 9/11 and nanoparticles. NANO magazine 2012;26:20–1.
[33] http://www.warhistoryonline.com/war-articles/atomic-bomb-hiroshima-eyewitness.html.
[34] http://it.wikipedia.org/wiki/Bombardamenti_atomici_di_Hiroshima_e_Nagasaki.
[35] Ron E, Preston DL, Mabuchi K, Thompson DE, Soda M. Cancer Incidence in Atomic Bomb Survivors. Part IV: Comparison of Cancer Incidence and Mortality Radiation Research 1994;137:98–112.
[36] http://www.toxicremnantsofwar.info/isafs-environmental-legacy-in-afghanistan-requires-scrutiny/.

CHAPTER 8

Food, Drugs and Nanoparticles

Contents

8.1 Introduction 163
8.2 Intentional and accidental contamination of food 170
8.3 Bovine spongiform encephalopathy and food 180
8.4 Vaccine contamination 187
References 193

8.1 INTRODUCTION

Hardly a day goes by without new nanoproducts being introduced and, even more commonly, would-be nanoproducts entering the market. So, creating an inventory is all but impossible and the reasons are numerous. Leaving aside the speed with which new products are released, there are products supposed to be "nano" that do not contain nanoparticles or have not been produced through Nanotechnology while, on the other hand, there are products that actually do contain particles, even nano ones, without being declared.

The amount of nanoproducts available on the market is already enormous, and increases every day. It is not always clear in what countries each product is available. Figure 8.1 shows the importance of nanoproducts in the different fields of applications, while Figure 8.2 presents the types of nanoparticles used in the different nanoproducts and their relative importance. Some producers deliberately omit information about the presence of nanoparticles in their products, as there is no worldwide specific regulation requiring the obligatory notification of all nanomaterials intentionally added to a product. The situation may look somehow paradoxical: as already mentioned, some companies claim the presence of non-existent nanoparticles, while others are shy about it, fearing a lack of acceptance by customers uncertain about the safety of nanoparticles. In any case, even when nanoparticles are reported correctly in the product's label, no quantities, size ranges and chemical final formulations in the products are declared [1].

A further possibility is that the manufacturer actually added nanoparticles but, possibly due to an insufficient knowledge of how nanoparticles

Case Studies in Nanotoxicology and Particle Toxicology
http://dx.doi.org/10.1016/B978-0-12-801215-4.00008-X

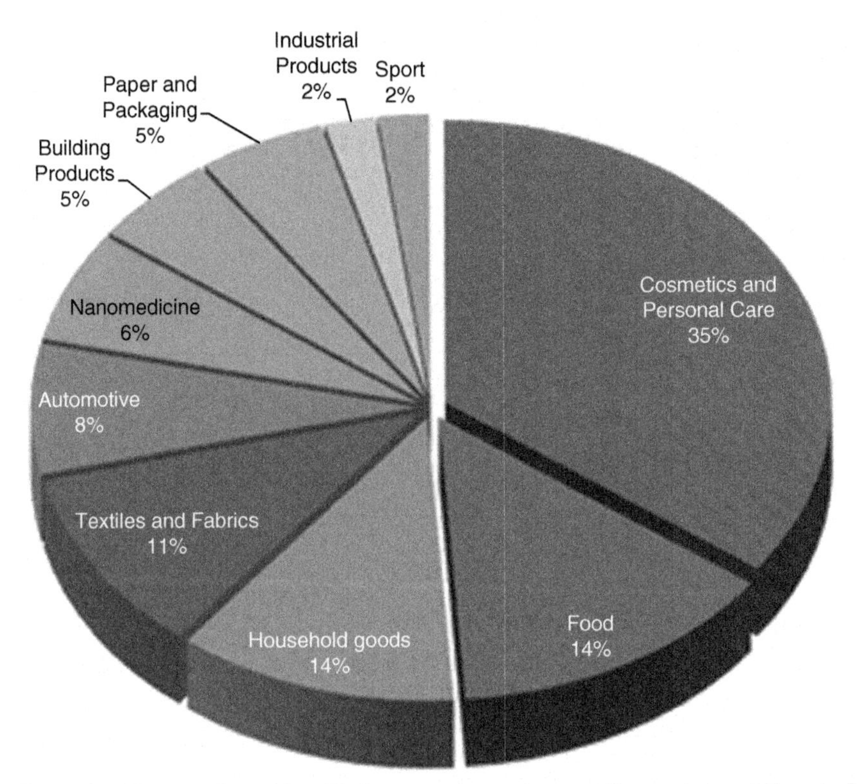

Figure 8.1 Nanoproducts identified per product-category. *(Source: Internal Report of INESE Project).*

behave, and, as a consequence, due to an incorrect handling, those particle aggregated, losing their original properties (Figure 8.2).

In the course of a project (INESE) we coordinated on nanoecotoxicity for the Italian Institute of Technology, we made a limited investigation on commercial nanoproducts. Our results obtained through our usual ESEM observations indicate that 31% of the products claimed as nano do not contain nanoparticles or only micron-sized aggregates (Figure 8.3).

Just as a couple of examples: in Figure 8.4 no silver, but only iron and titanium particles have been detected in the NanoCare NanoSilver Towel® (Korea). The unusual, undeclared presence of those small structures could be due to the industrial process used to manufacture those products and a consequent contamination. There is the possibility that the quantity of nanometric Silver added is below the instrument's sensitivity, but, if that is the case, the antibacterial property and activity are incorrectly estimated. A further example is reported in the NanoUP Toothpaste Au-Ag® (Japan)

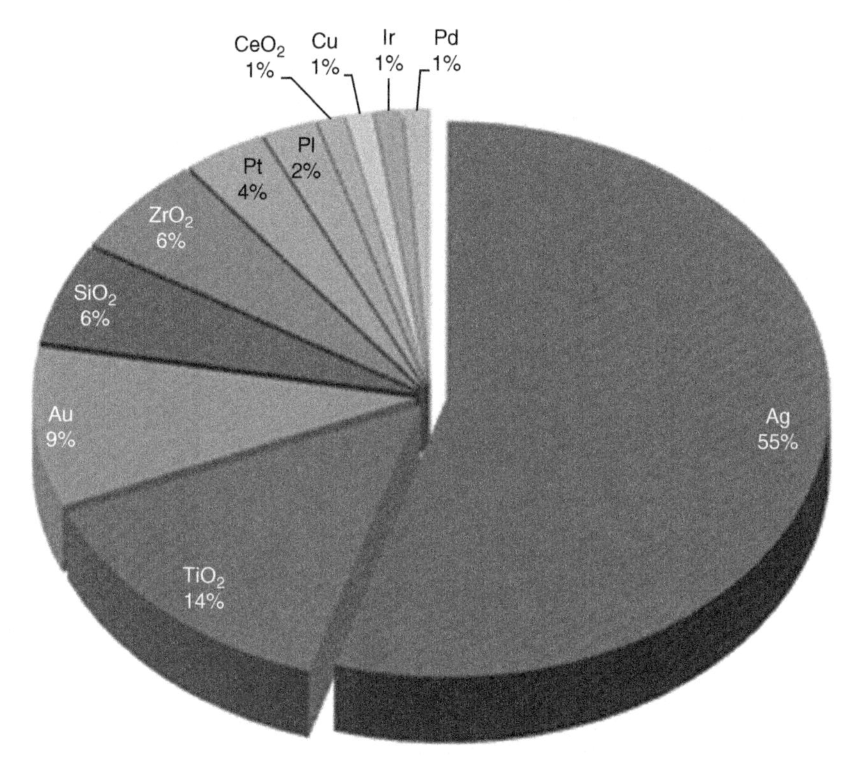

Figure 8.2 Chemical composition of nanoparticles in nanoproducts. *(Source: Internal Report of INESE Project).*

where gold nanoparticles have not been found (Figure 8.5) and a large amount of titanium-rich nanocontaminations has been identified in the NanoCare NanoSilver Beauty Soap® (Korea), where no silver nanoparticles have been found (Figure 8.6).

Additionally, in most of the samples, the nanoparticles intentionally introduced in the nanoproducts are hardly submicrometric or are even micrometric entities (Figure 8.6) and are not truly nanosized materials, as nano is bureaucratically understood to refer to size.

The following images show the presence of nanosized additives in products. Of course, there is the possibility that some nanoparticles were not seen by the FEG-ESEM especially those with a size below 10 nm and if present in isolated form. In many cases we identified them gathered together as aggregates.

Small aggregates of 100 nm silver nanoparticles have been found in the NanoCare NanoSilver Shampoo® (Korea), as illustrated in Figure 8.7. No

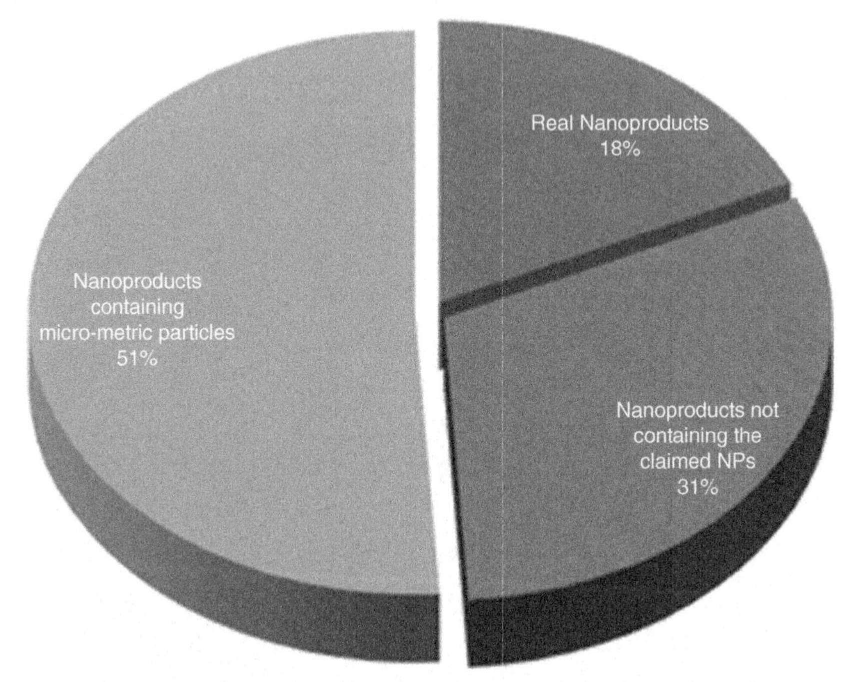

Figure 8.3 Results obtained by ESEM investigation of products claimed as nano *(Internal Report of INESE Project).*

significant research has ever been carried out regarding the effects of silver nanoparticles in repeated and very often long-lasting contacts with the skin.

In Figure 8.8, nanostructured agglomerates can be clearly seen in a paint additive containing TiO_2 nanoparticles dried at room temperature. With the ageing of the matrix, these aggregates can be released by the paint and

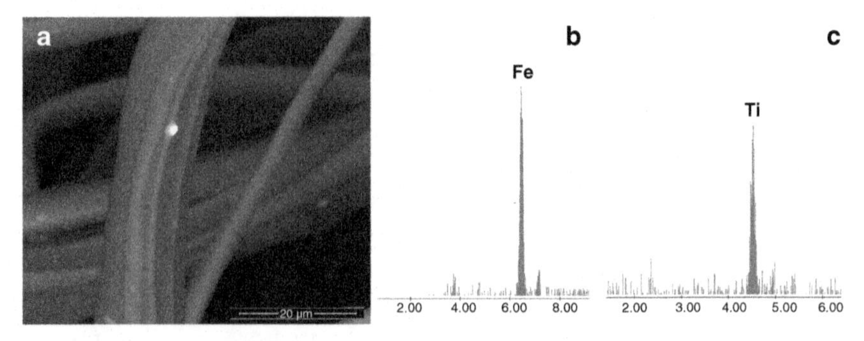

Figure 8.4 Image of a thread of a towel (a) where no nanoparticles of silver are detected but only iron- and titanium-rich contaminations have been found, as reported in the corresponding EDS spectra (b, c). There are no peaks of silver.

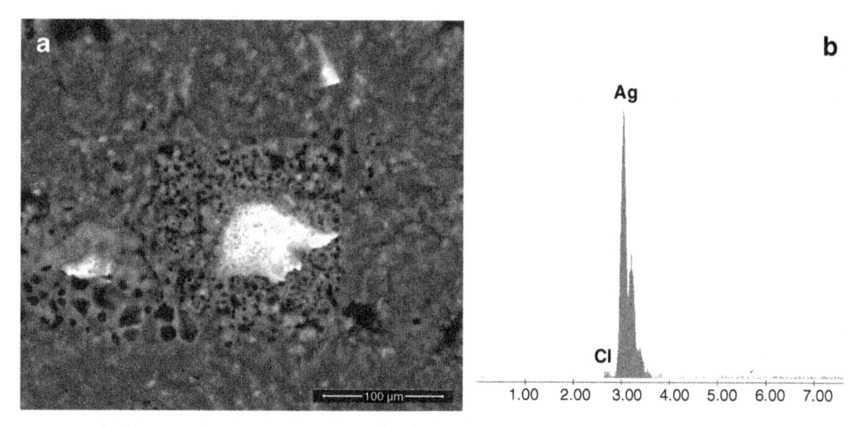

Figure 8.5 Low-magnification photograph of NanoCare NanoSilver Toothpaste (a) and EDS spectrum (b): silver micrometric debris have been detected.

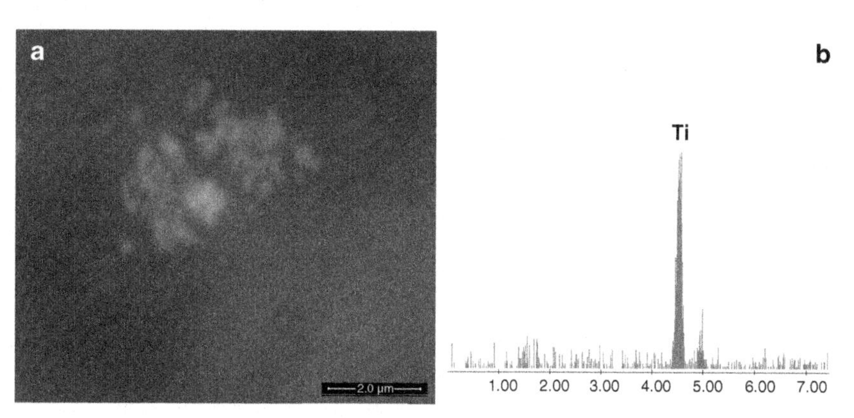

Figure 8.6 Image of NanoCare NanoSilver Soap (a) and EDS spectrum (b): a titanium-rich agglomeration of nanoparticles is recognizable.

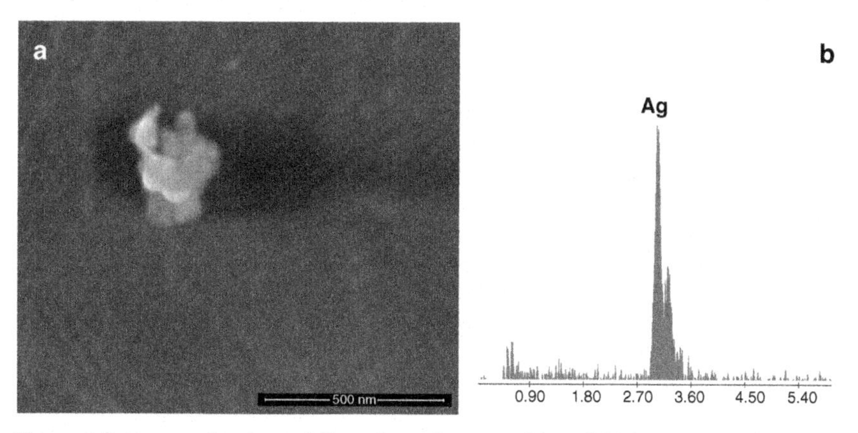

Figure 8.7 Image of a drop of NanoCare shampoo (a) and EDS spectrum (b): aggregated silver nanoparticles can be clearly seen.

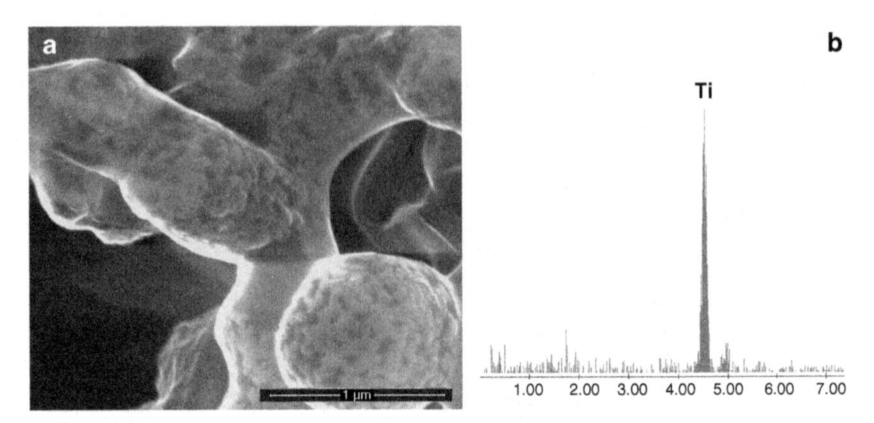

Figure 8.8 Image of paint additive enriched with clustered TiO_2 nanoparticles deposited on a paper filter (a): nanostructured agglomerations are clearly recognizable. EDS spectrum related to the aggregate (b).

contaminate the room where the persons live and breathe. Again, no exhaustive research on the subjects exists.

In conclusion, ESEM-EDS analyses reveal that only 18% of the samples analyzed confirm the presence of the nanosized particles claimed by the manufacturers.

These results suggest that a large part of nanoproducts available on the market could be considered "no-nanoproducts" and the information reported in the product's data sheet does not correspond to reality.

Figure 8.9 shows the content of titanium-dioxide nanoparticles in an anti-wrinkle cream (L'Oreal, France). The effect of smoothing the skin may be due, at least partially, to very thin wrinkles being mechanically filled by nanoparticles. As a matter of fact, the skin does not show any actual biologically observable improvement.

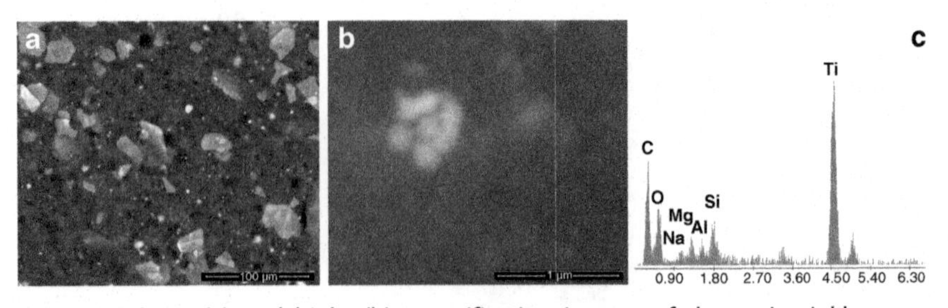

Figure 8.9 Low- (a) and high- (b) magnification images of the anti-wrinkle cream Revitalift: particles with a diameter < 250 nm, can be seen. EDS spectrum related to particles (c).

In our investigations we verified a few water filters used to sanitize drinking water, whatever the reason to use them on water distributed through controlled networks, as is done in most developed countries. As to "unsafe" water available elsewhere, their efficacy is dubious to say the least.

Nanosilver is more and more frequently used in many products mainly because of its well-known antibacterial properties: garments, air conditioners, washing machines, paints, toothbrushes and in the tap-water filters or the barrier pitchers to kill bacteria [2] in drinking water.

Silver has a bacteriostatic and, maybe, even a bactericidal action, but is also a well-known pesticide and so has been investigated by the EPA in the US: "Silver, a naturally-occurring element, is registered for use in water filters to inhibit the growth of bacteria within the filter unit of water filter systems designed to remove objectionable taste, odors, and color from municipally-treated tap water; these bacteriostatic water filters account for over 90% of its pesticidal use" [3], and ingesting a pesticide is something hard to justify. At present farmers can find commercially available nanopesticides whose biopersistence in the environment and in agricultural products is mostly unknown, with all the consequences linked, for instance, to microflora and microfauna.

Among others, we tested one of those commercially available filters (Brita, US) and found that silver nanoparticles had migrated into the water used for drinking (Figure 8.10).

Unfortunately, people who use those filters are not informed about the fact that they can ingest silver nanoparticles along with the water they drink and are aware only of the positive side of the issue, i.e., their water contains

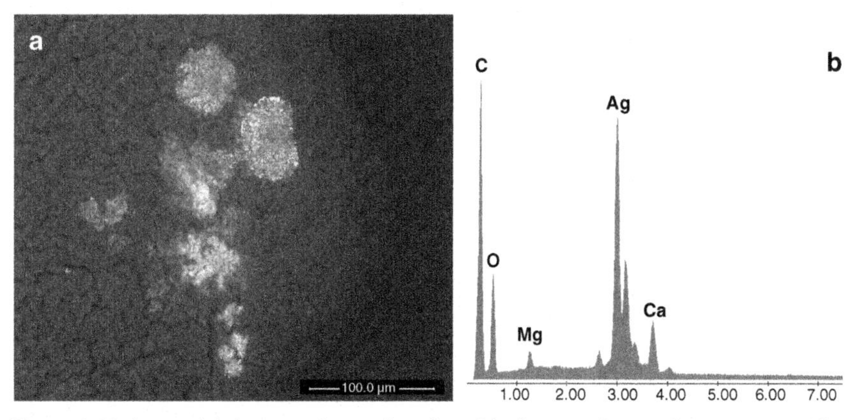

Figure 8.10 Image (a) shows carbon spherules with clusters of nanosilver nanoparticles (b) in a water filter for pitchers.

less living bacteria than regular tap water. Nor do most of them know that the chronic intake of silver can cause argyria [4] and that silver particles, like all non-biodegradable particles, can be the origin of nanopathologies.

One of the things we noticed in our investigations is that there is another discrepancy between what is declared in the label and what can be found in the products. Manufacturers who add nanoparticles to their products do not check if the status of nano is finally maintained. The tendency of nanoparticles to aggregate and the interactions with other components of the product may transform nano into microaggregates, with the consequences mentioned above.

In many cases, nanoparticle addition has no rational justification, and not only because they do not have the promised effect. Putting them in soaps or in washing machines, for example, is hardly more than a commercial gimmick with the side effect of having them end up in drain water that eventually and inevitably is released into rivers, lakes and oceans, poisoning plants and fish and entering the food chain. In Australia, the organization Friend of the Earth has already identified some hazards [5] in the nanoparticles of titanium dioxide and zinc oxide added to sunscreen that can react with the sunlight and harm the phytoplankton.

8.2 INTENTIONAL AND ACCIDENTAL CONTAMINATION OF FOOD

One of the reasons why engineered nanoparticles are used more and more in the preparation of industrial food is because they help protect it from bacterial contamination and allow it to enjoy a longer shelf life. But they are also used because of their capacity to enhance the food's look and, in some circumstances, its texture and even its taste. So, as a matter of fact, nanoparticles have deliberately entered the food chain.

Again, as a matter of fact, as occurs with virtually all nanotech applications regardless of the application field, no exhaustive study has ever been carried out to investigate the safety of those additions. In the best of cases (e.g., FDA regulations) the only burden producers are asked to bear is demonstrate that their products are not harmful, but the issue is still too complicated (and too expensive) to have protocols available and is still the object of debate from points of views that have nothing to do with science. So, little is actually done and the demonstrations of harmlessness are too often perfunctory. The consequence is that additions keep being incorporated in food for humans and for animals [6, 7].

Food packaging can also be treated with nanoparticles, from bottles to some kinds of wrappings, and in most cases nobody can vouch that those particles do not leach into food or beverages, concurring to an unintentional and in most cases disregarded contamination of food and drink [8].

Actually, nanofood does create some concerns, so that the Food and Agriculture Organization of the United Nations (FAO), some years ago, summoned worldwide-known experts to debate the problem. The report [9] we delivered discusses pros and cons and contains some simple criteria to be applied to obtain a safe(r) nanofood: high solubility of the nanocontent, no biopersistence, and bioavailability also of the possible transformed products. It is only obvious that business for the sake of business can hardly be advantageous if safety is the issue [10].

For the time being, waiting for more, easy to predict, nanotechnological additions, the consequence of particulate presence in food is just incidental pollution of environmental origin, and most, though not all, of it is represented by the fallout of the dust generated by combustions.

Fruit and vegetables are particularly exposed to that kind of pollution (see Chapter 6), inasmuch as, with the exception of what is grown in greenhouses, they do not enjoy any shelter from what falls from the sky. Fruits like, among others, oranges are usually peeled before being eaten and, therefore, the dust deposited on the skin is eliminated. Other fruits like, for example, cherries can be efficaciously washed, but vegetables like cabbages and other brassicaceae have a very rough surface where particles get easily trapped, and even a very careful wash can do very little. Paradoxically, a greenhouse-grown vegetable is more trustworthy than one grown in a more natural way. Figure 8.11 shows as an example what we found on a peach skin. Particles of titanium or titanium oxide, iron, manganese, silicates:

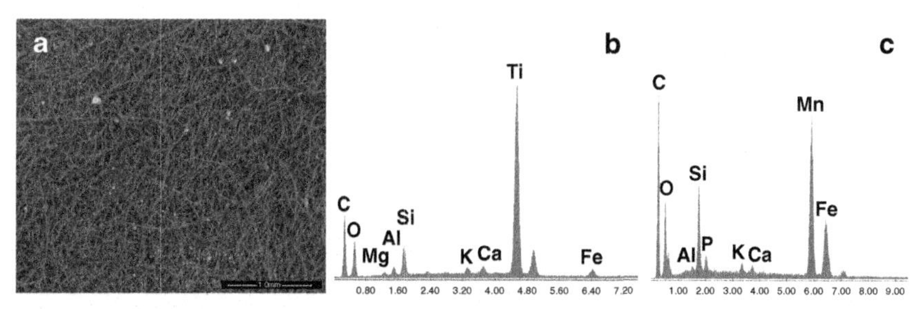

Figure 8.11 Image of peach skin with environmental particulate matter of different chemical compositions (a); debris of titanium-silicon-magnesium-potassium-calcium-iron (b); and particle of carbon-manganese-silicon-iron-phosphorus-calcium-potassium-aluminium (c).

inorganic, non-biodegradable debris without any nutritional value, but liable to contaminate the digestive system and the whole organism.

Many territories around industrial areas are heavily polluted and growing vegetables there is not advisable, to say the least. In some circumstances, the quality of the products grown there is so alarming that the local authorities had to forbid their farming. For instance, in some cases authorities forbid the consumption of vegetables grown around incineration plants, power stations or stainless-steel foundries both for humans and animals.

Cereals are usually milled without being previously cleaned and the dust deposited on the corns is then inevitably found in the flour. So, the products obtained with that flour keeps the environmental pollution and can increase the content of pollutants by the addition of other polluted compounds.

Figure 8.12 shows the debris identified in an Italian biscuit (Bistefani, Italy). In the small sample we analyzed, we identified three inorganic

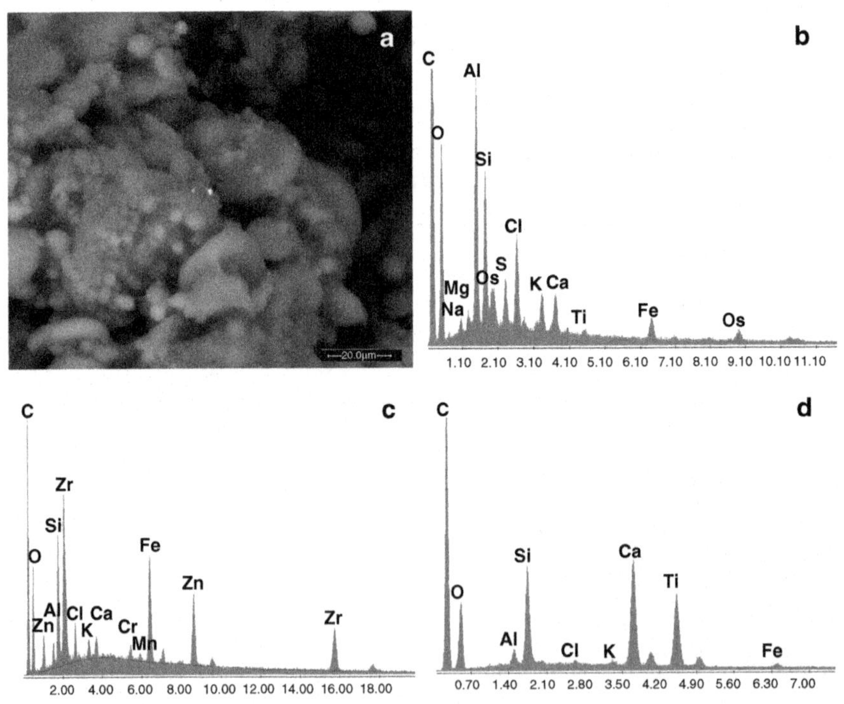

Figure 8.12 Image of an area of a biscuit where three different inorganic debris are visible (a). They are composed of aluminium-silicon-chlorine-sulfur-osmium-potassium-calcium-magnesium-titanium-iron (b), zirconium-silicon-iron-zinc-aluminium-chlorine-calcium-potassium-chromium-manganese (c), and silicon-calcium-titanium-aluminium-chlorine-potassium-iron (d).

debris containing, respectively, aluminium-silicon-chlorine-osmium, etc.; zirconium-silicon-iron-zinc-chromium, etc.; and silicon-calcium-titanium-etc.; to mention only the most-represented ones. It is obvious that that kind of food contains carbon, oxygen, sodium, chlorine, calcium, magnesium, potassium, but it is not as obvious, and certainly not healthy, that it contains elements such as osmium, zinc, chromium, etc., especially if they form non-degradable compounds.

Those pollutants could have been contained in the raw material (e.g., flour) and, in that case, it is likely that their presence is sporadic and their variety changes with every incoming consignment. But, as we saw more than once, it could also be due to the industrial process (e.g., wear of tools and machinery) and, if that is the case, procedures and all production instruments should be checked. In any case, the search for those pollutants is never done and does not belong to the received procedures of Good Manufacturing Practice [11, 12].

We analyzed a considerable number of industrial food products and the following are just a few examples: a homogenized children's food (Figure 8.13), an industrialized cold dessert similar to ice cream (Grand Soleil, Ferrero, Italy) (Figure 8.14), a chocolate snack (Mars, Mars Italia, Italy) and a piece of mortadella, a cold-cut pork meat. Mortadella is a very

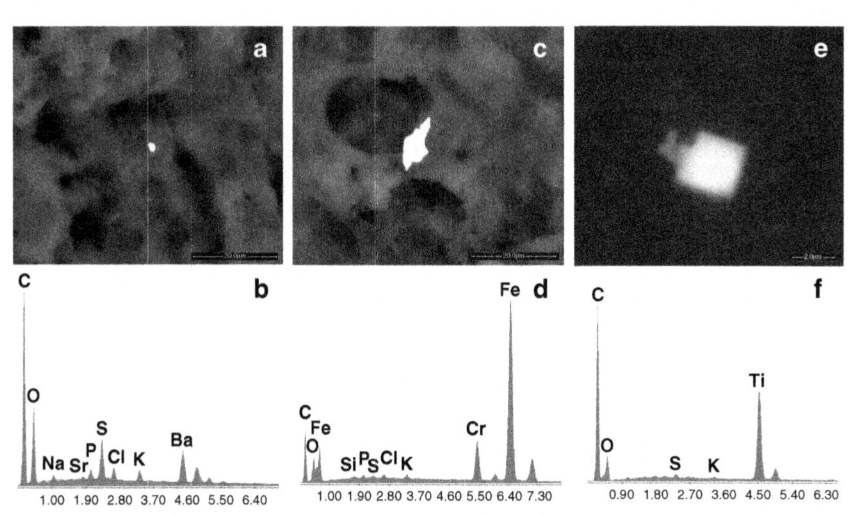

Figure 8.13 Images (a, c, e) show the debris found in a homogenized children's food. Some debris were identified as made of sulfur-barium-strontium-sodium-phosphorus-chlorine-potassium (b), iron-chromium (stainless steel),-silicon-phosphorus-sulfur-chlorine-potassium (d), and titanium-sulfur- potassium (f).

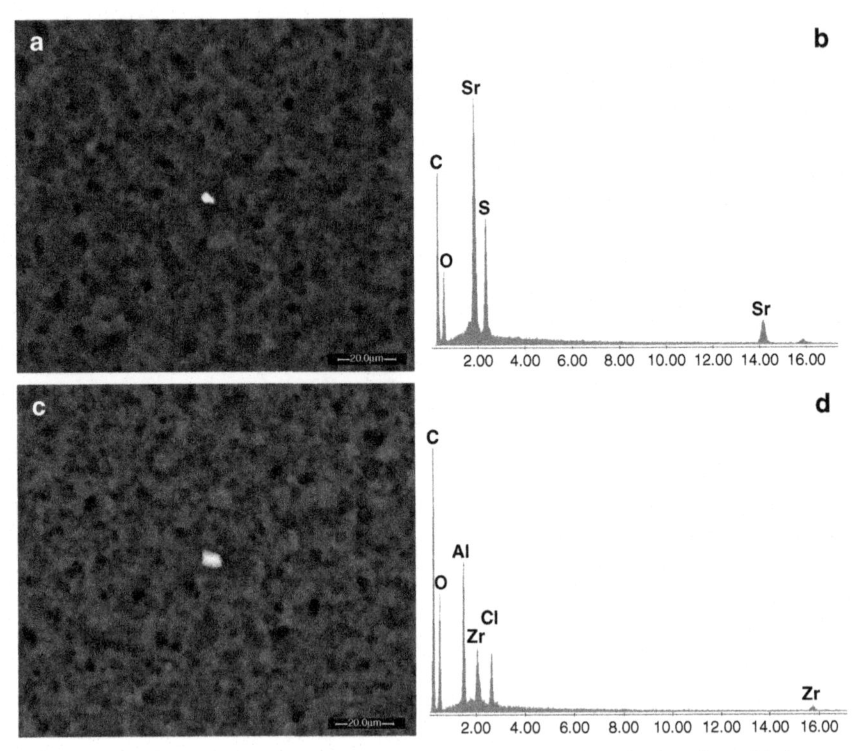

Figure 8.14 Images (a, c) show some micrometric debris identified in the cold dessert. They are composed of strontium-sulfur (b), and aluminium-chlorine-zirconium (d).

popular Italian cold cut, made by finely hashing frozen pork meat, and, to do that, stainless-steel blades are generally used. Having to cut great quantities of hard meat, they quickly lose their edge, so, for that reason, self-sharpening tools are used. It is obvious that the residues of the sharpening action end up in the final product and, in fact, many stainless-steel particles the size of some tens of microns can be found at the center of grey-brownish spots that are the reaction between fat and what results from the corrosion of that metal. In some samples we did not find any debris (Felsineo, Italy). All the other samples, with no exception, showed a particularly high presence of debris with different compositions: barium-sulfur-iron-chromium (the alloy of a sort of stainless steel), titanium (Figure 8.13), strontium-sulfur and aluminium-chlorine-zirconium (Figure 8.14), iron-copper-chromium (Figure 8.15), iron-chromium-manganese-nickel and iron-copper (Figures 8.16A and 8.16B). It is hard to tell the origin of some

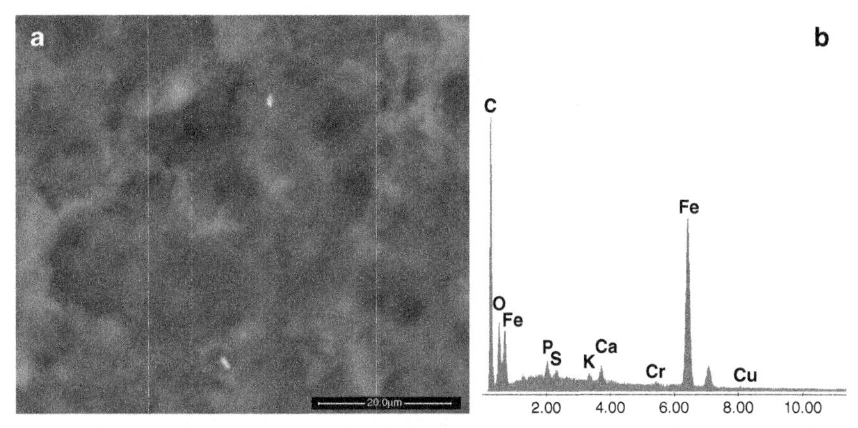

Figure 8.15 Image (a) shows two submicronic debris of stainless steel found in the chocolate crust of Mars. They are composed of iron-phosphorus-calcium-sulfur-potassium-chromium-copper (b).

debris. It is probable that some stainless-steel debris came from the wear of cutting knifes, blades or grinding tools. The only sample we found free from solid debris came from meat treated with the blades made of a particularly wear-resistant alloy.

A case that shows clearly how, probably unexpectedly for many people, pollution can travel is that of a loaf of bread we got from Tierra del Fuego, the southernmost tip of Argentina where there are no industries, car traffic is extremely scarce and no wheat is grown. In that bread we found particles of lead-calcium-chlorine-sodium-chromium, copper and gold. The presence of sodium-chlorine is understandable for the salt added to the bread, but it is difficult to explain the presence of lead-chromium or copper-gold

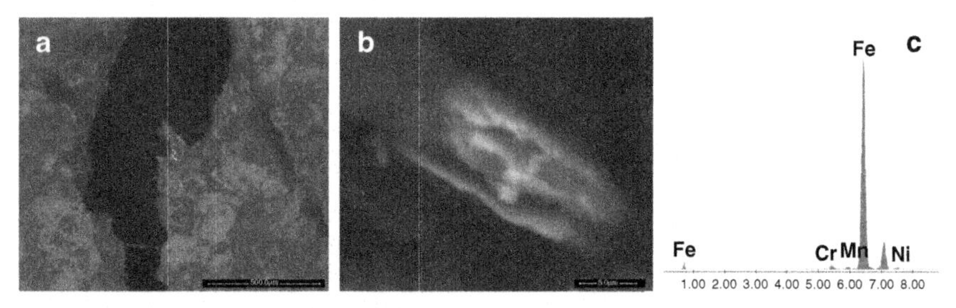

Figure 8.16A Images (a, b) show debris of a type of stainless steel identified in a type of mortadella. They contain iron-chromium-manganese-nickel (c).

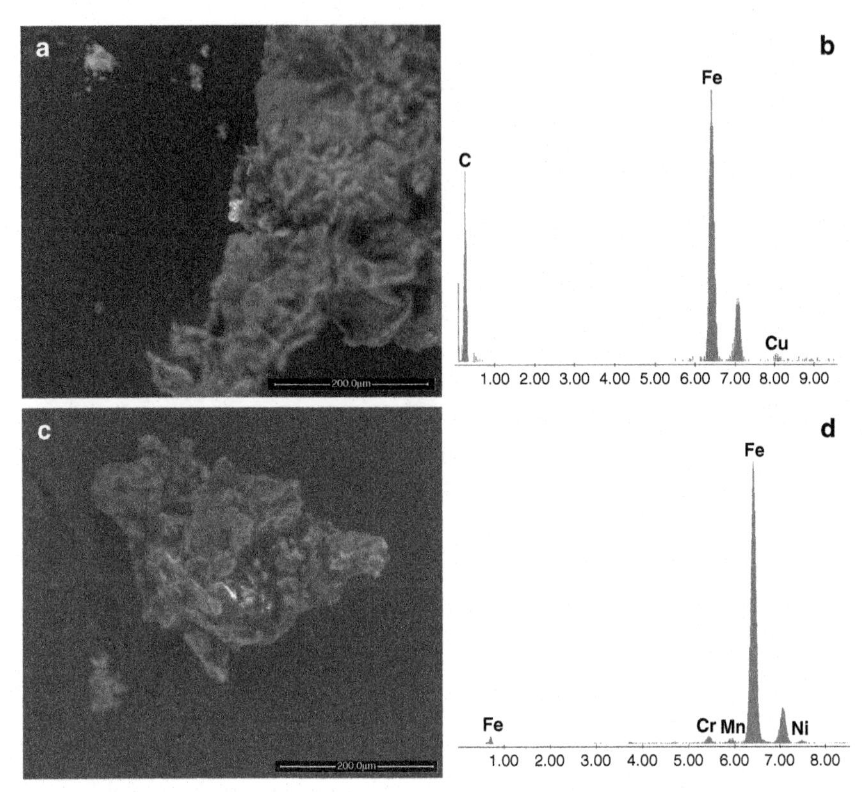

Figure 8.16B Images (a, c) show metallic debris embedded in the mortadella; they are composed of iron-copper (b), and iron-chromium-manganese-nickel (d).

(Figure 8.17). In that particular case, flour was imported from other territories and the pollution we found was imported as well.

In another sample of bread, Italian in that circumstance, we found tungsten carbide, a substance probably used to coat the grinding mills because of its anti-wear properties. If we guessed right, the non-biocompatible, biopersistent tungsten carbide we saw in the bread was there in spite of its anti-wear properties.

The presence of metal debris (e.g., nickel) we keep finding in many of the samples we check could explain some forms of food intolerance shown more and more frequently among the population.

Other unintentional, indirect food contaminations can come from non-stick cookware, products that, in a way, can be considered nanotechnological because of the addition of micro- and nanoparticles in their polymeric matrix.

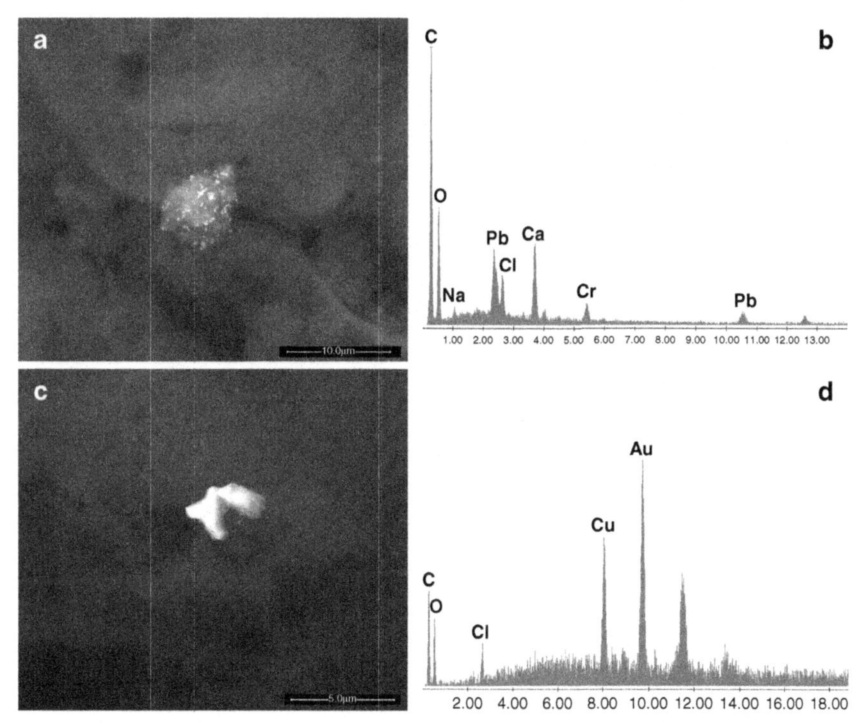

Figure 8.17 Images (a, c) show debris identified in a loaf of bread from Tierra del Fuego (Argentina). They contain lead-calcium-chlorine-chromium-sodium (b) and gold-copper-chlorine (d).

We had a chance to analyze some of those products, all manufactured by different companies, and found that all of them lose a more or less considerable part of the non-biodegradable, non-biocompatible particles embedded in their non-stick layer. Inevitably, those particles end up in the food. Figure 8.18A shows the debris remained attached to an adhesive tape simply by touching the non-stick layer of a pan surface. They are made of aluminium-silicon (c), but there are also metallic nanostructured debris present. While cooking, those particles are partially transferred to the food and are, inevitably, ingested. In another type of pan we found a similar release of debris but, in that case, the particles were composed of carbon-gold-zinc, etc., and of carbon-oxygen-silicon-barium-chromium-iron-copper, etc. (Figure 8.18B).

Once we analyzed a hamburger bought at a well-known fast-food restaurant, and in it we found silver particulate-matter aggregates. There may be more than one explanation to that undue presence, one possibility being

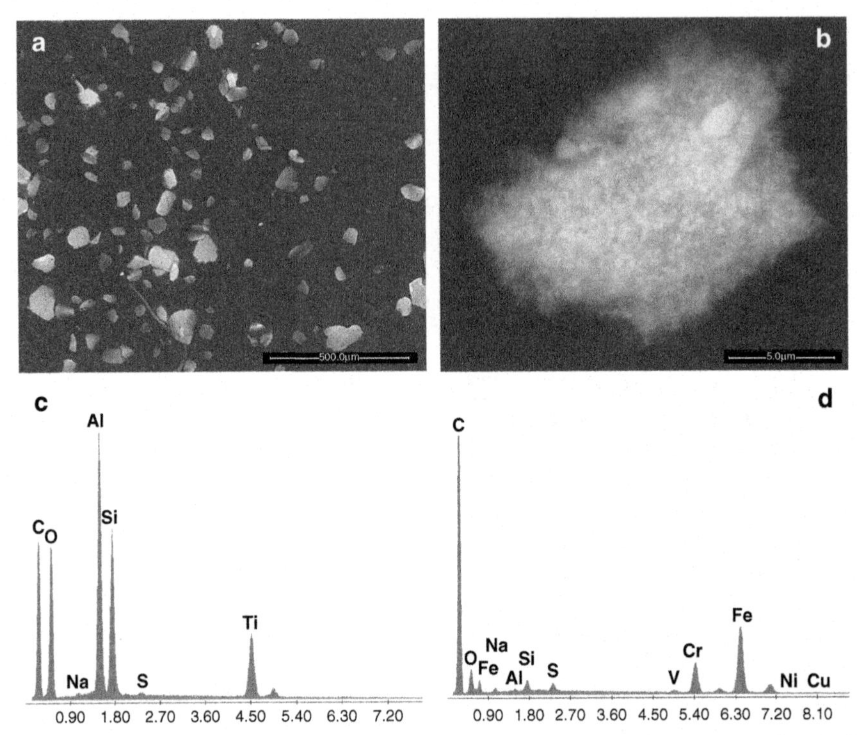

Figure 8.18A Image (a) shows the debris attached to adhesive tape; its main chemical composition is aluminium-silicon-titanium-carbon-oxygen-sodium (c). Image (b) shows a nanostructured debris containing particles of carbon-iron-chromium-oxygen-nickel-vanadium-copper-silicon-sulfur-aluminium (d).

that it was originally contained in the muscle of the animal and ended up there because silver is used as a pesticide, thus contaminating the hay the cattle were fed with. Without suggesting any unproven correlation, just as additional information and nothing more, we found very similar aggregates in a colon-cancer patient. As to hay, we had a chance to analyze a sample upon the request of a public prosecutor who was working on a bovine spongiform encephalopathy (BSE) case, and silver particulate was what we found there.

What is obvious is that if biopersistent particles enter into the food chain, animals and humans have a high probability to ingest them and their permanence in the body can trigger adverse biological reactions (Figure 8.19).

With increasing awareness regarding the dangers of eating contaminated food, organic food is growing more and more popular. The few organic food samples we analyzed with the aim of searching for particulate

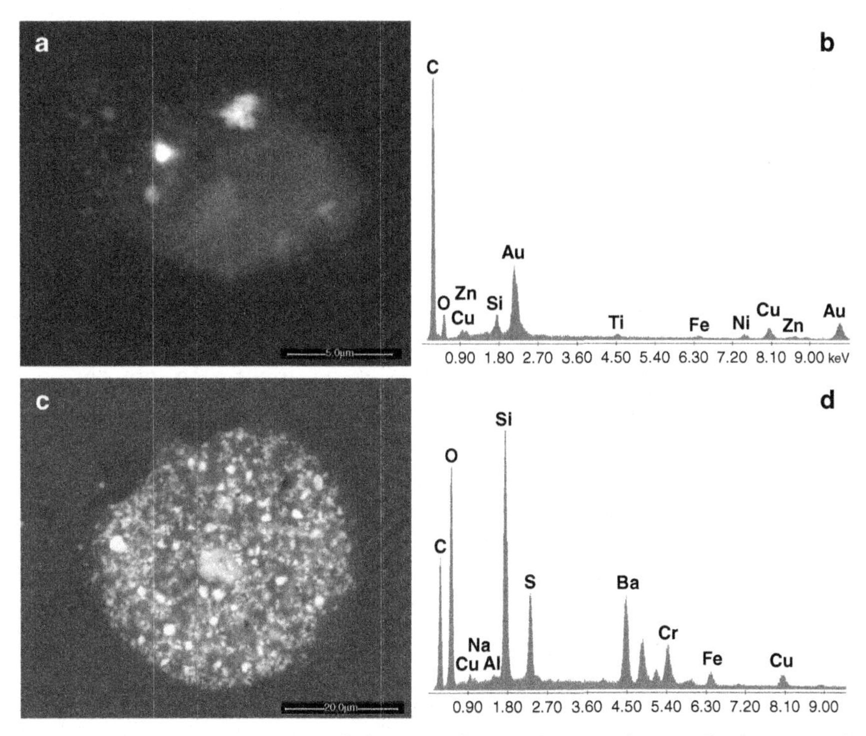

Figure 8.18B Images (a, c) show debris identified on the tape that touched a non-stick layer. Particles of gold-silicon-titanium-iron-nickel-copper-zinc (b) and an aggregate of submicronic debris of silicon-sulfur-barium-chromium-iron-sodium-aluminium-copper (d) were identified.

contamination did not show any significant differences compared to non-organic food. We can say the same thing regarding biodynamics. In both cases, the problems involving particulate contamination are ignored either by the law and by the producers. Organic and biodynamic productions are limited to the obligation to observe the steps of certain processes but the controls of the results are not particularly stringent. At least not as far as particulate contamination is concerned.

If food for human use may be polluted by micro- and nanoparticles, what is intended for animal feed shares the same condition. Years ago we acted as the consultants to a public prosecutor who tried to unmask an illegal traffic of additives for animal feed. In that case, the product was a compound used to increase milk production in cows. What we found was a ground mineral containing mainly bismuth, something obviously not bio-available to a cow's metabolism and far from beneficial.

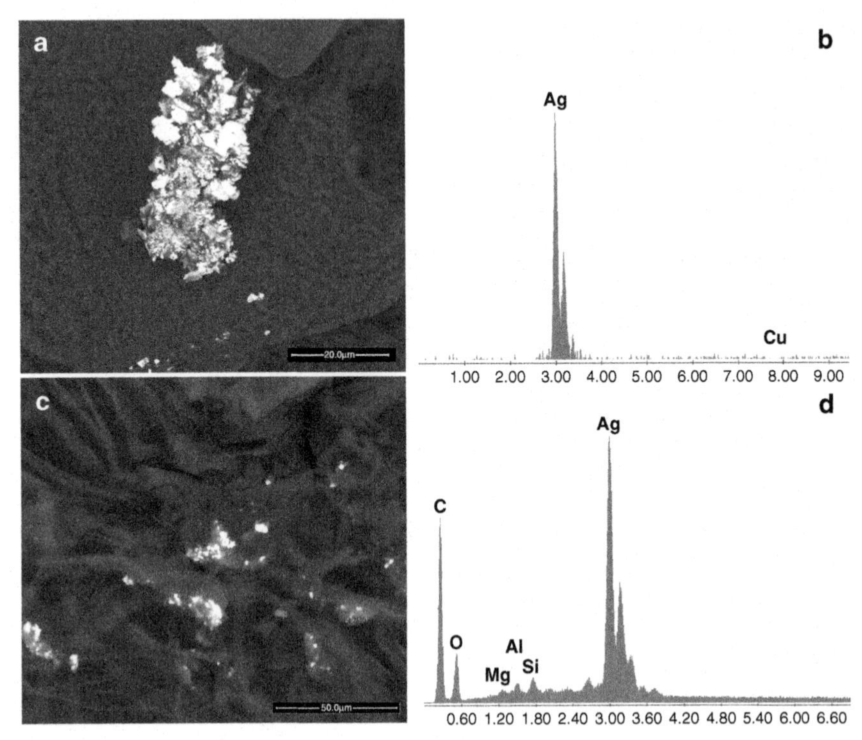

Figure 8.19 Image (a) shows a silver aggregate (b) found in an industrial hamburger. The image (c) shows also submicronic silver debris (d) identified in dried hay.

In the case of the shooting range of Quirra (see Chapter 7), for decades the local shepherds fed sheep and cattle with fodder grown under the pollution caused by the explosion of armament and by fuel for missiles. Controls are often disregarded.

8.3 BOVINE SPONGIFORM ENCEPHALOPATHY AND FOOD

Bovine spongiform encephalopathy (BSE) is a degenerative neurological disorder of cattle, but similar pathologies affect sheep (scrapie) and humans (Creutzfeldt–Jakob's disease). Kuru ("to shake"), very possibly a variant of the illness, was endemic in some tribal regions of Papua New Guinea until 1957, when ritual, funerary cannibalism was banned.

Gerstmann–Sträussler–Scheinker syndrome and fatal familial insomnia belong to the same group. Whatever the classification, all those illnesses are fatal. In fact, in the 1990s there was a sort of "epidemic" in Great Britain, and all governments took severe measures to prevent its diffusion, fearing the possibility of transmission to other species. The European Union banned the exports of British beef in effect from March 1996, thus creating an economic crisis involving the alimentary, cosmetic, biomaterial and medical fields.

So far, the nature of the pathogen, an agent that must be transmissible if the pathology is infectious as is officially declared to be, is not well understood and is still debated. The most widely accepted theory is that the agent is a modified form of a normal protein known as prion protein (PrP). This discovery, or theory, earned Dr. Stanley Prusiner the Nobel Prize [13, 14]. It is not within the scope of this book to discuss, criticize, accept or refute the prion theory. Instead, we limit ourselves here to a report in which we analyzed 20 samples of brain we received from the Veterinary Laboratories Agency of Scotland of cattle affected by BSE and our conclusions.

There are a few reasons why we were interested in analyzing those samples:

1. The simultaneous explosion of BSE cases in the UK did not follow the normal behavior and trend of an infectious epidemic. Its rapid disappearance as soon as the farmers stopped feeding their cattle with animal flour was, for instance, a matter worth considering more attentively than how it was done.
2. BSE is similar to Creutzfeldt–Jakob's disease (CJD) in humans, a pathology believed to be just as infectious. In 1972, years before that disease was known, Dr. Prusiner was in contact for a certain period with a patient affected by a form of dementia whose post-mortem revealed it to be CJD. Not aware of its alleged infectivity, Dr. Prusiner as well as other doctors took no precautions, but none of them developed CJD. Additionally, as far as we know, no cases were reported in literature of the transmission of the pathology among members of the same household or the staff of a hospital where a patient had been admitted.
3. The experiments performed to isolate the pathogenic agent involved heating the pathological brain tissue of animals affected by BSE at 600°C. The rats that received injections of the "cooked" BSE-infected cerebral matter developed a disease similar to BSE, and that event drove scientists to report that the material was infectious. However, we were not sure the sequence of events was correctly interpreted [15,16].
4. In contrast to other infectious agents, prions do not contain nucleic acids. According to theory, they are not living organisms but misfolded proteins that may replicate by transmitting a misfolded protein state, a condition that makes them extremely resistant to chemical and physical denaturation [17,18].

We think there is another possible and logical explanation to the phenomenon worth considering that has the peculiarity of being demonstrable. As always, we are open to discussion and documented criticism and are ready to accept the rational demonstration that we are wrong.

The FEG-ESEM and EDS analyses we performed on the brain samples showed the presence of foreign bodies. Figures 8.20, 8.21, 8.22, 8.23 and 8.24 show some of the debris we identified in a BSE-affected brain. The samples examined contained mostly metallic debris with very different, unexpected chemistry.

Those pieces of evidence allow us to explain some of the "prion" phenomena in a different way. As mentioned above, it is a fact that the so-called BSE epidemic in the UK disappeared very quickly when the practice of feeding cattle with animal flour was discontinued. Then it was supposed that the contamination was in the flour that had not been properly sterilized. We have reasons to believe that the flour was involved in the problem,

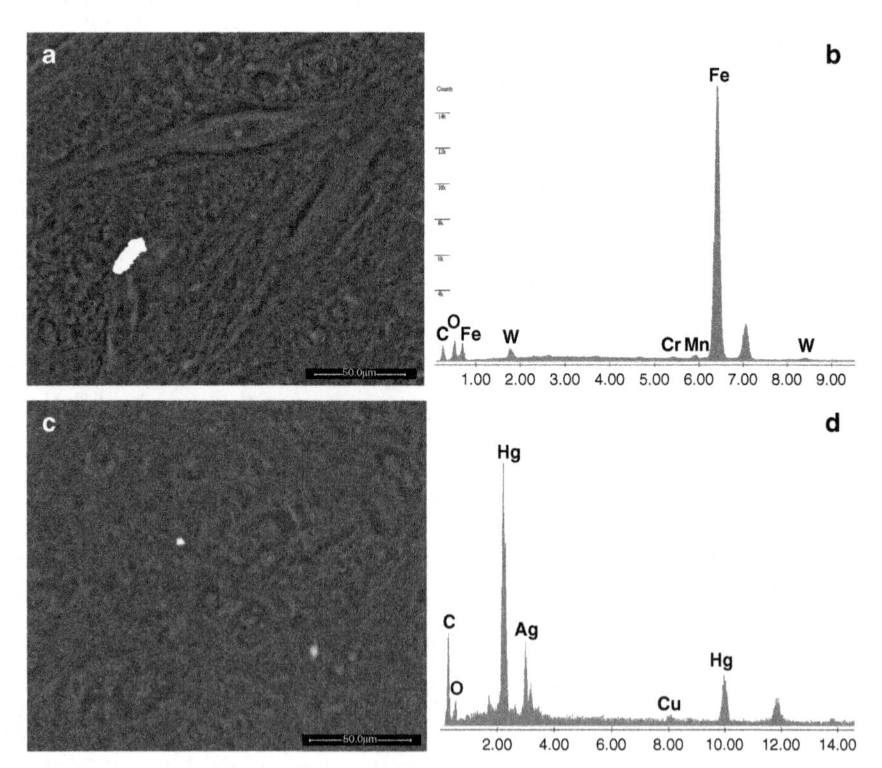

Figure 8.20 Images (a, c) show debris identified in a BSE-affected brain. They are composed of iron-chromium-manganese-tungsten (technical stainless steel) (b) and mercury-silver-copper (d).

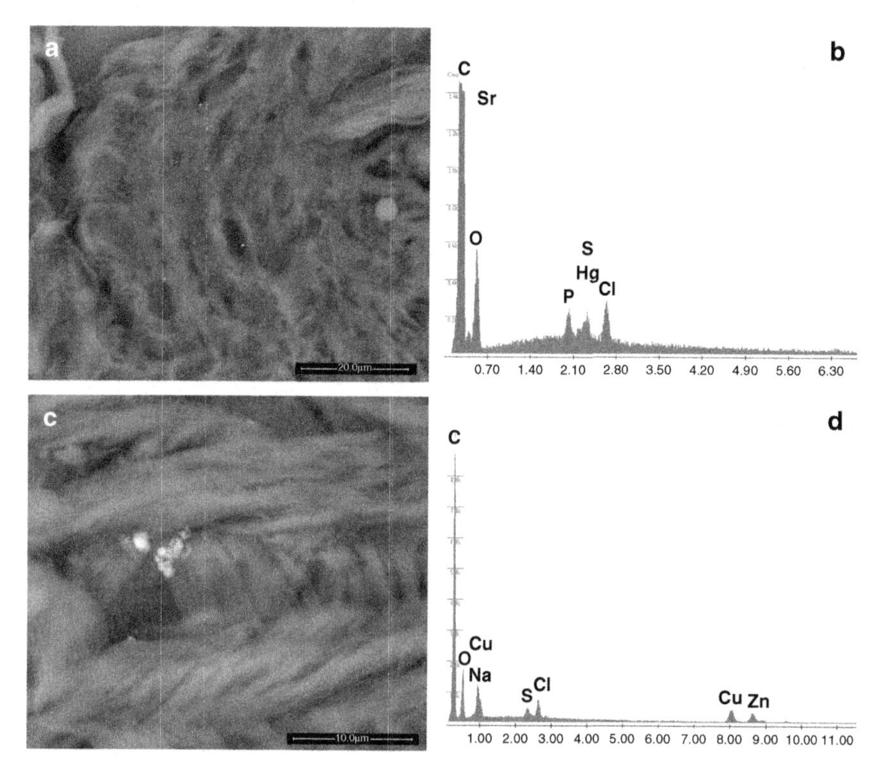

Figure 8.21 Images (a, c) show spherical nanosized debris (white dots) and a cluster, respectively, composed of mercury-chlorine-phosphorus-sulfur (b) and copper-zinc-sulfur-chlorine (d).

but that was only part of it. It must be considered that cattle cannot be fed with flour alone, given the difficulty in swallowing that dry, very coarse powder or granulate. So, it must be mixed with fluids that ease ingestion.

Water cannot be used, since it forms lumps. The only efficient way to prepare that feed is to mix the flour with oil. There is a real possibility that some

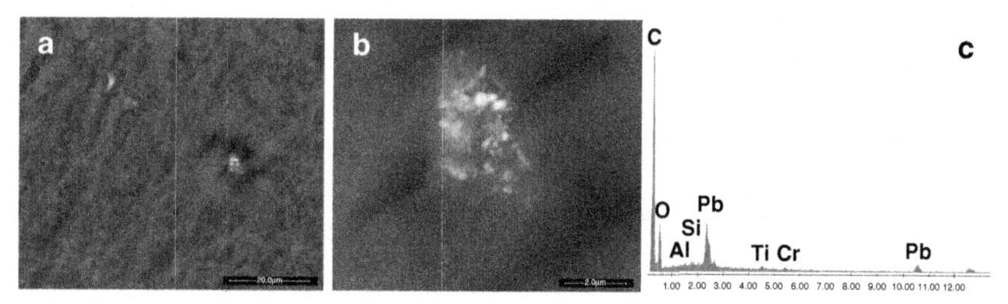

Figure 8.22 Images (a, b) show clusters of nanoparticles at different magnifications; they are composed of lead-silicon-aluminium-titanium-chromium (c).

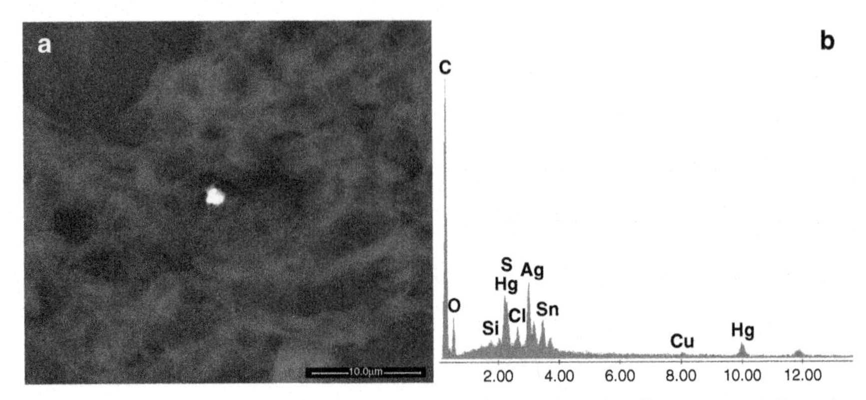

Figure 8.23 Image (a) shows a 800 nm-sized particle of sulfur-mercury-silver-tin-chlorine-silicon-copper (b).

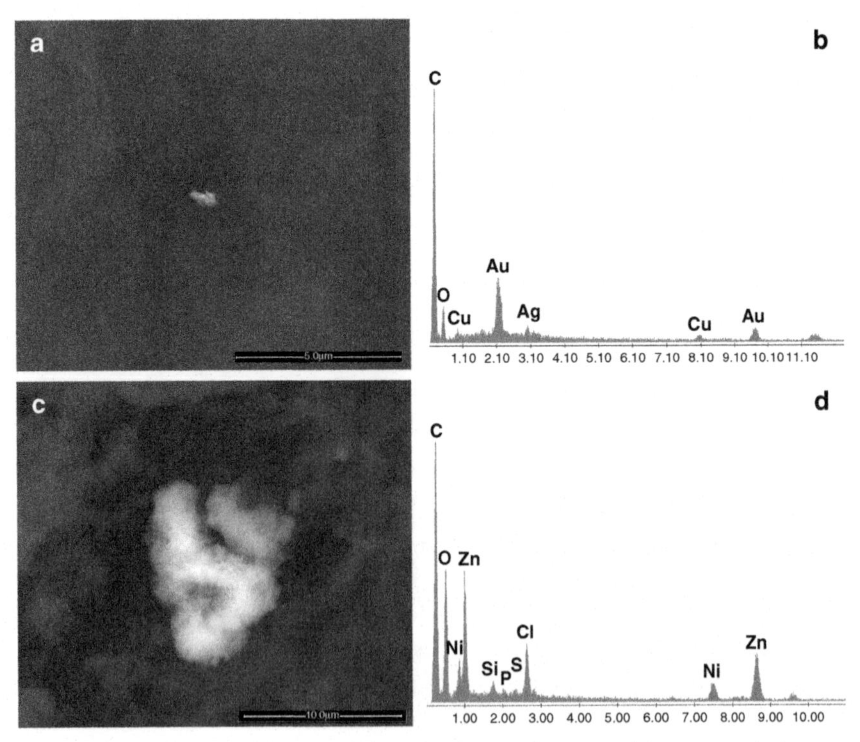

Figure 8.24 Images (a, c) show a debris and a cluster of nanoparticles, respectively composed of gold-silver-copper (b) and zinc-chlorine-nickel-silicon-phosphorus-sulfur (d).

farmers, in order to save money, used the exhausted oil of cars, a waste product that could be obtained, though not quite legally, without any cost and that contains particles generated by the mutual friction of mobile parts of engines.

If this hypothesis about preparing feed is true (we received a statement to that effect), numerous metallic particles from the friction phenomena and carbonaceous debris from the oil combustion were ingested with the flour. Once inside the body, due to their tiny size, they negotiated the digestive barrier and entered the blood flow, with a dispersion mechanism that is the same in all living beings. And the brain is one of the possible targets of those particles. Naturally those particles must have also traveled elsewhere, but the relatively short life of cattle did not allow time for the possible pathology to grow clinically manifest, yet they were immediately effective in the nervous system.

Debris with an oily layer coating gives them better interaction capability with cells (see transfection method in Chapter 3) and preserves them from corrosion and the ensuing release of metal ions. To better substantiate our hypothesis, 20 samples of BSE tissue were requested from the Veterinary Laboratories Agency of Scotland, were obtained and we analyzed them as we usually do. It must be added that the specimens were cut from the diseased brain tissues by a technician of the Veterinary Laboratories Agency without any prior specific investigation: namely, they were not selected under a nuclear magnetic resonance or with an ultrasound investigation selecting the areas presenting an inhomogeneity. The areas we were interested in were the denser or the calcified ones. In general, the interface between the normal and the diseased tissue contains more useful information. In this case, the sections we got were random. Nevertheless, 12 out of 20 samples contained foreign bodies, most of them metallic, as shown in Table 8.1.

The discussion of these results is not easy. Though unsorted and, therefore, not completely suitable for our investigation, more than half of the samples contained particles, most of which were made of calcium and of metals such as iron, iron-chromium-nickel, lead, titanium, etc.

These findings suggest different explanations for the results published in the current literature. The scientific consensus is that BSE prions are not destroyed by high-temperature procedures, meaning that even "well done," contaminated beef foodstuff remains infectious: an assumption that makes misfolded proteins pretty unique in their behavior.

What we found in the brains was metal particles that can resist high temperatures, and while the supposed 600°C is very hot for proteins, it is an easily bearable temperature for metal particles. In addition to that, those particles are foreign bodies, are pathogenic, are not degradable and

Table 8.1 List of elemental compositions of the debris identified in the BSE sections

Sample n.	Elemental	Compositions	of the	Debris		
1	FeCrNi					
2	Fe	Hg-Ag	Ni			
3	Fe					
4	AlSiCaFe	NaCl	CaFe	CaFeS		
5	MgCa	Si	Ca			
6	MgSiFe	Ca	PbBa	Fe		
7	Fe	Ca				
8	Fe	AuCuAg				
9	FeNi	FeZn	PbTiCr	FeTiMg	AlSi	BiCl
10	Fe	ZnClNiSi				
11	Fe	NiFe	HgAgSnCu			

are certainly there, without the need for the hypothesis of the existence of other pathogens with peculiar features.

The experiments aimed at demonstrating that brain tissue affected by BSE can trigger the same pathology in other animal species did not consider that that tissue could have contained particles like those we found and that the temperature they underwent left those particles unaltered.

The neurological symptoms the animals affected by BSE showed can be easily explained considering the elemental composition of the foreign bodies we identified in the brain samples. Most of them are metallic and, obviously, are good electrical conductors. For that reason, the electric field generated by the nervous structure can be altered by the presence of foreign electric conductors adjacent to it and in touch with other nervous tissue. They can also damage the myelin sheath of neurons, inducing a demyelination. The local damages caused by that condition can be expressed peripherally as tremors, abnormal posture, loss of coordination and difficulty in standing.

If metal particles cannot be degraded, their oil coating can, which can interfere with the local proteins. That biointeraction can be the cause of the formation of what is called "protein corona" (see Chapter 3). The interaction is driven by the physical-chemical surface characteristics of the debris but also Van der Waals forces can unfold the proteins, thus altering their three-dimensional morphology. Such a mutual action binds the protein in an irreversible way and enzymatic proteases are unable to destroy the link, a behavior similar to the one attributed to prions.

All that is by no means to say that prions do not exist or do not have the characteristics reported in literature: it is simply an invitation to take other possibilities into account.

Though lacking the prescribed evidence based on large numbers, we think it reasonable to suspect that the presence of similar debris can cause other neurological disorders such as Alzheimer's and Parkinson's diseases. The difference could be due to the morphology, size and chemical compositions of the particles captured by the brain. Targeted autopsies supported by nanopathological analyses could verify this hypothesis. Heavy metals have long been in the list of suspects in some neurological illnesses, but, to our knowledge, their particulate form has never been taken into consideration.

8.4 VACCINE CONTAMINATION

Few drugs are the object of controversies and disputes as much as vaccines, including the actual validity of the premarket trials that particularly the new ones undergo; the safety of the chemicals added; the appropriateness of their qualitatively-different, concentrated administration to newborn babies and to soldiers; their real efficacy; their side effects; the information released by producers and those printed in the leaflets enclosed in the package; and the reliability of the medical literature in most cases sponsored by producers are just a few of the points generally discussed and, unfortunately, unscientific opinions and lobby interests too often overwhelm objectivity.

It is not our intention to participate in any way in those discussions nor to take sides. Besides, nothing of that falls within the scope of this book. We will not go beyond reporting some of the results obtained by analyzing 27 different vaccines, all limited to one sample. This means that the results are valid only for those 27 individual specimens (with the exception of the last one, where we analyze 5 specimens); any extrapolation is not ours but its scientific legitimacy is left to the reader's mind.

In an investigation carried out for a German anatomical pathology institute we analyzed a tissue with a granuloma that was formed in a child's arm short after the injection of a vaccine [19]. In that tissue we found many debris containing aluminium. After that we decided to analyze other vaccines to check if inorganic particles were there.

It is our duty to declare that the first 19 of a total of 27 (at the time of writing this book) (see list of Table 8.2) vaccines were analyzed for the graduation thesis of Ms. Francesca Sola of the University of Parma (Italy). All 27 presented some anomalies in their composition, as we found particulate matter that should not be present in a drug destined to be injected in the human body and definitely not in a newborn baby. The quantity of each vaccine we analyzed was 20 microl deposited on a cellulose filter.

Table 8.2 List of some of the vaccines analyzed with their indications

Vaccine	Production technique	Indication	Al content declared	Producer
Inflexal V	inactivated virus	anti-flu	No	Berna Biotech
Vaxigrip	inactivated, split virion	anti-flu	No	Sanofi Pasteur MSD
Anatetall	contains anatoxin	anti-tetanus	Yes	Novartis Vaccines and Diagnostics Srl
Tetabulin	recombinant subunits	anti-tetanus	No	Baxter AG
Infanrix (DTPa) paediatric	contains anatoxin	diphtheria-tetanus-acellular-pertussis	Yes	GlaxoSmithKline S.p.a.
Stamaril Pasteur	attenuated 17 D strain	yellow fever	No	GlaxoSmithKline S.p.a.
Thypherix	Vi polysac-charide	Salmonella typhi	No	GlaxoSmithKline S.p.a.
Priorix	attenuated	measles-mumps-rubella-varicella	No	GlaxoSmithKline S.p.a.
Engerix-B	recombinant subunits	hepatitis B	Yes	GlaxoSmithKline S.p.a.
Varilrix	attenuated	varicella	No	GlaxoSmithKline S.p.a.
Dif-Tet-All	contains anatoxin	diphtheria-tetanus	Yes	Novartis Vaccines and Diagnostics Srl
Menjugate-kit	conjugate	meningococcal group C	Yes	Novartis Vaccines and Diagnostics Srl
Focetria	inactivated	H1N1-flu	No	Novartis Vaccines and Diagnostics Srl
Gardasil	recombinant	cervical cancer	Yes	Merck & Co., Inc.

All vaccines names are trademarked

It is good pharmaceutical practice to add the smallest possible number and quantity of ingredients to any active principle. That is because very few substances, if any, can be considered absolutely free from side effects, because side effects may vary even much from subject to subject and because mixing components means adding possible reactions that are more and more difficult to predict and may come totally unexpected. In the case of vaccines, the situation can be even more complex when they are administered in combinations as happens as a rule with infants and soldiers. In any case, every component, regardless of its concentration and quantity, should be mentioned in the datasheet and in the patient's leaflet if only to prevent as much as possible allergic reactions and, needless to say, nothing notoriously harmful should be contained in any drug.

In the past thimerosal, chemically ethyl(2-mercaptobenzoato-(2-)-O,S) mercurate(1-) sodium, was usually added to vaccines [20] as a preservative. Due to its toxicity, it was replaced by aluminium hydroxide, a chemical with debatable advantages over thimerosal in terms of toxicity.

The images that follow (Figures 8.25, 8.26, 8.27, 8.28, 8.29, 8.30) present a selection of the debris identified in the above-mentioned vaccines. No vaccines, and no injectable solutions, at that, may contain particulate matter. Infanrix contained titanium particles (Figure 8.25); the presence of debris of iron-chromium and nickel (stainless steel), besides aluminium, were detected in Anatetall (Figure 8.26), while debris of copper-lead-tin-zinc were identified in Inflexal (Figure 8.27).

Stamaril showed particles of chromium-silicon-copper-iron-magnesium-sulfur-potassium-calcium in a substrate of aluminium and saline

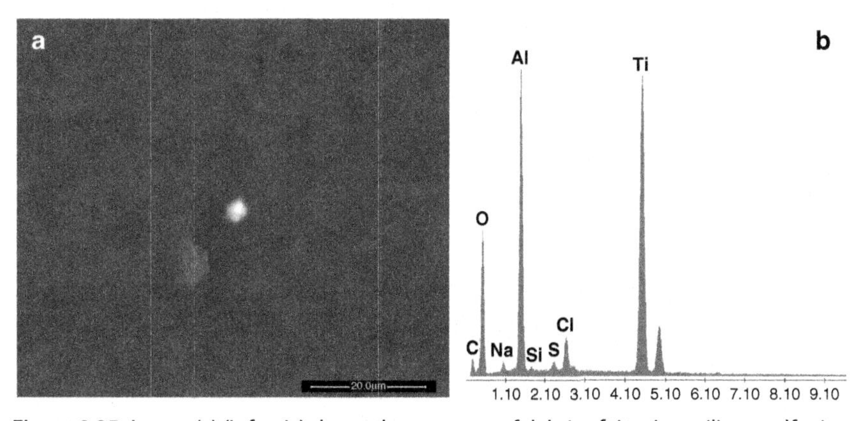

Figure 8.25 Image (a) (Infanrix) shows the presence of debris of titanium-silicon-sulfur immersed in an environment of aluminium diluted in saline solution (sodium-chlorine) (b).

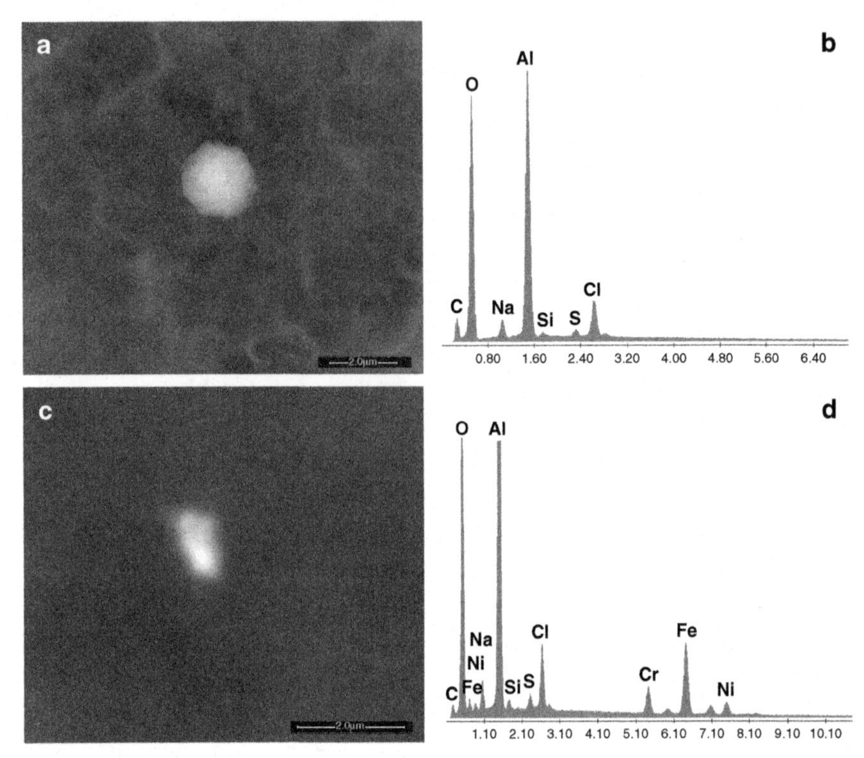

Figure 8.26 Images (a, c) (Anatetall) show 2-micron-sized debris of aluminium-silicon-sulfur immersed in a saline solution (sodium-chlorine) (b) and iron-chromium-nickel-silicon sulfur with aluminium in saline (d).

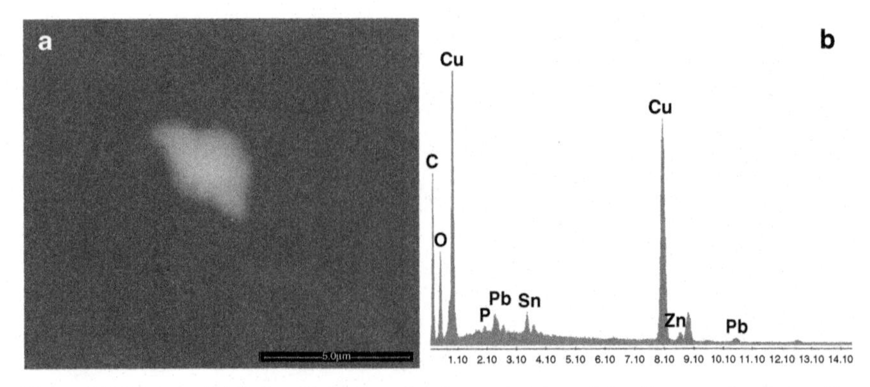

Figure 8.27 Image (a) (Inflexal) shows a 4-micron-sized particle of copper-lead-zinc-tin-phosphorus (b) identified in the Inflexal vaccine.

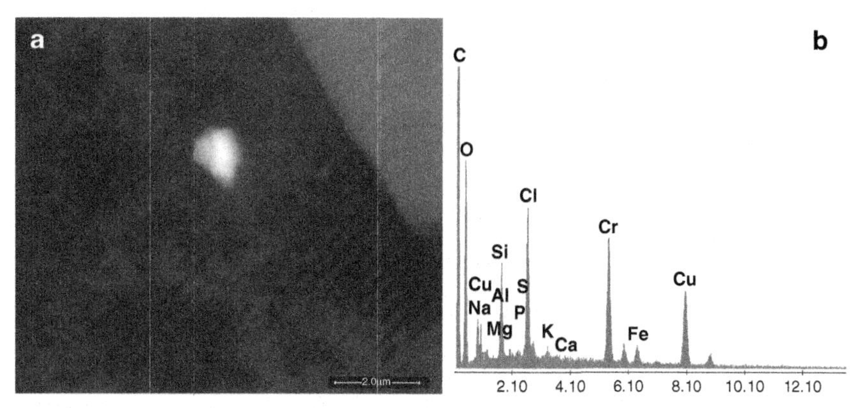

Figure 8.28 Image (a) (Stamaril) of a 1.8-micron-sized debris identified in the Stamaril vaccine. It contains chromium-silicon-copper-iron-magnesium-sulfur-potassium-calcium (b).

(Na-Cl) (Figure 8.28). Thypherix showed aggregates of tungsten nanoparticles (probably tungsten carbide) and calcium (Figure 8.29). In a substrate of aluminium we identified a particle of copper-iron (Figure 8.30 (a, c)) and nanosized particles of lead-bismuth in Gardasil. The EDS spectrum is the result of the subtraction of the signal coming from the particles from the aluminium substrate.

The presence of solid, non-biodegradable and non-biocompatible foreign bodies in vaccines was unexpected to us and, excluding the absurdity of any introduction done on purpose, the most immediate explanation

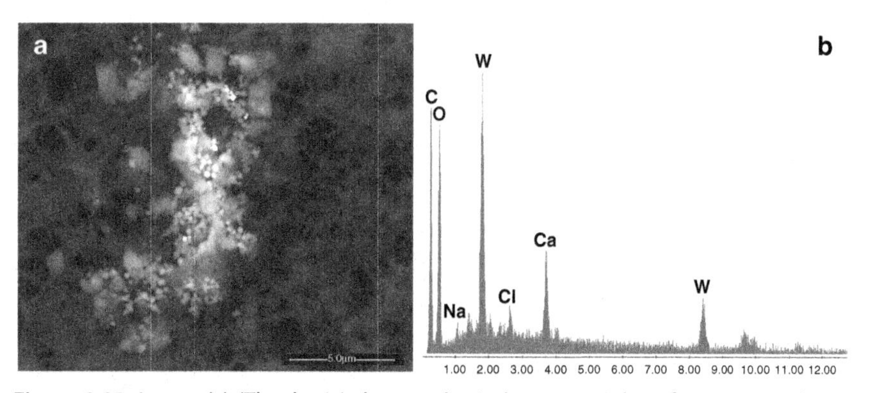

Figure 8.29 Image (a) (Thypherix) shows spherical nanoparticles of tungsten-calcium in a saline solution (b).

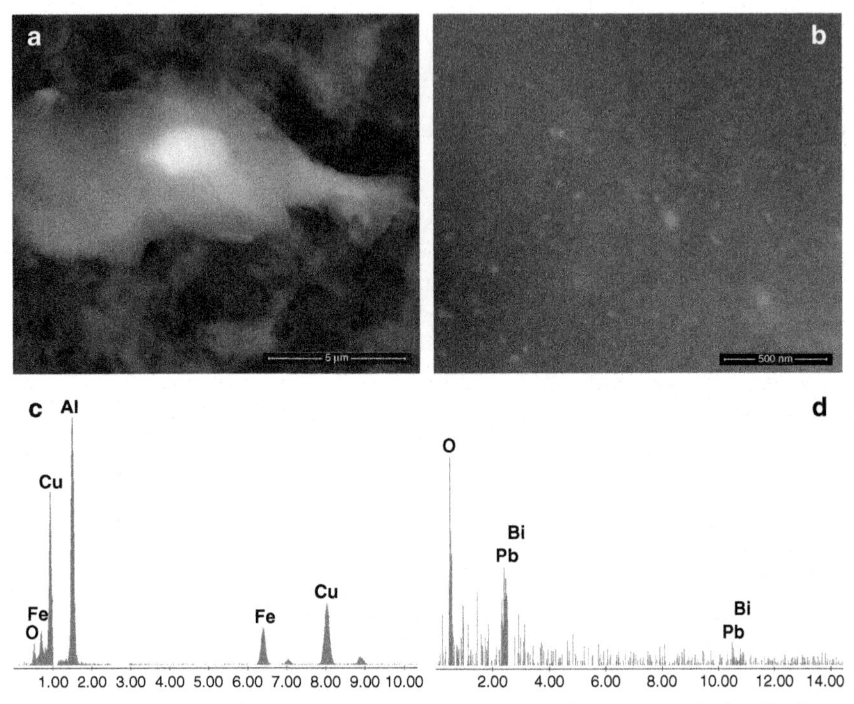

Figure 8.30 Images (a, b) show the particulate matter, also nanosized, identified in a drop of Gardasil vaccine. The micron-sized particle in (a) is composed of aluminium-copper-iron (c), while the nanosized one (b) is composed of lead-bismuth (d).

to those presences is an incidental contamination occurred during the preparation of the product or contained in one or more of its many ingredients. Taken for granted that that is most likely the case, unfortunately we could not conduct any investigations in the factories because of the lack of cooperation by producers and the indifference on the part of health authorities.

If, given the presence of non-biodegradable particles, the formation of a common local reaction (e.g., a granuloma) at the vaccine inoculation site, a reaction often seen in animals, can be understandable, but correlating the administration with systemic adverse effects including neurological ones may need some not-yet-available or, at least, not-scientifically-indisputable explanations. The new pieces of evidence found during our research could offer indications for new, in our opinion indispensable, investigations.

REFERENCES

[1] Internal Report of the Project INESE (Impact of Nanoparticles in Environmental Sustainability and Ecotoxicity) supported by the Italian Institute ofTechnology 2006-2013.
[2] www.manstore.co/purificup-nanosilver-membrane-filter-tap-water
[3] www.epa.gov/oppsrrd1/REDs/factsheets/4082fact.pdf
[4] https://www.princeton.edu/~achaney/tmve/wiki100k/docs/Argyria.html
[5] http://emergingtech.foe.org.au/sites/default/files/FOE_nanotech_food_report_low_res.pdf
[6] Mahler GJ, Esch MB, Tako E, Southard TL, Archer SD, Glahn RP, Shuler ML. Oral exposure to polystyrene nanoparticles affects iron absorption. Nature Nanotechnology 2012;7:264–71.
[7] Nikonov IN, Folmanis YG, Folmanis GE. Iron nanoparticles as a food additive for poultry. Doklady Biological Sciences 2011;440(1):328–31.
[8] Duncan TV. Applications of nanotechnology in food packaging and food safety: barrier materials, antimicrobials and sensors. J. of Colloid and Interface Science 2011;363(1):1–24.
[9] FAO REPORT (FAO/WHO Expert Meeting on the Application of Nanotechnologies in the Food and Agriculture Sectors: Potential Food Safety Implications, MEETING REPORT, Ed. Food and Agriculture Organization and WHO. 2010; 1-103 (http://www.fao.org/docrep/012/i1434e/i1434e00.pdf)
[10] Report of the Danish Ministry of Environment, Environment Protection Agency, Supplementary Survey of Products on the Danish Market Containing Nanomaterials, Part of the "Better control of nano" initiative 2012-2015, Environmental Project No. 1581, 2014.
[11] Gatti A, Tossini D, Gambarelli A, Montanari S, Capitani F. Environmental Scanning Electron Microscope and Energy Dispersive System investigation of inorganic micro- and nanoparticles in bread and biscuits. Critical Reviews in Food and Nutrition 2009;49(3):275–82.
[12] Charbonneau JE. Investigation of foreign substances in food using scanning electron microscopy-x-ray microanalysis. Scanning 1998;20:311–7.
[13] Prusiner SB. Novel proteinaceous infectious particles cause scrapie. Science 1982;216:136–44.
[14] Prusiner S. Some Speculations about Prions, Amyloid, and Alzheimer's Disease. New England Journal of Medicine 1984;310(10):661–3.
[15] Mad cow disease: Still a concern.MayoClinic.com (CNN). 10/2/ 2006. Retrieved 2009-06-20 (http://www.riversideonline.com/health_reference/Infectious-Disease/ID00012.cfm)
[16] Bovine Spongiform Encephalopathy – "Mad Cow Disease" Fact Sheets. Food Safety and Inspection Service. March 2005. Retrieved 2008-04-08. (http://www.cfsph.iastate.edu/FastFacts/pdfs/bse_F.pdf)
[17] Oesch B, Westaway D, Wälchli M, McKinley MP, Kent SB, Aebersold R, Barry RA, Tempst P, Teplow DB, Hood LE, Prusiner. A cellular gene encodes scrapie prp 27-30 protein. Cell 1985;40(4):735–46.
[18] Meyer RK, Mckinley MP, Bowman KA, Braunfeld MB, Barry RA, Prusiner SB. Separation and properties of cellular and scrapie prion proteins. Proc. Natl. Acad. Sci. USA 1986;83(4):2310–4.
[19] Hansen T, Klimek L, Bittinger F, Hansen I, Capitani F, Weber A, Gatti A, Kirkpatrick JK. Mastzellreiches Aluminiumgranulom, Der Pathologe, 29(4):311-3.
[20] http://www.fda.gov/BiologicsBloodVaccines/SafetyAvailability/VaccineSafety/UCM096228

CHAPTER 9

Occupational Cases

Contents

9.1 Introduction	195
9.2 Printers and nanoink	196
9.3 Cases of spontaneous pneumothorax	200
9.4 Working pollution in nanotechnology laboratories	204
References	208

9.1 INTRODUCTION

Throughout this book we have tried to make it clear that, once inorganic, non-biodegradable particles have been captured by an organ, there are no known ways to get rid of them. At least not in a significant enough way. That does not mean that they do not exist: simply, that we do not know them. In the discussion dedicated to the firefighters' sweat in which we found particles, we reported our observations about particle elimination. According to what we saw, traces of elimination do exist, but we are not aware of the efficiency of the mechanism nor of the origin of those particles nor if their chemistry and/or other characteristics have a specificity with some particular tissues. Without prejudice to accepting evidence to the contrary, our opinion is that the efficiency is modest and that the particles we found in the sweat were not embedded in any organ but came from fluids like the blood or, less probably, from fatty tissues.

In other circumstances and more than once, we found particulate matter in the seminal fluid (see Chapter 6), and that is further evidence that it can somehow leave the body. Again, we do not know what part of the organism it comes from and, in this case, we cannot expect great efficiency.

So, given the big difference between what goes in and what possibly can go out, a small but continuous introduction of particles in the body may, in some respects and with due differences, correspond to a much more substantial one concentrated in a very short time.

In a working place, in the majority of cases, with continuity, with a high frequency, for a relatively long fraction of the day, one is subjected to interference with the same type of low-dose pollution. The visible difference with a great, one-shot, invasion of particulate pollutants is that, unlike what happens, for instance, in soldiers exposed to explosions, in the former case the pathology can take a long time to manifest itself, and, fortunately, in many circumstances,

Case Studies in Nanotoxicology and Particle Toxicology
http://dx.doi.org/10.1016/B978-0-12-801215-4.00009-1

the time is so long that there are no clinically detectable signs for many years or, perhaps, ever. Obviously, checking an indoor environment, as many workplaces are, is much easier or, at least, more possible, than doing the same thing outdoors. So, if a variety of protective measures can be taken for people working, for example, in a factory or in a laboratory, primary prevention, i.e., not producing pollution, is the only efficient way to protect people outdoors.

However, as always happens, the first step to the solution of a problem is being aware that the problem exists. The next, immediately following, is being willing to solve it. Often, and for many reasons, eyes and brain stay stubbornly closed.

9.2 PRINTERS AND NANOINK

In its earlier versions, toner, the powder used in printers, photocopiers and fax machines, was a relatively coarse mix of carbon powder, iron oxide and some organic compounds. As customers demanded more and prints grew more sophisticated, in color and with higher and higher resolution, new substances and smaller and smaller particles down to nanometric size were constantly introduced in the composition. It is no surprise that, that very volatile dust can be dangerous for people who work in the vicinity of the machines (e.g., people using a computer with a printer placed besides it), and, in particular, for those who work in the maintenance of the equipment.

New toners and inks prevent free dispersion in the environment more than older ones. Their cartridges are sealed and the operations of replacing them are safer than in the past.

We were given to analyze some biopsic samples of a patient who had been busy for about 20 years cleaning and repairing printers and fax machines and, unfortunately, no one had taught him to take the most necessary precautions, from using a nose-and-mouth mask to wearing dedicated clothes and disposable gloves.

The collection of symptoms the subject reported were invalidating: he was troubled by an extremely frequent pollakiuria (his bladder had become very small and overactive), and he also suffered from insomnia, loss of short-term memory, numbness and soreness of the arms and legs, tremor, difficulty in walking and handling objects and jerky movements. Ecography showed abnormal areas in the pancreas, in the kidneys, in the liver and in the spleen with the liver, and the spleen was also enlarged. X-rays showed abnormal areas in the inferior lobes of both lungs whose nature was not diagnosed.

The biopsy of the bladder we analyzed was exceptionally contaminated by particles, many of which were unusually small in size, down to a few tens of nanometers. Apart from some particles, what we found was very homogeneous: Mercury-based particles also containing lead-tin, chlorine-sulfur and iron-chromium, respectively. It is interesting to observe the threadlike shape of some of those particles (Figure 9.1 a, b, c, d, e, f) whose origin is hard to tell.

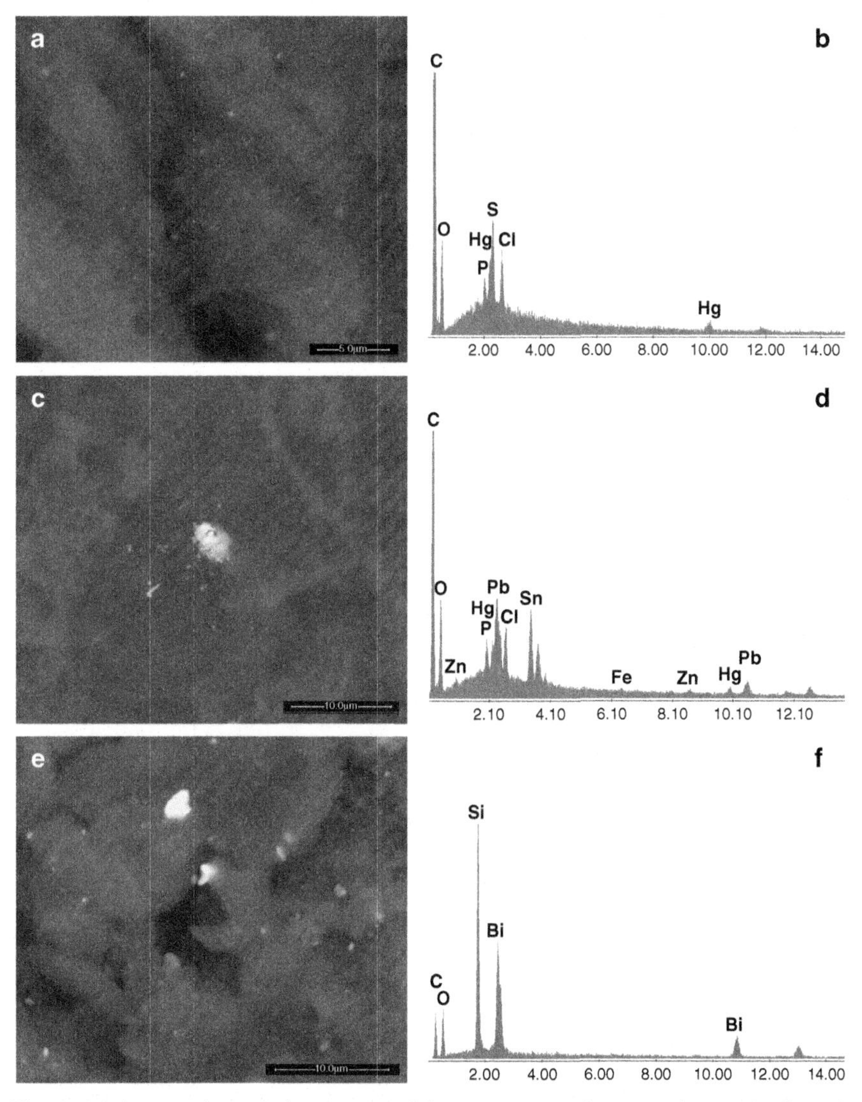

Figure 9.1 Images (a, c, e) show some of the numeorus micron- and nanosized particles identified in the bladder biopsy. They are composed respectively of: debris sulfur-mercury-chlorine-phosphorus (b), lead-tin-mercury-chlorine-phoshorus-zinc-iron (d) and silicon-bismuth (f).

Table 9.1 Summary of particles identified in the bladder biopsy of a worker

Analysis n.	Size in μm	Chemical composition
1	1.5	C, O, Cl, S, Ba
2	0.1-0.5	C, S, O, Hg, Cl, P
3	0.1-3	C, Pb, O, Sn, Cl, Hg, P, Zn, Fe
4	3	Cr, C, O, S, Cl, Hg, Fe
5	wire	C, Hg, Cl, O
6	4	C, Fe, Cr, O, Cl, Ni, Hg, S, Si
7	100	Si, C, O, Cl, S, Ti
8	0.1-1	Si, Bi, O, C

Table 9.1 shows a selection of the particles identified in the bladder biopsy and their elemental chemical compositions. The presence of lead-tin-mercury and other mercury-based particles was ubiquitous and their size was homogenous (0.1-0.5 micron).

In any case, particles that size cannot be attributed to a "normal," municipal, environmental origin and a reasonable explanation we found for them was linked to the patient's job, even if, as will be decribed later, we could not find the same compositions in the toner samples we analyzed. A possibility we could not prove is that the domestic environment the subject lived in or, in any case, an environment much frequented by him contained those pollutants. The last possibility, at least in our opinion, is that the patient had used for a long time some products in which that particulate matter was present. As is understandable, some of the foreign bodies we detected could hardly matter with the patient's working activity and were almost certainly of environmental origin (mercury, lead, tin, silicon, bismuth, etc.).

We could not verify the presence of organic substances like, for instance, alkylphenolthoxylate, an endocrine disrupting chemical banned in 2003 (EC Directive 2003/53/CE), as our technique is not suitable to do so.

To complete this short description of the case, we had a chance to see that the patient had rather violent, immediate reactions, vomiting in particular, whenever even very small quantities of toner dust were brought in the room we were in: a reaction typical of multiple chemical sensitivity. We tried to introduce some toner with the subject being totally unaware and the reactions were the same.

Trying to give a satisfactory explanation to the case, we analyzed six different samples of toners, five black and one magenta, toners for printers and photocopiers to which the patient probably had been exposed (1) Toner Ricoh FT 5351, 2) Toner Olivetti, 3) Toner Ricoh Katun FT

6750, 4) Toner Ricoh Aficio FT 450, 5) Toner Developer Ricoh, 6) Katun Aficio Color 3506).

As soon as we started this very limited research of ours, we realized that the chances of finding something that would put us safely on the trail of what we were looking for was, to say the least, remote. Not only was the variety of products currently available at the time of the analyses enormous, but their commercial life was in many cases pretty short and what belonged even to a recent past and could have been the cause of the disease we were investigating on was actually impossible to come by. It must be added that the patient had started to work 20 years before he gave us his samples. In any case, it may be of some interest to report some of our findings.

Besides carbon particles that were the base to all products, the first sample contained a great number of particles a little below and a little above 1 micron in size. Silicon and vanadium were the prevalent elements, but titanium and zirconium were also present.

Toner number 2 showed very small, often nanometric, iron particles either isolated and in clusters inside larger carbon particles. Outside, iron-barium particles were found.

The third specimen was similar to the second, with isolated and clustered iron-chromium-nickel (stainless-steel) small particles within and without the carbon matrix.

Titanium, tin, antimony and silicon 1-3 micron particles, were found in the next sample, all of them outside the carbon matrix that contained some-tenths-of-micron particles of iron and silicon. 1.5-micron particles of iron, silicon, zinc, nickel, copper and manganese were inside the carbon particles of sample 5, while much larger ones (60-80 microns) were outside.

The particles observed in the last specimen, the magenta toner, had a size that ranged between slightly more than 1 micron to 10 microns whose prevalent composition was titanium and iron accompanied by a lesser presence of aluminum, magnesium and silicon.

To the best of our knowledge, though black carbon has been extensively studied, in-depth medical investigations on the impact produced by those powders on the whole organism and their possibility to be moved to a fetus with the ensuing consequences have never been carried out. The presence of certain non-biodegradable and non-biocompatible particles, many of which are sized below 1 micron, is an element of considerable difference from carbon alone.

As to the case of our patient, in the six toners we analyzed we could not find the chemistry we had seen in his biopsy sample, but the already-mentioned large variety of products available since toner came into use and their often ephemeral presence in the market made our quest the equivalent to looking for a needle in a haystack.

So an unexceptionable explanation to the unusual syndrome troubling our patient is still missing. Nevertheless, leaving the widest possible room open to discussion, considering also the enormous quantity of particles detected, we believe that the occupational origin is the most likely.

9.3 CASES OF SPONTANEOUS PNEUMOTHORAX

We had a chance to participate in research carried out by the University of Modena (Italy) on 201 cases of spontaneous pneumothorax to which 20 post-mortem control cases were added of subjects that died not always from but always with pulmonary pathologies, excluding pneumothorax [1,2].

In 14 of the 201 cases we found a condition of pneumoconiosis, a restrictive lung disease caused by the inhalation of dust, known to be, with few exceptions, of occupational origin. More specifically, what we found could be better defined as a pleural coniosis, i.e., a collection of inorganic, non-biodegradable and non-biocompatible particles at the pleural level.

That condition, absent in the 20 control lungs, induces fibrosis with distortion of the pulmonary juxta-pleural structures and with the consequent formation of fibrous plaques, blebs, blisters and chronic giant-cell inflammation. This situation can lead to the rupture of the pleura and pneumothorax.

In all 14 cases we considered, the presence of the particles found in the specimens could be easily explained by the occupations of the patients: mechanic, hairdresser, ceramic-tile worker, mason, welder, etc.

In all cases we checked as much as possible the places where the subjects had worked, finding particulate matter compatible with what we had found in the biological samples where we had identified the dust most probably responsible for the necroses and lesions of the tissue that induced the pneumothorax they all had suffered from.

Figures 9.2, 9.3, 9.4, 9.5, 9.6 present some examples of debris identified in the surgical lung lesions of workers affected by spontaneous pneumothorax who were operated on. The EDS spectra clarifying the chemistry make the causes of the spontaneous pneumothorax clear.

Figure 9.2 shows images of the lung tissues where numerous debris (white dots) are entrapped. The chemical composition clarifies that they are ceramic debris (silicon-aluminum-magnesium-sodium-phosphorus-sulfur-potassium-calcium-iron). The patient worked in a ceramic-tile

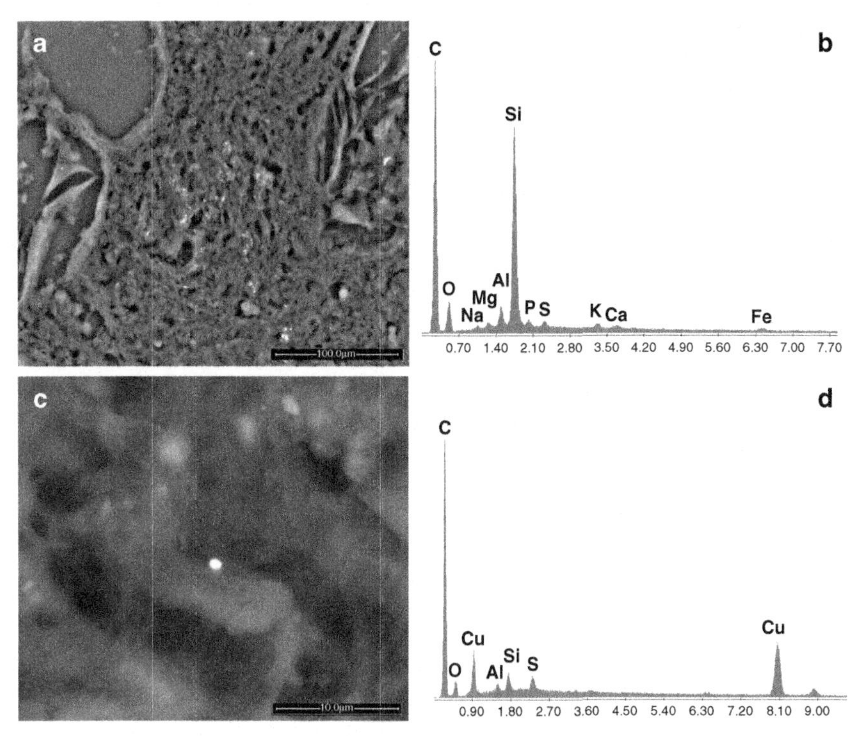

Figure 9.2 Images (a, c) show a ceramic worker's lung where numerous debris (white dots) are visible. The chemical composition is silicon-aluminum-magnesium-sodium-phosphorus-sulfur-potassium-calcium-iron (b) and copper- silicon-aluminum-sulfur (d).

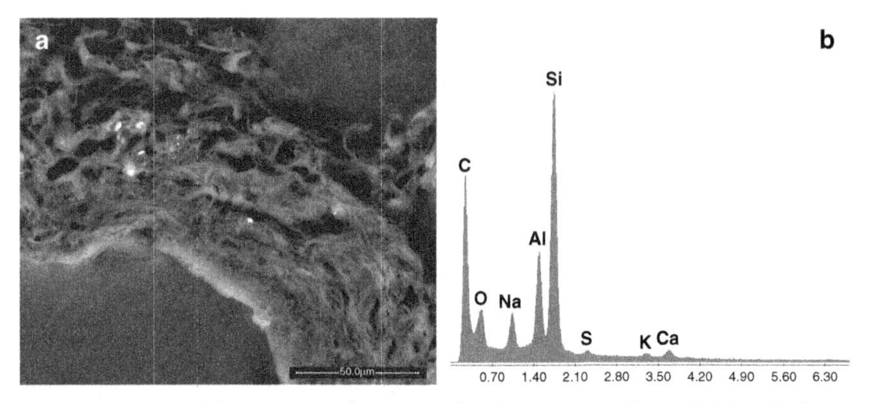

Figure 9.3 Image (a) presents a lung sample of a mason, where debris of silicon-aluminum-magnesium-sodium-sulfur-potassium-calcium (b) are visible.

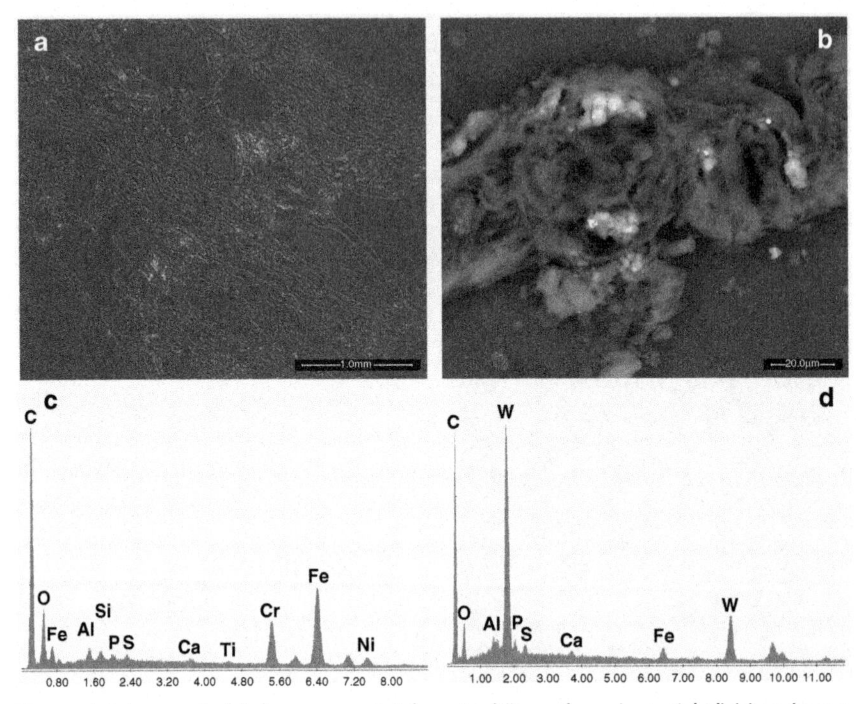

Figure 9.4 Images (a, b) show some stainless steel (iron-chromium-nickel) (c) and tungsten-aluminum-iron-phosphorus-sulfur-calcium debris (d) in a welder's lung. Note the sperical shape of the particles.

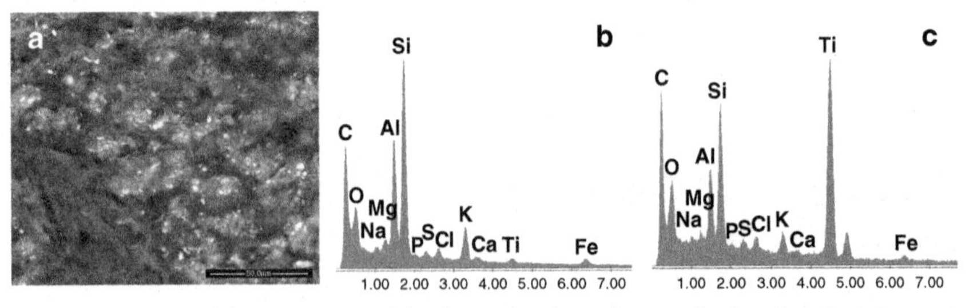

Figure 9.5 Image (a) presents a wide dissemination of ceramic dust in a hairdresser's lung tissue; silicate dust containing titanium (b, c) is also visible.

industry. In the sample we also identified a toxic spherule of copper-aluminum-silicon-sulfur which can be related to ceramic glaze. The series of lantanides also present in some debris confirm an occupational source of pollution.

Figure 9.3 is related to the ceramic dust identified in a mason's lung tissue who was also affected by silicosis. The debris are composed silicon-aluminum-sodium-sulfur-calcium-potassium, and the chemical composition

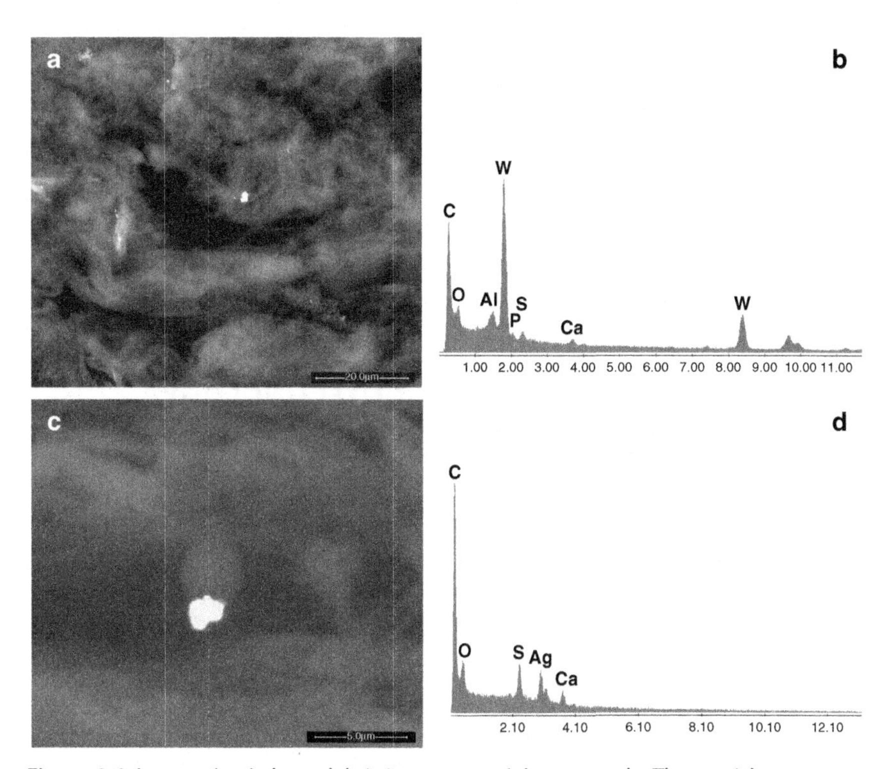

Figure 9.6 Images (a, c) show debris in a surgeon's lung sample. The particles are composed of tungsten-aluminum-phosphorus-sulfur-calcium (b) and silver-sulfur-calcium (d).

can be related to the brick and plaster (CaS) dust, probably calcium-sulphate, that when is bound to two molecules of water is gypsum.

Figure 9.4 presents metallic debris of stainless steel (iron-chromium-nickel) found in the lungs of a welder.

Figure 9.5 is related to a hairdresser who was also affected by tuberculosis (TB). The lung sample was full of ceramic particles also containing titanium. The analyses of the hair products did not show the presence of ceramic materials, but the narrow working space showed the presence of the plaster dust and of the paint (titanium dioxide). The limited workspace could have facilitated transmission from an infected customer. Certainly the presence of ceramic, non-biodegradable dust in the lung could have depressed the immunosystem that was not able to react to a bacterial invasion.

Figure 9.6 presents a special case: a surgeon who developed a pneumothorax, but during the surgical operation an aneurism of the aort was identified. Two of the debris identified were composed of tungsten-aluminum-phosphorus-sulfur-calcium and sulfur-silver-calcium. A possible explanation of the rarity of these findings in a person who seems not exposed to a particular environmental pollution could be in the use of a saw and vessel-welding devices that have tungsten carbide tools. No occupational explanation was found for the 2-micron-sized particle of silver-sulfur.

Though the particles found were different, they behaved as what they are, foreign bodies, and the final effects were similar in all the subjects examined.

9.4 WORKING POLLUTION IN NANOTECHNOLOGY LABORATORIES

With increased frequency, nanotechnology laboratories are being introduced everywhere, and as is only obvious, the people working there make and handle nanoparticles, with all this may entail, volatile and pathogenic as they are. When the phenomenon started, much to our surprise, we noticed that very often the people employed in those laboratories did not realize the criticality of what they were working with and very few precautions, if any, were taken. Fortunately, today much has changed, including, at least in part, the chemistry of the particles, and in more than one circumstance, though not always, the awareness achieved is, in a way, reassuring.

As part of the European Project DIPNA (FP6-NMP-2006-032131) we took part in, a specific project aimed at verifying nanopollution in nanotech research laboratories and industries.

One of the first things we did was to create a wearable sensor (Pollution, Italy) capable of collecting dust suspended in the air through a battery-powered device, and use a 4-l/min pump (EGO, Zambelli, Italy) to collect dust on cellulose filters; we also used adhesive tape to collect the dust deposited on furniture. The sensor was worn on the coat of the people working in the laboratory who could easily carry the device, and mainly through that simple equipment we carried out investigations in 12 laboratories where nanoparticles were produced and handled:

- 4 in Spain (Colloidal preparation, Dry and wet)
- 1 in the USA (Dry-Laser)
- 1 in Germany (Dry-Laser Ablation)
- 1 in The Netherlands (Dry in glass coating)
- 2 in Denmark (Dry and wet, dry for food)
- 1 in France (Dry-Laser)
- 2 in Italy (Dry and wet nanotubes)

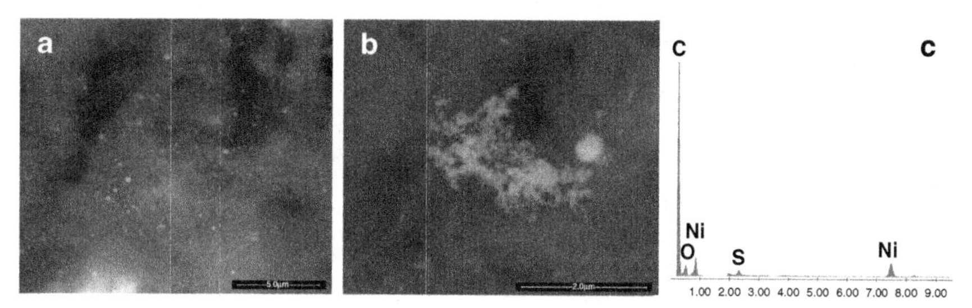

Figure 9.7 Images (a, b) show carbon nanotubes with nickel nanosized contamination (c).

In all of them we measured pollution in the areas where the nanoparticles were generated and in other rooms. We also collected dust simply touching some surfaces with small adhesive carbon discs.

In a laboratory where carbon nanotubes were produced, we found 0.1-1-micron nickel spherules attached to the carbon structures (Figure 9.7), whose origin we had no chance to check (a catalyst contamination or a cross-contamination due to other activities?). If those nanotubes were meant, for example, to make conductive and high-strength composites or were used for energy-storage and energy-conversion devices or for sensors [3,4], probably the risk for human health was minimal or there was no risk at all. But, if they were used to make hair dye or other cosmetics, they represented a direct risk for their consumers. For this reason, it is mandatory to check the final product for any involuntary contamination and to assess the risk their use involves.

In one of the laboratories checked, besides the contribution due to environmental air pollution (Si-Al- based particulate matter), we found microsized calcium-based particles and calcium-based aggregates of nanoparticles. In addition, we detected the presence of metallic nanoparticulate contaminants (Au, Zn, Ag, Fe-Cr-Ni, Ni, Ti-Fe-Mn, Ti) either inside the laboratory and in the offices. The sensor located in the center of the laboratory collected large (18 μm in diameter), elongated, bismuth-based, crystalline particles certainly not generated in the laboratory, since no bismuth belonged to the composition of the products. Also, on a laboratory-staff coat a comparatively large amount of cadmium-based crystals (Figures 9.8A and 9.8B) with an acicular and/or platy habit was found and is worth noting: the crystals ranged from about 5 μm to nanoscale. The presence of sulfur in the EDS spectrum of these cadmium-based crystalline clusters may suggest that they were greenockite (CdS) crystals, a mineral once, before being aware of its toxicity, commonly used as yellow pigment in various industrial applications and now used in some electric batteries. No one in the laboratory could guess the origin of that contaminant.

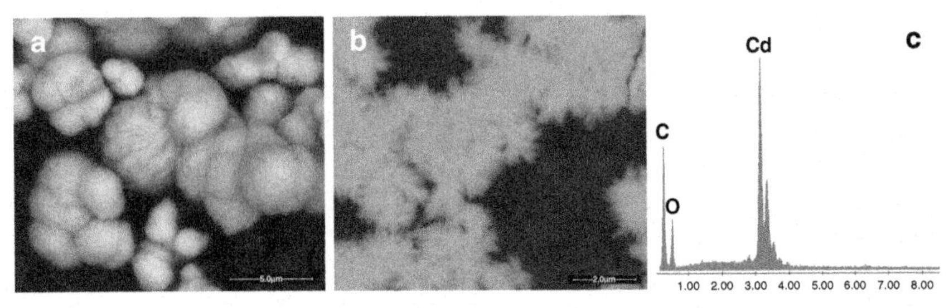

Figure 9.8A Images (a, b) also present nanometric-sized particles of chlorine-zinc-copper-tin-cadmium (c) identified on a lab coat after 48 hours of use.

The detection of gold-based particles in different locations, such as the secretary's office, of a laboratory where those particles were produced suggests that the lab coats may act as carriers for the dispersion of the particles actually produced there and of particulate pollutants.

In another laboratory where different nanoparticles were produced in different shifts, we clearly identified traces of the particles they had made: zirconia, platinum, lead, etc. (see Figure 9.9 (a, b, c, d)). Particles lingering in the laboratory rooms are a source of cross-contamination affecting the next project. But despite the filtering measures taken, particulate pollution coming from outside the building can also be found, most likely, as already mentioned, carried in by garments, since nanoparticles tend to adhere virtually to any object. So, a window left open in an office, even removed from the laboratories, can let in pollutants eventually contaminating the whole production.

Single and clustered cerium-titanium nanoparticles and larger iron-based spherules were captured by our carbon discs used to check a laboratory

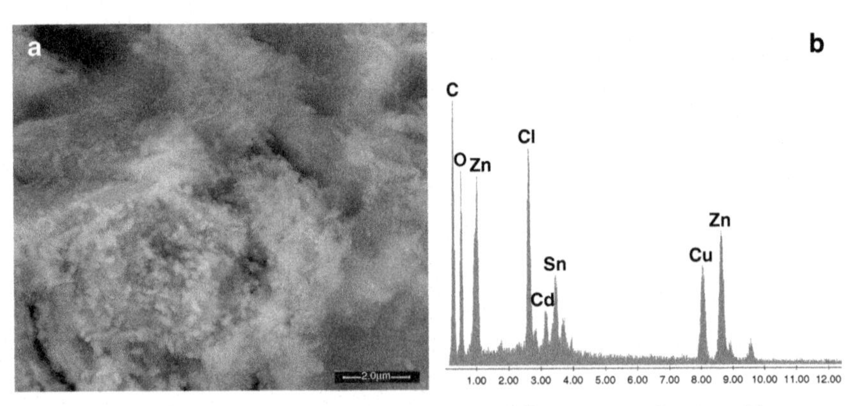

Figure 9.8B Images show crystals of cadmium at different magnifications. The nanometric size of some of the crystals is clearly visible.

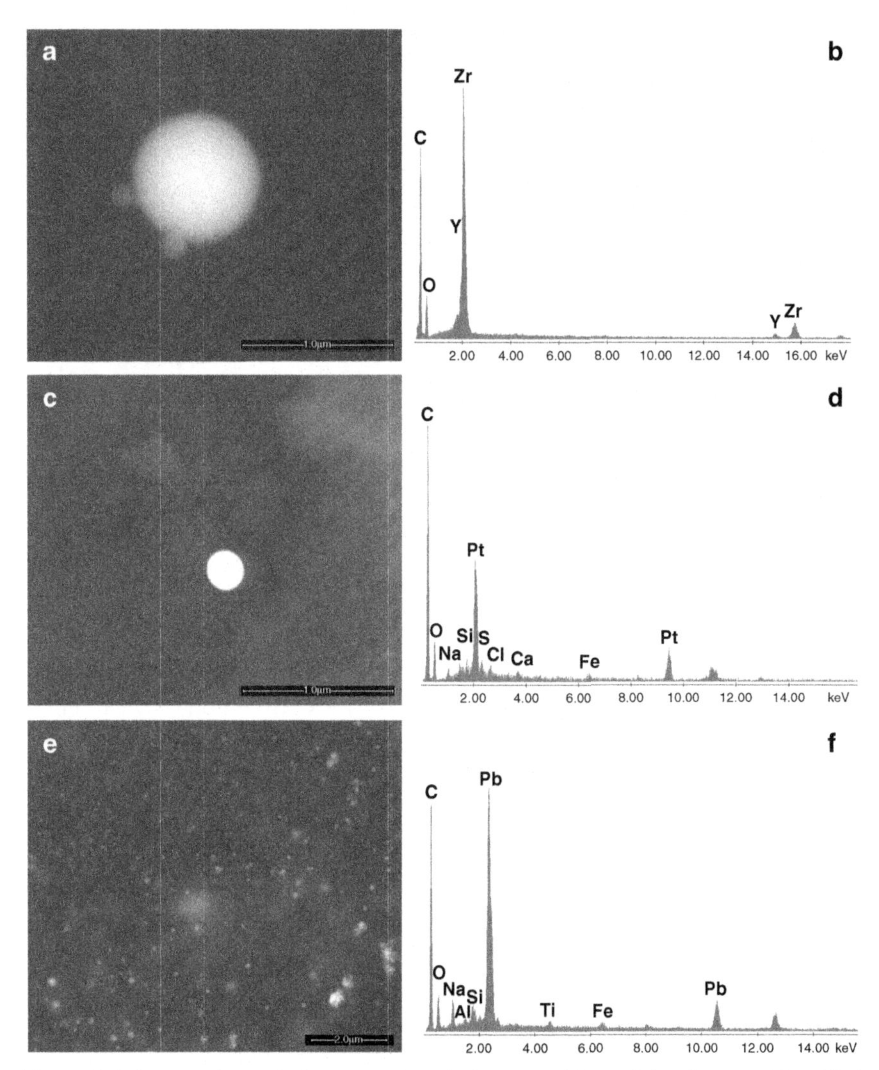

Figure 9.9 Images (a, c, e) show the chemical composition of the dust we identified in the filter of an air pump. Traces of the production clearly remain in the laboratory: zirconia (b), platinum (d), lead (f).

where dry fibers were produced. While the former were cross-contaminants coming from the internal production, the latter, of evident combustion-origin, came from outside.

In another circumstance, we found iron nanoparticles contaminated by copper, while in a different laboratory we found 0.1-1-micron nickel spheres on the carbon nanofibers produced. We are not aware of how firmly those particles adhere to the fibers. Nor do we know if the aging of the

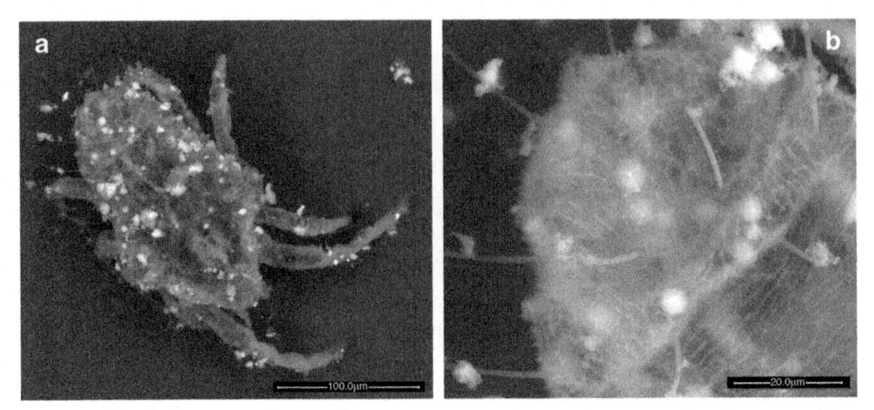

Figure 9.10 Images of an acarus at different magnifications that entrapped the environmental pollution of its habitat.

product results in the detachment and release of the nickel particles into the air or, as is the case when carbon fibers are used for food packaging, if food is contaminated.

Because of the peculiarities of nanoparticles, their being so mobile, their strong tendency to adhere to other particles as well as to virtually all objects and surfaces including skin and hair and the ease with which they can enter the human organism, nanotechnology laboratories are potentially dangerous places for people who work there. So careful and meticulous precautions must be taken to protect workers and accurate controls must be taken on a regular basis of the indoor environment.

In one of the filters we used to collect the air pollution we also found an acarus that entrapped part of the laboratory pollution. It is not possible to know if this pollution induced some harmful health effects (Figure 9.10).

These pieces of evidence must persuade manufacturers and policymakers to activate specific measures to protect the workers of these industries or laboratories.

REFERENCES

[1] Barbolini G, Reggiani Bonetti L, Gatti A, Capitani F, Gambarelli A. Le basi morfologiche del pneumotorace spontaneo primitivo, in Il pneumotorace spontaneo: dalla genetica al trattamento chirugico. Publ. by Athena, Audiovisuals Modena Italy 2005; 51-62.
[2] Barbolini G, Gatti A, Murer B. Pleura, Trattato di Istopatologia. Publ. Piccin Nuova Libraria Padova (Italy). 2006; Ch.8.4:1081–98.
[3] Katz E, Willner I. Biomolecule-functionalized carbon nanotubes: applications in nanobioelectronics. Chem Phys Chem 2004;5(8):1084–1104.
[4] De Volder M, Tawfick SH, Baughman RH, Hart AJ. Carbon Nanotubes: Present and Future Commercial Applications Science. 2013; 339:535–38.

CHAPTER 10

Miscellaneous Cases

Contents

10.1 Impact of smoking 209
10.2 Diabetes, chronic-fatigue syndrome and other pathologies that could be explained from a different point of view 216
10.3 Other possible effects of nanoparticle exposure 228
References 229

10.1 IMPACT OF SMOKING

Despite the awkward, in some instances ironically or naively heroic, attempts of the tobacco industry to deny the evidence, and in contrast with some "scientific" and "medical" literature of the past, it is clear that smoking increases the risk of contracting lung cancer, besides a list of other pathologies. Among other agencies and medical societies, the existence of a strong link between oncologic diseases and the breathing of organic, toxic compounds coming from the combustion of tobacco has been ascertained by the World Health Organization and the International Agency for Research on Cancer. According to them, the culprits are the carbonaceous by-products generated not only by tobacco leaves, but also by paper, graphite additives, and flavor/aroma correctors, etc. Though unexceptionable, evidence-based findings convinced us that something escaped from this evaluation, and what escaped can supply hints to explain other effects related to tobacco smoking. In fact, smoke is indicated as responsible for a number of diseases, first of all cardiovascular ones, which, at least in appearance, do not have anything to do with cancer.

In the tobacco strips we analyzed and identified micro- and nanosized inorganic particulate matter that follows the incineration of the leaf. That does not mean that tar, nicotine and all the many chemicals contained in tobacco smoke are not pathogenic agents or are of lesser importance, but it means that the scenario is somewhat more complex than commonly believed.

As seen in many circumstances and described throughout this book, vegetables grown outdoors collect dust falling from polluting sources, and tobacco leaves are no exception.

The debris deposited on tobacco leaves is characteristic of the environmental pollution of a certain place and indicative of a certain period. For example, in a previous work [1] we identified in some Bosnian cigarettes

Case Studies in Nanotoxicology and Particle Toxicology
http://dx.doi.org/10.1016/B978-0-12-801215-4.00010-8

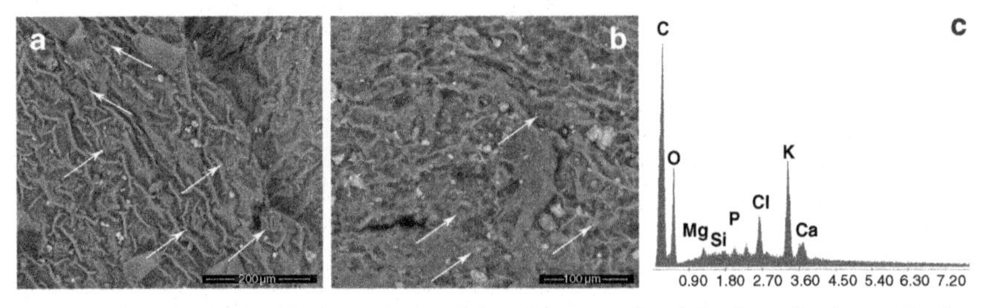

Figure 10.1 Images (a, b) show the surface of a strip of a dehydrated tobacco leaf at two different magnifications with the environmental debris (c) deposited. The arrows indicate the stoma of the tobacco leaf.

the war pollution, recognizable, among other characteristics, by the presence of dust containing uranium. It must be explained that during the long siege of Sarajevo, all the local industries were closed for lack of energy, but the tobacco factory continued to work, since drying the leaves and making cigarettes requires very low energy consumption. In such a case, the risk of smokers is also that of breathing a special environmental pollution for many months or years.

Tobacco leaves are not cleaned before being worked, so the pollution accumulated during their growth remains there. When the leaf is dehydrated, it shrinks to a very small volume; water evaporates but particles stay where they are and concentrate (Figure 10.1). This means that smokers do not inhale just the known toxics of tobacco but also a concentrated version of the environmental pollution.

The temperature of a cigarette ember can reach 1,100°C and, at that condition, some particles (tin ones, for example) can melt. During their inhalation, though, with the rapid decrease of temperature they solidify again, taking a spherical morphology. It can also happen that, during melting, an aggregation with other particles occurs, leading to the formation of clusters that, immediately after, deposit either inside the mouth mucosa or along the respiratory system or inside the lung alveoli. Particles made of higher-melting elements remain unchanged, but heat makes them more reactive in the biological environment where they end up.

If the particle size is submicronic, there is a high probability that they can cross the lung barrier and enter the blood stream. This possibility opens the door to other pathological effects of smoking such as, for instance, an impaired peripheral blood circulation that can lead to tissue necrosis. These particles, being naturally hard, can scratch the endothelium of the blood vessels inducing vasculitis with the ensuing reparation and calcification phenomena, and, over time, also cause the separation of the elastic layers that occurs in the arteries when the hemodynamic loads on the vascular wall exceed the adhesive strength between layers. The consequence is aneurysm or pseudoaneurysm, particularly in areas characterized by turbulent flow. But those particles can also occlude some small vessels, either because of their size or because of their thrombogenicity, thus being the agents of an altered vascularization or a necrosis.

Of course, in general, the particles produced by tobacco smoking and entering the organism are in no way different from any other similar particles and can be the origin of the same pathologies.

It is also true that, after years of smoking, the accumulation in the alveoli of tar and particulate foreign bodies, commonly covered with a thin carbonaceous coating that could also contain traces of pesticides or other substances used in tobacco cultivation, can physically impair the lung functionality, particularly in regards to CO_2-O_2 exchange. Also, some forms of asthma can be reasonably due to a reaction against particles, either because of their shape, size or chemistry or a combination of those

Table 10.1 List of tobacco strips with the cigarette brand and origin

	Cigarette brand	Country of purchase
1	Merit (Altria Group Inc.)	Italy
2	Diana (Altria Group Inc.)	Italy
3	Muratti (Altria Group Inc.)	Italy
4	Philip Morris (Altria Group Inc.)	Australia
5	Marlboro Red (Philip Morris Int.)	Italy
6	Marlboro (Philip Morris Int.)	France
7	Camel (Reynolds Tob. Comp.)	Italy
8	MS filtro (BAT)	Italy
9	MS Club Oro (BAT)	Italy
10	821 (BAT)	Italy
11	Fortuna (Altadis)	Italy
12	Aura Extra (Fabrika Duhana Sarajevo)	Bosnia & Erzegovina
13	Drina King Size (Fabrika Duhana Sarajevo)	Bosnia & Erzegovina
14	Aura Lights (Fabrika Duhana Sarajevo)	Bosnia & Erzegovina
15	Drina Lights (Fabrika Duhana Sarajevo)	Bosnia & Erzegovina
16	Aura SuperLights (Fabrika Duhana Sarajevo)	Bosnia & Erzegovina
17	Sumer Star – King Size (Iraq)	Iraq
18	Sumer – King size (Iraq)	Iraq
19	Natural Spirit (Santa Fe Nat. Tob. Comp.)	United States
20	Yuma Organic (rolling tobacco)	France
21	Pueblo Organic (rolling tobacco)	Germany

Table 10.2 Selection of chemical composition and size of particles identified on the surface of the tobacco strips analyzed

Cigarette brand	Chemical composition *	Size & morphology
Natural tobacco leaf (standard)	K, C, O, S, P, Ca, Si, Mg K, C, O, Ca, Cl, S, P, Mg	
Marlboro	Cu, C, Zn, O, Fe, Ti, S, Au, Si, Al, Cr	5 μm (spherules)
MS filter	Ti, Fe, C, K, O, Cl, Ca, P, S, Si, Al, Mg	3 μm
	S, C, Ba, K, O, Ca, Cl, Sr, P, Al, Mg, Fe	7 μm
	C, O, Ca, P, K, Nd, Ce, La, Si, S, Al, Mg, Cl	3 μm
	C, Pb, K, O, Sb, Ca, Sn, Cl, Si, P, Al, Mg	2 μm
	C, Fe, O, K, Si, Ti, Al, Cl, Ca, S, P, Mg	0.5-3 μm
Aura Extra	W, Ca, C, O, K, Al, Mg, Cu	0.8 μm
	Sn, C, Si, Pb, O, Al, Na, P, Fe	30 μm
	K, Ca, Al, C, O, Cl, Mg, Si, S, P, Nd, Ce, Fe, La, Th	7 μm
	Al, C, Bi, O, S, Si, Na, Fe, Cu	0.2 μm
	Zr, Si, K, C, Ca, O, S, Mg, Cl, Al, Fe	7 μm
Drina King Size	C, O, K, Si, Th, U, Ca, P, Al, Zr, Y, Mg, S, Fe, V, Ce,	3 μm
	Zr, Si, Hf	7,5 μm
Aura Super Light	K, Ca, C, Nd, Ce, O, Mg, P, Cl, La, Pr, Th, S, Si, Al, Fe	20 μm
Sumer Star – King Size	P, O, Ce, Nd, C, La, Si, K, Ca, Th, Al, Cl, U, Mg	10 μm
Philip Morris Australia	C, Ca, P, Ce, Nd, K, La, O, Si, Cl, S, Th, Mg, Fe, Al	4 μm
Natural Spirit	C, O, Ca, K, P, Si, Ce, Al, Pb, La, Fe, Mg, Cl, Na	2 μm
	Ti, Fe, O, C, Mg, Si, K, Ca, Al, S	10 μm
Yuma Organic	Ti, Fe, Mn	8 μm
	Fe, O, Si, Al, C, Mg, Ti, Mn	5 μm (spherules)
Pueblo	P, Ce, Nd, C, O, La, Pr	0.4-1.4 μm
	Fe, O, Si, Al, C, Mn	2.5 μm (spherules)
	Fe, Mn, Ti, Si, Al	5 μm
	W, Co	0.2-1 μm
	Pb	

* Note: Elements are listed starting from the one that showed the highest peak in the EDS spectrum.

factors. Depending on their chemical composition, some particles can be degraded, e.g., by corrosion and the corrosion products can be further pathogens. As already described elsewhere, the smallest particles can enter cell nuclei and damage the DNA [2]. Larynx, pharynx, esophagus, oral cavity, nasal cavities and sinuses share the same risk with the lungs. A major fraction of the totality of the particles introduced into the alveoli is likely to enter the blood circulation [3–7], thus increasing all risks typical of particles, from hypercoagulation to all other nanopathologies, including, for instance, some forms of leukemia [8–16]. All the authors who noted those phenomena indicated in the carbonaceous and organic content of smoke the cause of those diseases, but none of them took the ever-present inorganic particles into account.

With our usual method we examined 21 different types of tobacco cigarettes. Table 10.1 reports some of the most interesting debris identified and their chemical composition.

The natural elemental composition of tobacco can include carbon, oxygen, potassium, sulfur, phosphorus, calcium, silicon, magnesium or carbon, oxygen, potassium, calcium, chlorine, sulfur, phosphorus and magnesium (Figure 10.1(c)). The presence of other elements is due to pollution. (Table 10.2)

Figure 10.2 shows some particles deposited on a strip of Pueblo tobacco leaf. Some of them are occluding a stoma, not allowing the leaf respiration. The EDS spectrum of the 4-micron-sized whiter debris on the right (not considering the basic tobacco composition) is composed of aluminium-phosphorus-strontium-silicon-lanthanum-cerium. Figure 10.2(c) concerns two spherical debris found in a smoker's lung affected by cancer. Their composition is cerium-lanthanum-neodymium-praseodymium-iron-chlorine. In this analysis we did not consider the debris of inorganic materials surrounded by carbonaceous compounds, since it is difficult to get a clear picture of the carbon of the leaf substrate with the carbon of the debris.

The analyses on the different tobacco leaves showed the interesting presence of elements such as strontium, hafnium, gold, lanthanum, cerium, neodymium, praseodymium, thorium and uranium.

Figure 10.3 presents a selection of spectra among those analyzed with very particular chemical compositions of the debris identified in some cigarettes. We found debris of tungsten-cobalt, lead and silicon-zirconium-yttrium-aluminium, etc. In this last case, the probable origin of this contamination was likely the ceramic-material industry. The presence of yttrium

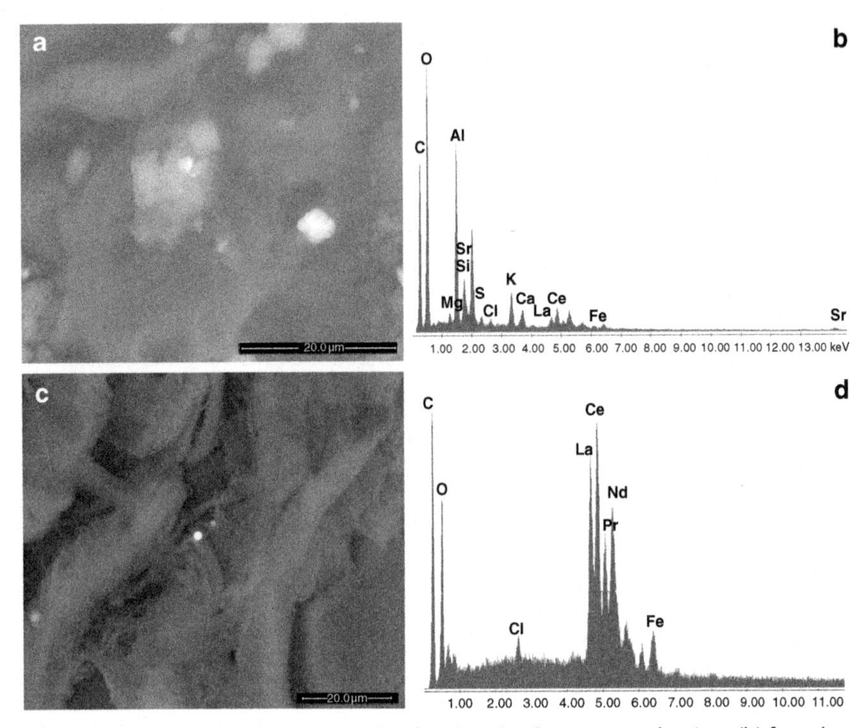

Figure 10.2 Image (a) shows a particle of cerium-lanthanum-neodymium (b) found on a strip of tobacco. Particles with similar elements (d) were identified in a surgical sample of a smoker's lung (c).

with zirconium indicates a probable doping of zirconium-silicon with yttria (yttrium oxide). This debris must be also weakly radioactive.

The case of cigarettes is a very particular one, since pollutants are voluntarily inhaled.

As is the case with many other pathogens, particulate matter in smoke can have different impacts on different subjects. Besides the variability that differentiates one individual from another, the same brand of cigarette, if manufactured with tobacco coming from different parts of the world, will result in products with potentially different health hazards. Also, the blending of different tobaccos coming from different cultivation areas will lead to a more or less contaminated product. This is also valid for organic tobacco (see Natural Spirit and Pueblo) contaminated by toxic pollutants as well as non-organic products.

It is obvious that what has been said about tobacco and its particulate pollution is true for any other leaf intended to be smoked.

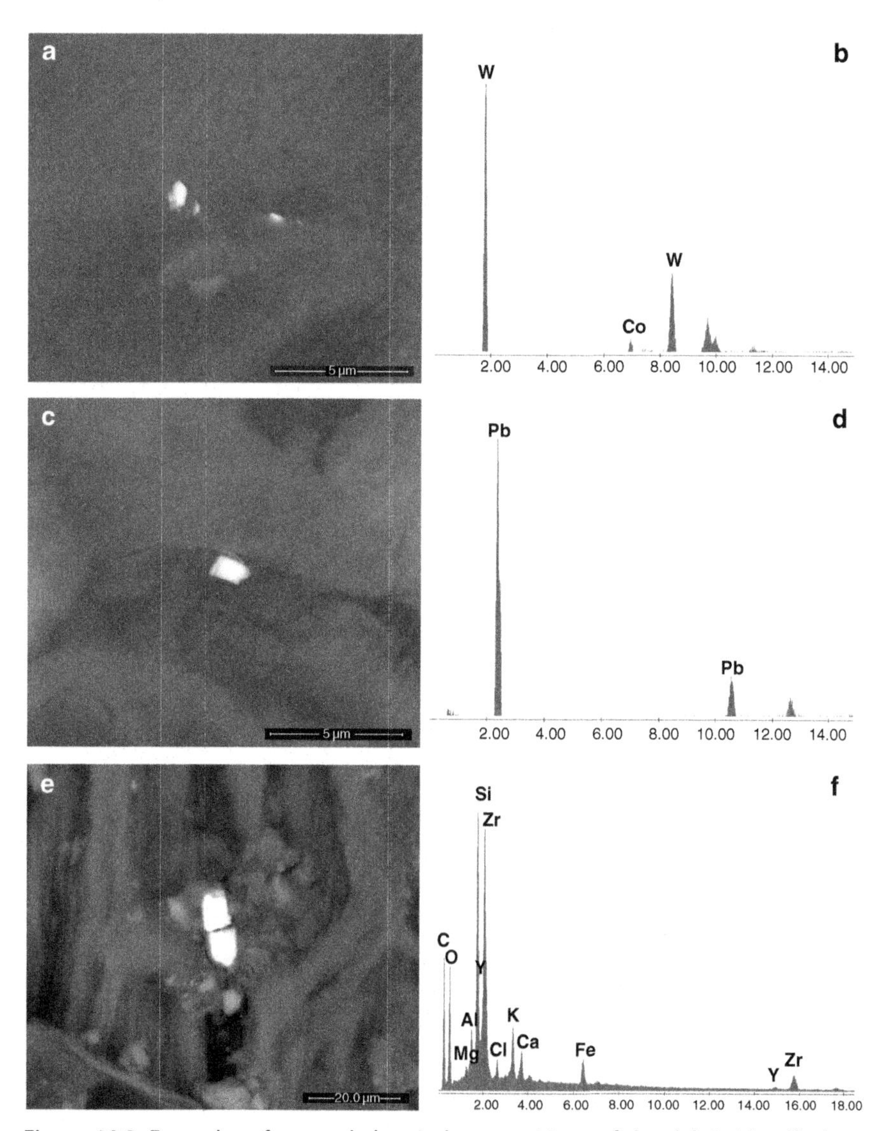

Figure 10.3 Examples of unusual chemical compositions of the debris identified on some tobacco strips (a, c, e). Debris of tungsten-cobalt (b), lead (d) and silicon-zirconium-yttrium (f) are clearly identified.

Besides primary prevention, i.e., cultivating tobacco in certainly hard-to-find pollution-free territories, if technically possible, one of the solutions to mitigate or even eliminate the consequences of dust pollution on tobacco leaves is cleaning them before being dried.

10.2 DIABETES, CHRONIC-FATIGUE SYNDROME AND OTHER PATHOLOGIES THAT COULD BE EXPLAINED FROM A DIFFERENT POINT OF VIEW

The paucity of research funds and the lack of economic interest by the pharmaceutical industry toward pathologies involving a relatively small number of subjects are mainly responsible for a long and far-from-exhaustive list of what are called "orphan diseases." Vasculitis and cryoglobulinemia, for instance, two illnesses of interest to nanopathology, are among them [17], but the National Organization for Rare Disorders in the US registers a list of about 6,500. Sharply rising pathologies like, among others, multiple chemical sensitivity (MCS) are either ignored or summarily dismissed and treated as of psychiatric origin.

We are convinced that some of them, in particularly among the new ones or the ones on the increase, could be caused by submicronic particles of once-rare or non-existent compositions and, in any case, never so present in the environment and in food. Rarity or uniqueness could also be due to particular, personal exposures impossible to repeat. Verifying the hypothesis of a particulate origin can be done by looking for foreign bodies in the blood or, if the pathology is compatible with that origin, in biopsies or surgical samples when they are available. We developed a new method of blood investigation for patients affected by cryoglobulinemia and leukemia and the results look impressive. Figure 10.4 shows the appearance of the precipitated cryoglobulines and the identification of the foreign bodies with their EDS spectrum. Inside the blood-serum precipitates we could identify many foreign bodies: Among them clusters of silver-aluminium-gold-chlorine-copper nanoparticles (Figure 10.4(b)), or aluminium-silicon-iron.

As already mentioned elsewhere, in some positive cases, when possible, anamnesis work with a patient specifically addressed to find and characterize the particulate pollutants that may be present in the biological sample can be of great help to trace the source and, if the patient is still exposed to it, eliminate it. That is the basis of a protocol needing specialized personnel that personalized medicine must adopt. Interdisciplinary work involving medical doctors, physicists, biochemists and proteomics/metabolomics experts could solve what is now the mystery of cryoglobulinemia. The mere elimination from the blood of this "pollution" and the elimination of the exposure could contribute to the improvement of the symptoms and perhaps of the disease.

At the time of writing we are working on a project involving chronic myeloid leukemia. Preliminary results indicate that in all the four fractions

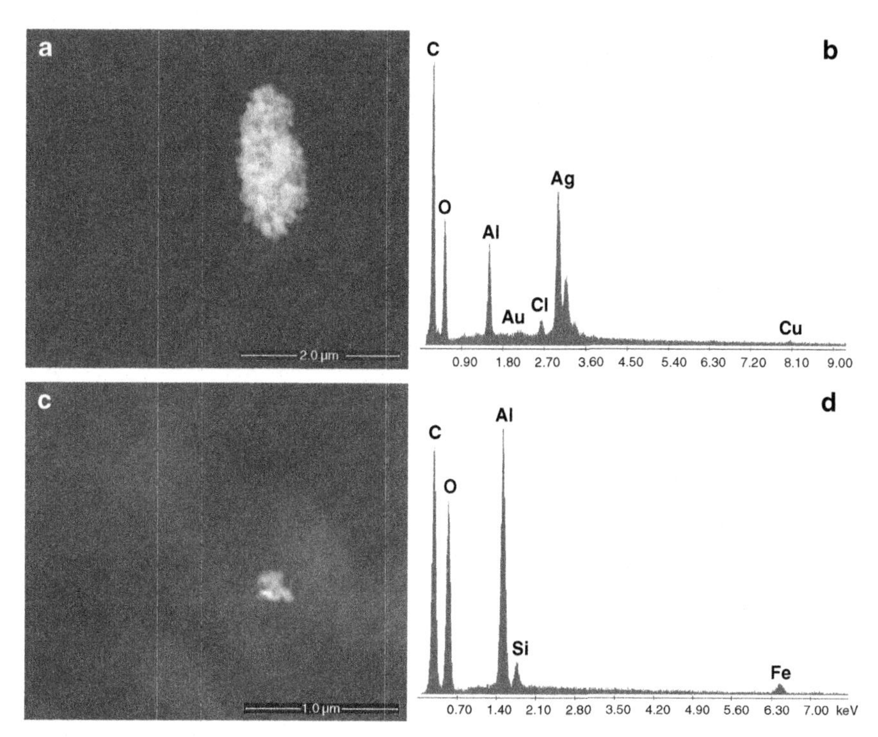

Figure 10.4 Images (a, c) show some examples of the particles identified in two cryoprecipitates. The first contains a silver-aluminium-chlorine-gold aggregate of nanoparticles (b), and the second contains aluminium-silicon-iron (d).

of blood we are considering (red cells, white cells, platelets, plasma), there are foreign bodies that, according to their chemistry and superficial energy, interact in different ways with the different components to which they remain anchored. Figure 10.5 shows particles of micro- and nanosized foreign bodies detected in different blood fractions in patients affected by chronic myeloid leukemia.

We have reasons to believe that chronic-fatigue syndrome (CFS) and post-traumatic stress disorder (PTSD) could be the consequence of exposure to an important, acute, submicronic pollution. The invasiveness of submicronic particles and their wide dispersion in the body can be responsible for the symptoms that patients complain of. All organs, tissues and cells are stressed by those undue presences for which they have no efficient elimination mechanisms.

Many of those who were present during the collapse of the Twin Towers in New York and were exposed to the pollution that ensued showed symptoms that were classified as belonging to PTSD. They included "intrusive memories,

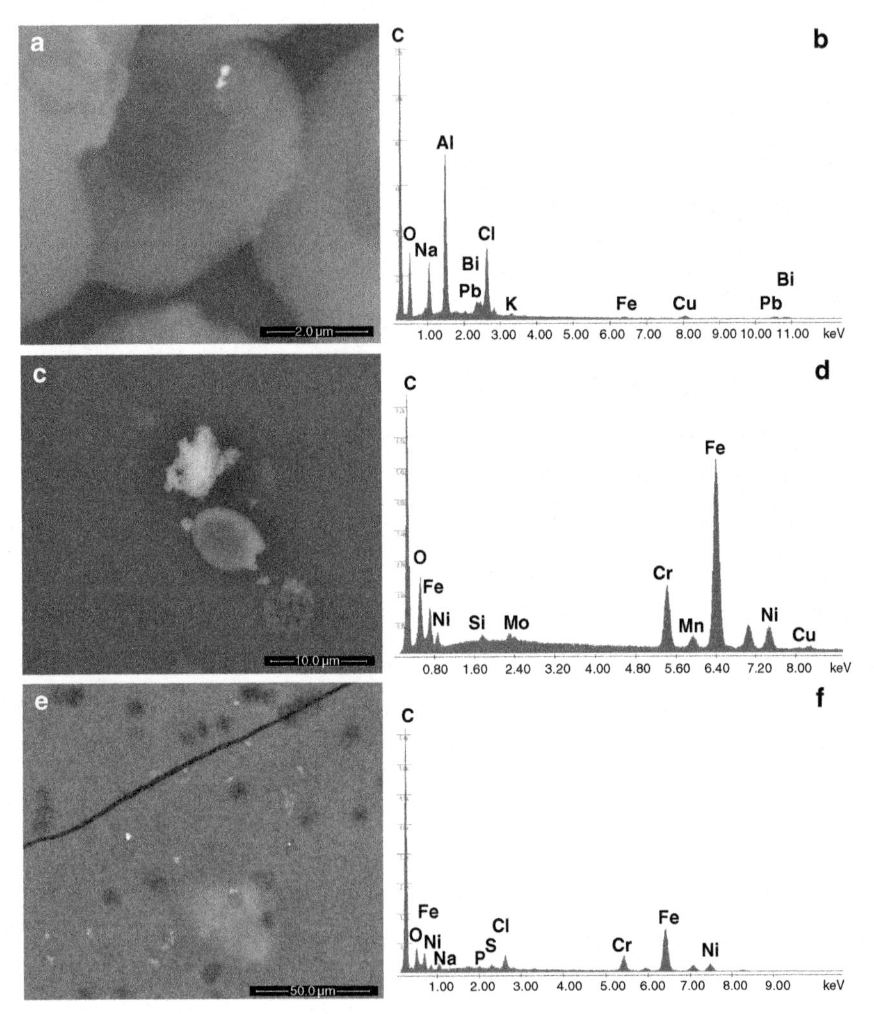

Figure 10.5 Images show the presence of foreign bodies attached to the blood components: aluminium-bismuth-lead (b) particle attached to a red cell (a), iron-chromium-nickel-copper-silicon-molybdenum (d) debris in contact with white cells (c), and nanosized stainless steel (f) debris entrapped in the plasmatic fraction (e).

avoidance, negative changes in thinking and mood, or changes in emotional reactions" [19]. We agree that after having been present in a disaster or having been the victim, for instance, of a physical assault or other crimes, these symptoms can appear, but in the case of 9/11 there was more. For example, some firefighters told us [20] that they were suffering from acute breathing problems, such as asthma; eye, nose and throat irritation; nausea, headaches, vision impaired by flashes and flickers; and that some of those symptoms were worsening.

They reported that they were suffering also from stress and anxiety related to their poor health and to their uncertain future rather than to the spectacle of which they had been the witnesses. We believe that all these symptoms can have a physical source.

In addition to the situations already discussed in this book, in two cases of eye cancer we had a chance to observe we discovered that tiny debris had entered the retinal circulation. Such presences can occlude some blood vessels or, during their passage, can briefly interrupt the retinal circulation or enter into physical contact with the retina, causing the vision symptoms the firefighters suffered from. Figures 10.6, 10.7 and 10.8 show details of eyes explanted for fibrous flogosis (Figures 10.6 and 10.7) and for a squamous-cell carcinoma (Figure 10.8). In all samples there is the presence of metallic debris (lead-tin and stainless steel).

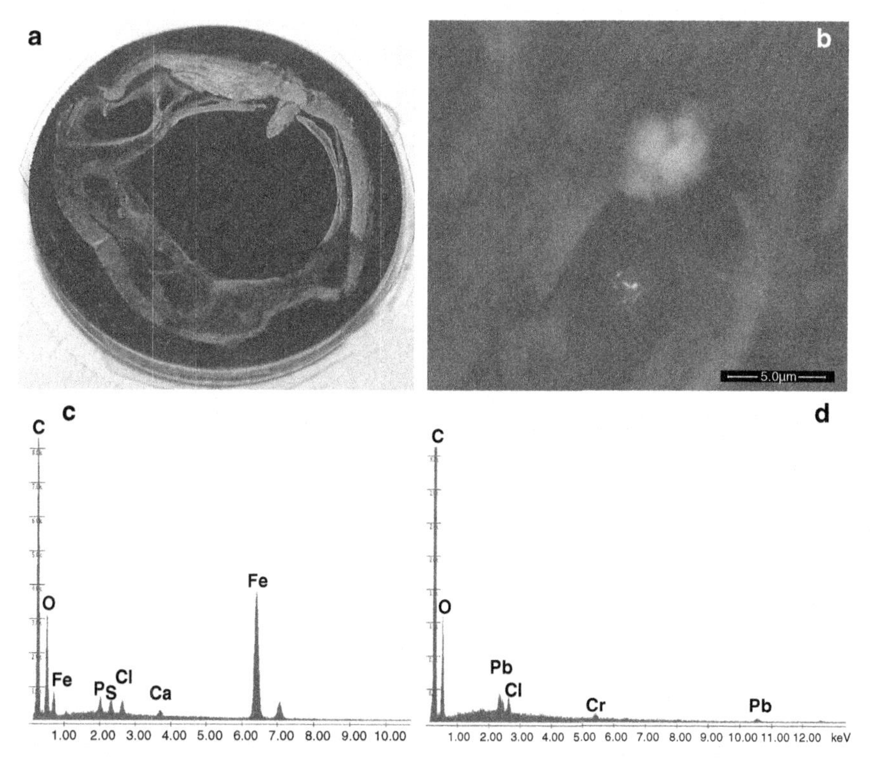

Figure 10.6 Images (a, b) show a section of an explanted eye affected by fibrous flogosis at two different magnifications. The EDS spectra show some of the chemical compositions of the foreign bodies identified: iron precipitates (c) and lead-chromium-chlorine particulate matter (d).

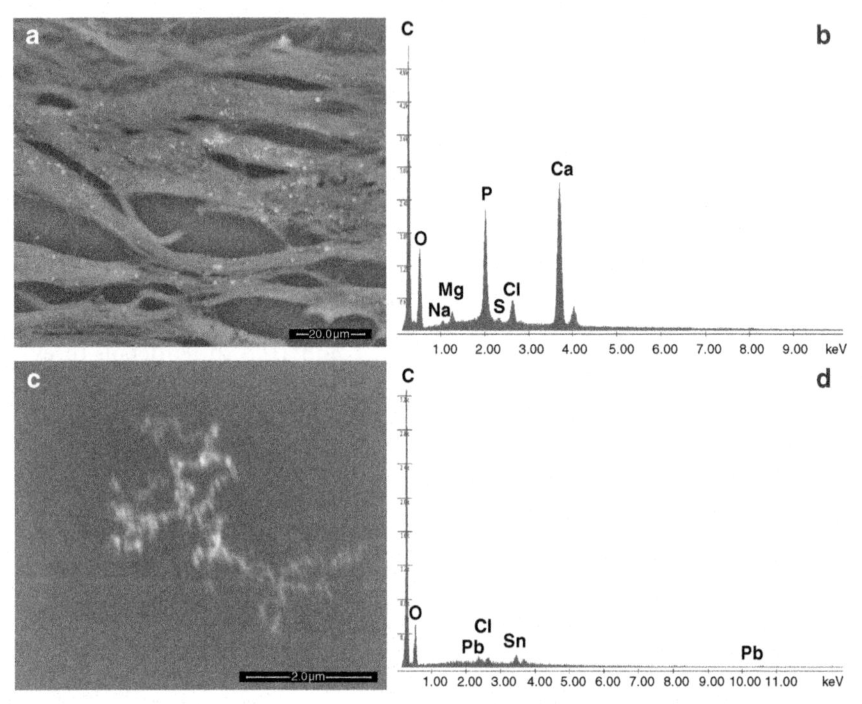

Figure 10.7 Images show an area of the eye affected by fibrous flogosis completely covered by calcium-phosphorus spherules (a,b) and a cluster of lead-chlorine-tin nanoparticles (c,d).

The 9/11 health registry [21] reports that "environmental exposure or attack-related stress reduced fetal growth in some women," and that "child development may be more influenced by maternal mental health than by direct effects of disaster-related pre-natal stress" [22,23]. We do not intend to disprove the psychological explanation, but we proved through electron-microscopy analyses of actual biological specimens that particulate matter can pass from mother to fetus and observed the consequence of that phenomenon. Blaming stress is certainly much easier and much less expensive than investigating other hypotheses, but that may require ignoring some verifiable, objective data.

We also suspect that type-I diabetes could be induced, certainly among other causes, by the physical presence of particles captured by the pancreas and triggering a foreign-body inflammation that depresses beta-cell functionality. We could not verify our hypothesis, since we never had a chance to get suitable samples of that organ, but we think that the possibility is worth pursuing, if only to show it is wrong.

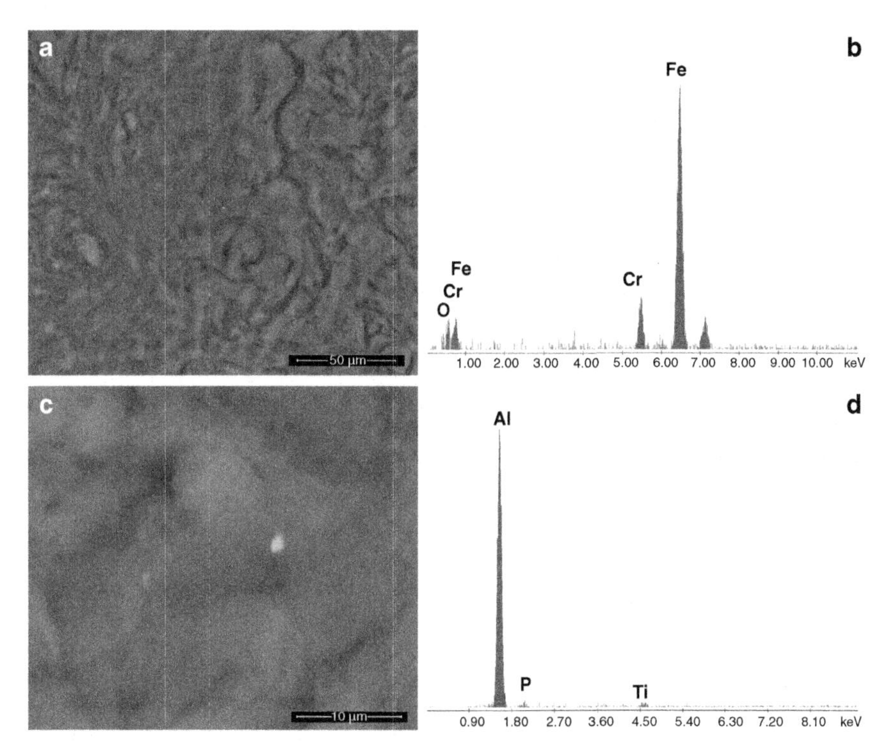

Figure 10.8 Images (a, c) of a soldier's eye tissue affected by squamous-cell carcinoma. Stainless-steel (iron-chromium) (c) and aluminium-titanium-phosphorus (d) particles are present.

In any case, in our opinion, getting rid of particles whenever it is possible is an excellent prevention measure that, among its advantages, does not include the use of drugs. At the time of writing we are working on a system to clear the blood from particles in patients affected by leukemia, but, of course, the same system can work on any subject even if there are no connections with leukemia. People who inhaled a large quantity of dust could immediately benefit from the treatment; this is true in some circumstances, for example, with soldiers or firefighters.

We experienced very few cases of patients who could not find any doctor willing to issue a diagnosis related to the symptoms they suffered and/or to prescribe a therapy. One of them, a particularly singular one, had some filaments growing in his mouth. Since they would not burst independently, his mouth mucosa was irritated, so the patient broke the tissue himself to extract them, in our presence, and we analyzed them. Figure 10.9 shows a filament at two different magnifications with its chemical composition and also shows a stainless-steel particle embedded. The chemical composition

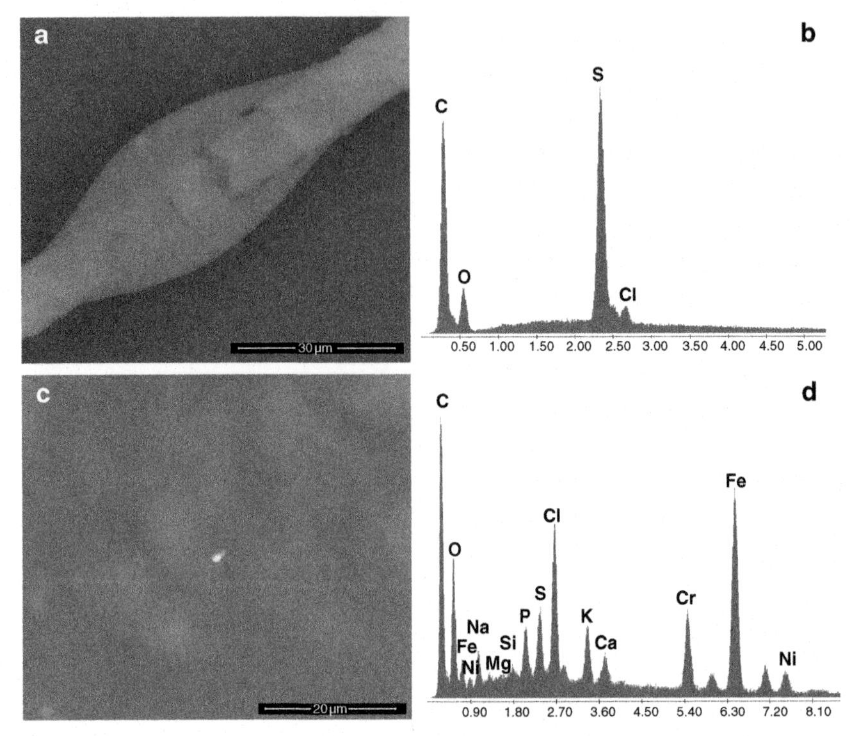

Figure 10.9 Image (a) shows a filament extracted from the patient's mouth. It is composed of carbon, sulfur and oxygen (b). The nanosized debris (c) is composed of carbon, iron, chlorine, oxygen, chromium, sulfur, potassium, phosphorus, nickel, sodium, magnesium, silicon, calcium and nickel (d).

of the filament is similar to a polysulfone (carbon-oxygen-sulfur), but the contribution of the biological tissue (carbon-oxygen-phosphorus-sulfur-chlorine) can bias the analysis. In those filaments we could detect further debris. The debris of Figure 10.9(c) is probably stainless steel and also contains other elements belonging to the tissue where it was entrapped, making identification very difficult. We have no explanation to give nor do we have opinions and the case remains a mystery to us.

Sometimes, in the course of a visit, the patient reports a long collection of symptoms, many of which "make sense," i.e., fit within the frame of a certain pathology. A few of them, however, in some cases apparently without importance, are hard or impossible to understand if the approach is the "usual" one and, be it consciously or unconsciously, the doctor ignores them. In our experience, some, or, sometimes all, of those disregarded signs can be very important: so important that, in some circumstances, they can change the diagnosis, allow to identify the source of the pathology, suggest the correct therapy, if any, and remove the patient from the actual origin of his/her troubles. That is the case of many particle-induced illnesses. By characterizing the debris, it is possible, or easier, to understand the route they took, from their source to their final target, a set of information that is often the key to fill the gaps in a puzzle; otherwise, even in the best of circumstances, incomplete and unsatisfactory results are often the case.

We repeated throughout this book that ours are works in progress and much is still to be explained. One of the phenomena that we find hard to fully understand is the frequent presence of calcium-phosphorus spherules in cancerous tissues. Histopathologists classify them as calcified areas resulting from an inflammatory process, without other more in-depth analysis.

In those formations, the calcium/phosphorus ratio is often 1.67 as in hydroxyapatite, a mineral that cannot be formed in spherical shape at 37°C. So, we suppose that the formation could be catalyzed by enzymes, but we do not know for sure. In the past some scientists thought that those spheres were nanobacteria, but others, including us [24], are convinced that the growth of those tiny balls is carried out by successive depositions of concentric layers driven by a diffusion and local saturation phenomenon as can be guessed from the regularly-stratified shells visible in a sectioned spherule (Figure 10.10).

A possibility is that, for some reason, for instance, with the presence of a foreign body, the local cell metabolism is damaged and the balance of calcium and phosphorus is altered without any possibility to eliminate it by exocytosis.

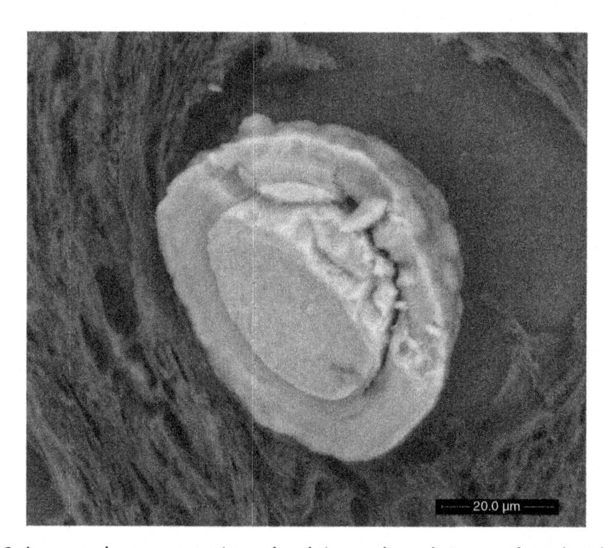

Figure 10.10 Image shows a sectioned calcium-phosphorus spherule identified in a thyroid cancer. Different layers are clearly visible that witness to the different growth steps.

There is an accumulation of the compound and, as soon as the local saturation concentration is reached, it precipitates on or inside a cell organelle. The cell cannot survive the attack to its organelles and dies, leaving only its inorganic, no more biodegradable content. If that is the case, the phenomenon is a very simple chemical one and could play an important role in carcinogenesis.

We note that some proteins such as albumin or fetuin-A, present in the blood, are avid binders of calcium and calcium-phosphate and if they bind a nanoparticle, they become biopersistent, acting as seeds for further crystallization and anomalous calcification [24].

Other precipitates such as calcium-phosphorus-zinc that we observed in a child's brain (see Chapter 5) as well as some iron-phosphorus or sodium-phosphorus precipitates we detected in other circumstances seem to obey to similar mechanisms. Figures 10.11 and 10.12 show examples of calcium-phosphorus precipitates identified in cancer tissues (Figure 10.11, kidney and Figure 10.12, breast). As can be seen, the calcium-phosphorus spherules

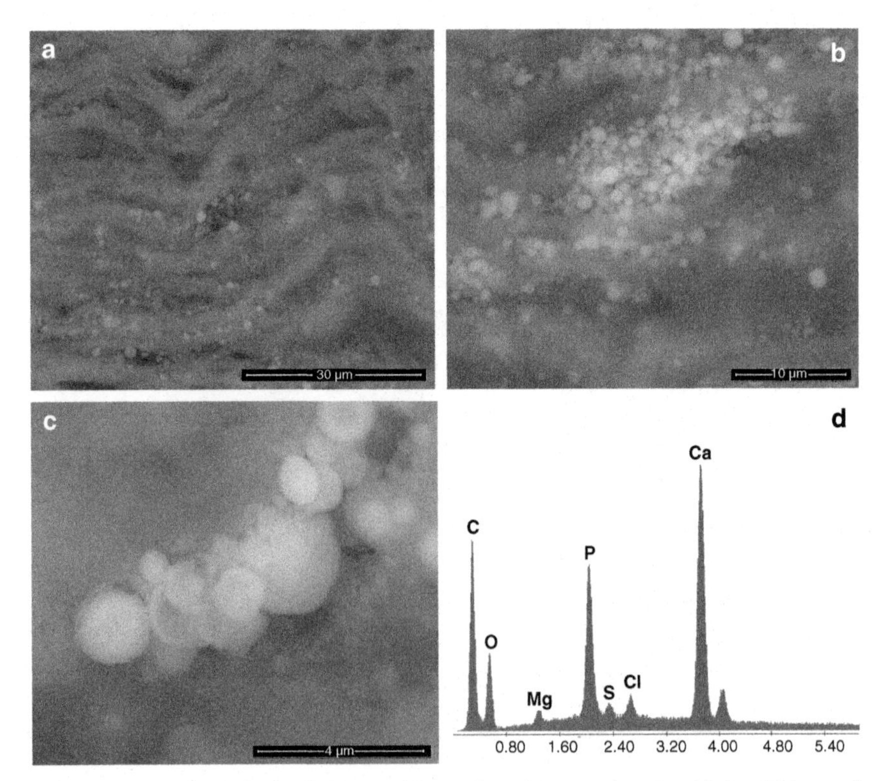

Figure 10.11 Images (a, b, c) show calcium-phosphorus spherules (d) in a kidney affected by cancer.

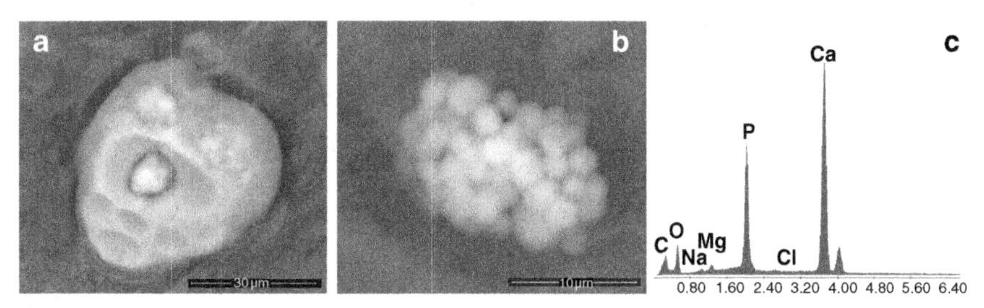

Figure 10.12 Images (a, b) show large calcium-phosphorus spherules (c) in a breast cancer.

invade all the tissue. In Figure 10.12(c) the internal shells of calcification growth are well visible. In contrast, Figure 10.12 shows the evolution of these spherules that aggregate to a coalescence, forming a wide area of calcification well visible under X-rays.

In the course of our nanoecotoxicity tests, after feeding earthworms with cobalt nanoparticles, we analyzed their digestive tract and found similar calcium-phosphorus spherules (Figure 10.13). We wonder if this "calcification" is a universal biological reaction to some particular stimulus?

In cancer tissues we also often find iron-phosphorus precipitates that we think of endogenous origin (see Figure 10.14), but also in this case we have no clear explanation to offer.

In a brain disease (pathology lacking a diagnosis) we found many areas of the frontal lobe with zinc-phosphorus-calcium precipitates whose origin remained unknown (see Figure 10.15). Similar precipitates of sodium-phosphorus found in cancer tissues continue to be of unknown source (Figure 10.16). Figure 10.17 show the EDS spectra with

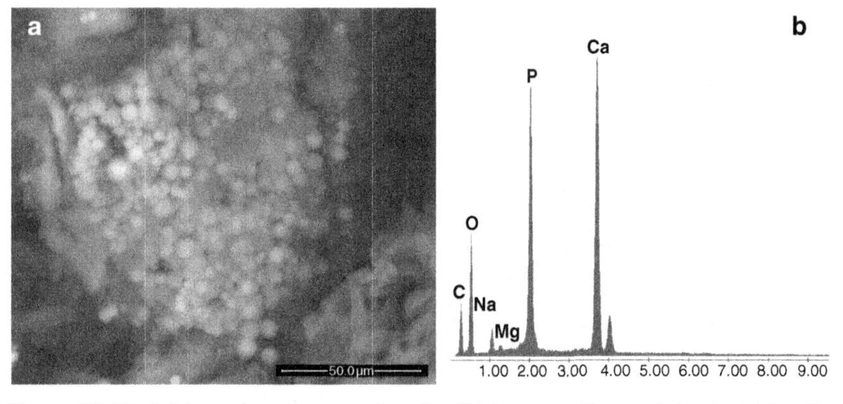

Figure 10.13 Calcium-phosphorus spherules (b) in an earthworm's body (a) (section before clitellum) fed with cobalt nanoparticles.

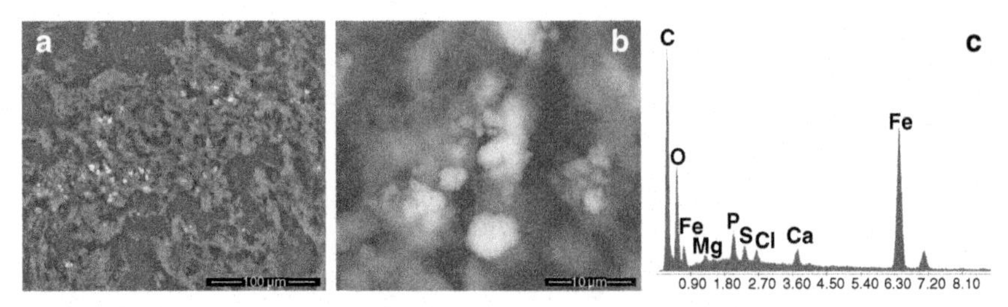

Figure 10.14 Images (a, b) show iron-phosphorus precipitates (c) in the kidney of a patient affected by kidney-and-pancreas carcinoma.

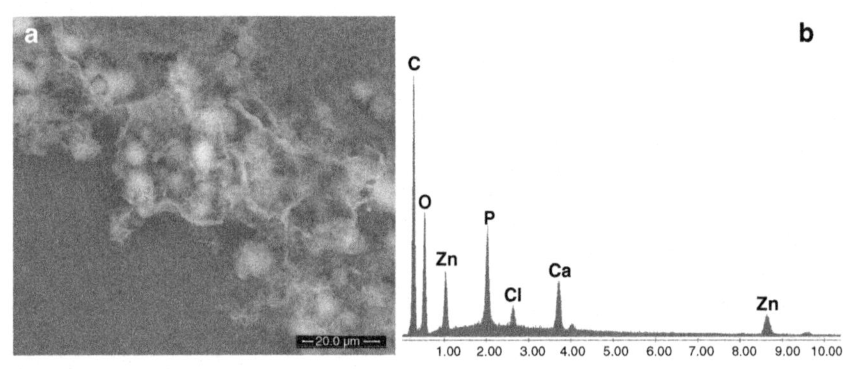

Figure 10.15 Image (a) shows zinc-phosphorus precipitates (b) in brain tissue with encephalitis.

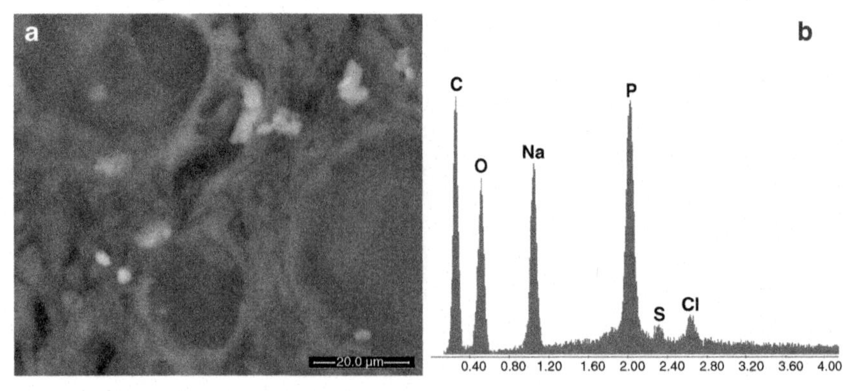

Figure 10.16 Image (a) shows sodium-phosphorus precipitates (b) in a thyroid affected by carcinoma.

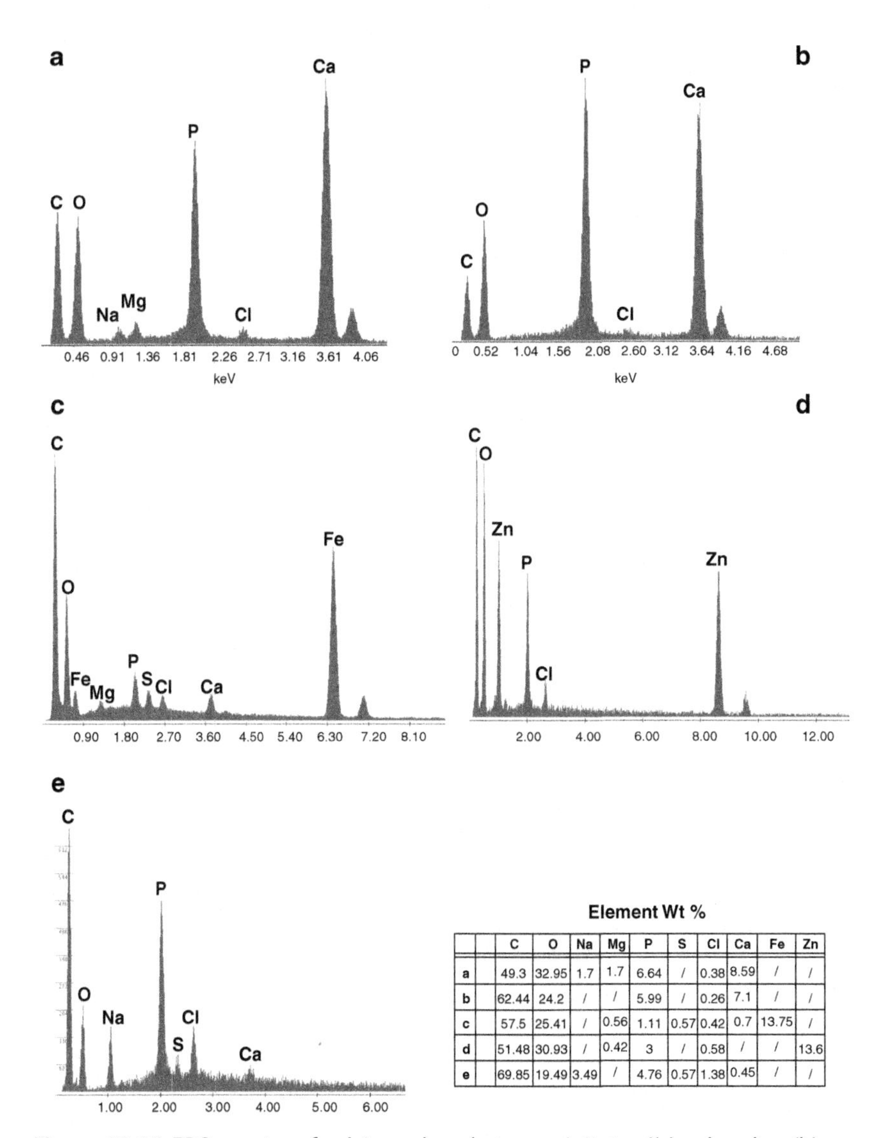

Element Wt %

		C	O	Na	Mg	P	S	Cl	Ca	Fe	Zn
a		49.3	32.95	1.7	1.7	6.64	/	0.38	8.59	/	/
b		62.44	24.2	/	/	5.99	/	0.26	7.1	/	/
c		57.5	25.41	/	0.56	1.11	0.57	0.42	0.7	13.75	/
d		51.48	30.93	/	0.42	3	/	0.58	/	/	13.6
e		69.85	19.49	3.49	/	4.76	0.57	1.38	0.45	/	/

Figure 10.17 EDS spectra of calcium-phopshate precipitates ((a) spherules, (b) crystals), iron-phosphorus precipitates (c), zinc-phosphorus precipitates (d) and sodium-phosphorus precipitates (e). The semi-quantitative analyses of every spectrum for every element are shown.

the semi-quantitative analyses of the different typology of precipitates: calcium-phosphorus, iron-phosphorus, zinc-phosphorus, and to sodium-phosphorus.

We are convinced that these precipitates are part of the carcinogenesis mechanism or, somehow, though we do not know how, related to it. We are also convinced that our pieces of evidence, if studied with oncologists and molecular biologists, could shed light on cancer, something still a great deal mysterious.

10.3 OTHER POSSIBLE EFFECTS OF NANOPARTICLE EXPOSURE

At this point in the book, we hope it is clear that, as repeated above, much of nanopathology is a work in progress and that questions are far more numerous than are answers. If one goes back with one's mind to certain events from the past that have never been clarified, it is inevitable to suspect that micro- and nanoparticles may be in some way involved. Hiroshima and Nagasaki are among them and have been briefly discussed in Chapter 7.

Another case worth attention is the "romantic" curse of the Pharaohs, according to which those who violate the Egyptian monarch's tomb are hit by mysterious disease. It is no surprise that bacteria and radiations were included among the possible culprits, as usual without any solid evidence against them. Now, we would like to add another possible suspect: of course, particles.

In 1922, 3,246 years after the young pharaoh Tutankhamun was laid in his tomb in the Valley of the Kings, the archaeologist Howard Carter and George Herbert, 5th Earl of Carnarvon, violated his nearly intact sepulchre, a room probably intended for someone else as it was not so grand as Egyptian royal tombs used to be. The justification for such an uncommon modesty is the untimely, unexpected death of the Pharaoh. One year later in Cairo, not having yet turned the age of 57, the Earl died short after having accidentally cut a mosquito bite on his face while shaving. Much later in London, in 1939, Carter died of lymphoma at the age of 64. Twelve other people who had worked with Carter and Herbert in the tomb died in the space of 24 years and one after 60 years.

Nothing particularly unusual in all that. Nevertheless the "theory" of a curse, also regarding the opening of other tombs and the death of people who had been somehow involved in the works, started to circulate.

Leaving aside at least dubious curses and with the lack of evidence of immediate deaths that never seem to have occurred, it would be interesting to know if those people showed symptoms such as, for example, chronic fatigue and lymph-gland swelling.

Tutankhamun's tombs had been excavated by stonecutters in limestone rock and the production of dust was inevitable. Just as inevitable was the crumbling of over 32 centuries into tinier and tinier dusts specks of plaster and fresco paintings and of some of what was contained in the small, never-aired rooms. So, the people engaged there in the 1920s must have inhaled those particles, something that is far from beneficial. In most cases, nanoparticles are rather slow acting, but, again in most cases, they are very efficient.

In the recent past, when, sometime after having caused explosions inside them, US soldiers entered the caves and tunnels of Tora Bora (Afghanistan), one of the strongholds of the Taliban and the suspected hideout of Al-Qaeda leader Osama bin Laden, they wore special protections and used respirators. Yet that place was very extensive and, as far as we know, very efficiently ventilated. That wise behavior means that the possibility of getting in touch with nanoparticulate matter, though not particularly publicized, is given very serious consideration.

REFERENCES

[1] Gatti AM, Montanari S. Nanopathology: The health impact of nanoparticles. Singapore: PanStanford Publishing Pte. Ltd; 2008. Calcium-Phosphorus.6:231-37.
[2] Gatti AM, Quaglino Daniela, Sighinolfi Gian Luca. A Morphological Approach to Monitor the Nanoparticle-Cell Interaction. Int.J.l of Imaging 2009;2(S09):2–21.
[3] Nemmar A, Hoet PHM, Vanquickenborne B, et al. Passage of inhaled particles in to the blood circulation in humans. Circulation 2002;105:411–7.
[4] Gatti AM, Montanari S, Monari E, et al. Detection of micro and nanosized biocompatible particles in blood. J. of Mat. Sci. Mat in Med. 2004;15(4):469–72.
[5] Gatti AM, Montanari S. Retrieval analysis of clinical explanted vena cava filters. J. of Biomedical Materials Research: Part B 2006;77B:307–14.
[6] Gatti AM, Montanari S, Gambarelli A, et al. In-vivo short- and long-term evaluation of the interaction material-blood. Journal of Materials Science Materials in Medicine 2005;16:1213–9.
[7] Gatti AM, Rivasi F. Biocompatibility of micro- and nanoparticles: Part I in liver and kidney. Biomaterials 2002;23:2381–7.
[8] International Agency for Research on Cancer. IARC monographs on the evaluation of carcinogenetic risks to humans. Tobacco smoke and involuntary smoking, vol. 83. Lyons, France: IARC Press; 2004.
[9] Schwartz L, Guais A, Chaumet-Riffaud P, et al. Carbon dioxide is largely responsible for the acute inflammatory effects of tobacco smoke. Inhal Toxicol. 2010;22(7):543–51.
[10] Abolhassani M, Guais A, Chaumet-Riffaud P, et al. Carbon dioxide inhalation causes pulmonary inflammation. Am J Physiol Lung Cell Mol Physiol. 2009;296(4):L657–65.
[11] Stayner L, Bena J, Sasco AJ, et al. Lung cancer risk and workplace exposure to environmental tobacco smoke. Am J Public Health 2007;97(3):545–51.
[12] Sasco AJ, Secretan MB, Straif K. Tobacco smoking and cancer: a brief review of recent epidemiological evidence. Lung Cancer 2004;45(Suppl 2):S3–9.
[13] Boffetta P, Hecht S, Gray N, et al. Smokeless tobacco and cancer. Lancet Oncol. 2008;9(7):667–75.
[14] Besson H, Renaudier P, Merrill RM, et al. Smoking and non-Hodgkin's lymphoma: a case-control study in the Rhône-Alpes region of France. Cancer Causes Control 2003;14(4):381–9.

[15] Polesel J, Talamini R, La Vecchia C, et al. Tobacco smoking and the risk of upper aero-digestive tract cancers: A reanalysis of case-control studies using spline models. Int J Cancer 2008;122(10):2398–402.
[16] Vollset SE, Tverdal A, Gjessing H. Smoking and deaths between 40 and 70 years of age in women and men. Ann Intern Med 2006;144(6):381–9.
[17] https://www.rareconnect.org/en/community/cryoglobulinemia
[18] https://www.rarediseases.org/rare-disease-information/rare-diseases
[19] http://www.mayoclinic.org/diseases-conditions/post-traumatic-stress-disorder/basics/symptoms/con-20022540
[20] http://www.nyc.gov/html/doh/wtc/html/residents/residents.shtml
[21] http://www.nyc.gov/html/doh/wtc/html/residents/know.shtml
[22] Ohlsson A, Shah PS. the Knowledge Synthesis Group of Determinants of Preterm/LBW births. Effects of the September 11, 2001 Disaster on Pregnancy Outcomes: A Systematic Review. Acta Obstetricia et Gynecologica Scandinavica 2011;90(1):6–18.
[23] Harville E, Xiong X, Buekens P. Disasters and Perinatal Health: A Systematic Review. Obstetrical & Gynecological Survey 2010;65(11):713–28.
[24] Young JD, Martel J. The rise and fall of nanobacteria. Scientific American 2010;1:52–9.

CHAPTER 11
The Future of Nanotechnologies

Contents

11.1 Introduction 231
11.2 Can nanomedicine solve the unsolved problems of medicine? 233
11.3 The success of nanosilver 238
11.4 Analysis of the end of the life cycle of nanoproducts 240
11.5 Prevention and systems of prevention 243
11.6 Present and future 243
References 246

11.1 INTRODUCTION

This side of human psychology has been discussed in more than one religious text: Man and his ultra mundane projections are often prone to succumb to the temptation of challenging God or, in one way or another, to take His place. A naive temptation, an impossible mission, to be sure, be one a believer or not, if only because we know just a tiny fraction of the rules governing the universe that hosts us, but this is how man's brain works. Nanotechnologies are just the latest temptation with the possibilities they offer to construct new matter, a matter with unwanted properties that seems to bend to the will of its creator. The obvious stumbling block standing in the way of our demiurgic claim is that man does not create anything, not going beyond reassembling what already exists. But, that is something that transcends the limits and the scope of this book. In any case, new nanotech materials are growing more and more available, are influencing economy and society and can do that in ways that are either unpredictable or whose prediction may be, at least in some respects, unwelcome.

Though progress is something desirable, most steps forward in technological progress have their downside and often problems manifest only after a relatively long time. So, curbing enthusiasm and observing all novelties with a critical eye could be advantageous to avoid falling into errors such as those relating, for example, to asbestos, to tetraethyl lead, to fluorine-carbides, to some drugs and to a long list of materials greeted and welcomed with keen excitement, guaranteed as harmless by the vast majority of scientists and then disowned because of the disasters they proved to be responsible for. Prudent experimentation means spending time, and time

Case Studies in Nanotoxicology and Particle Toxicology
http://dx.doi.org/10.1016/B978-0-12-801215-4.00011-X

means money. But, often that time is not wasted, and indeed, it becomes an investment, preventing mistakes that can be extremely expensive, if money is the only element of judgment.

The industrial areas where nanotechnologies are not applied or are not expected to find applications, are very few, and one of the greatest expectations concerns medicine.

In spite of a certain lack of modesty and of admirable progresses, medicine is still far from being able to solve a considerable number of problems. Among them (cold, hay fever, alopecia...), cancer, a multifaceted pathology is on the increase, and many of its varieties, whose origin is still largely unknown and a matter of opinion at least as much as it is of science. It is only natural that, when the origin of an illness is unknown, therapies are inevitably addressed to symptoms.

Be as it may, with researchers busy trying to unravel the mysteries of cancer, diagnosticians and therapists are very attentive to emerging technologies, and rightly so.

If on one side the race to new nanotech applications is too fast, on the other it is evident that progress is not fast enough and the approaches we have used so far are not fully satisfactory. In any case, due to its enormous, still unexplored possibilities, Nanotechnology may be a powerful weapon worth trying both to diagnose a pathology and to treat it.

As has been already pointed out more than once, in the case of medicine, and in all other cases, at that, we ought not to be influenced by haste and by the eagerness to recover the money invested and see gains multiplied.

If it is hard to induce investors in the medical industry to abide by the rules of prudence, particularly when some non-impossible, though yet mostly undefined, negative consequences may be expected to occur not immediately, but after a significant amount of time, institutional regulators are there to act as the custodians of wisdom. It is only fair that users, consumers and also investors are made aware of what the real concerns are, so as to avoid health problems, besides legal and moral responsibilities, and, as to investors, financial losses. For easy-to-guess reasons, medicine is the most ticklish subject and must be treated with great care.

Administering preparations containing nanoparticles without being certain of what their lot is going to be is something that, in the long run, may cause trouble, and even the simple, uncontrolled disposal of nanotechnological products at their life-end may present more than one contraindication. So, it is essential to legislate on the basis of all current scientific knowledge without neglecting anything and always using prudence and honesty as a guide. Health and the environment, two factors that are closely related, are things on that we cannot afford to act upon lightly.

11.2 CAN NANOMEDICINE SOLVE THE UNSOLVED PROBLEMS OF MEDICINE?

To the question of whether nanotech-based medicine can solve unsolved health problems, the answer is a non-committal yes and no. What is sure is that nanomedicine, like any other approach, will never solve all problems.

In some picturesque contexts nanoparticles are sometimes called "magic bullets" for the ability they have to cross any physiological barriers and, if handled the right way, hit exactly the mark aimed at. Nanotechnologists construct them to interact with the wanted targets, cells or organelles or even molecules at nanolevel, and do that with the highest accuracy. But, on closer inspection, this ability can also have some negative aspects.

When nanotechnologists thought it was a brilliant thing to inject iron-oxide nanoparticles into an organism for imaging purposes or for cancer therapy, they might have been only superficially aware of the defense mechanisms the body stands against nanosized particulate matter and, because of that, they neglected to pay due attention to the basic concept according to which all injectable/implanted materials must be compatible with the organism either immediately or over time. And biocompatibility of nanomaterials is not the same as that of what medicine has been accustomed to using for a very long time. Assuming that the introduction of something into the organism has only the result of getting the desired effect is at the very least naive. All drugs, without exceptions, have side effects, and, not too rarely, some of those effects can outweigh the desired ones. In that regard, nanotech products are not different. Among the problems inherent in nanoparticles with the characteristics of those we study is that we know very little about them, and, in particular, we largely ignore how much and, what is more important, if, we can get rid of them. As a matter of fact, what we saw as far as elimination is concerned does not look particularly encouraging, and so is what we saw in terms of tissue reactions. Their extreme invasiveness, their strong capability to interact with organs, tissues and cells, and their possible biopersistence are key-points to be carefully considered whenever a new nanoproduct is being created and whenever an advance in nanomedicine is envisioned.

Our opinion is that nanotechologies can be very helpful to medicine if used wisely, and the first concern must be to be able to remove what has been introduced in the body after it has done its job. "Point-of-care" devices and extremely miniaturized "lab-on-chips" could represent a bright future, provided we can tether them, maintain complete control of them and be sure of all their effects, particularly long-term ones. Diagnostic imaging, cancer therapy, regenerative medicine and the preparation of new drugs through the synthesis of specific nanoparticles and with targeted delivery are all subjects in which nanotechnologies can play a very important, even crucial, role and we must be careful not to fail, because a failure, perhaps dictated by haste, could close prospects of enormous interest.

As briefly mentione above, one of the applications concerns the synthesis of multifunctional nanoparticles endowed of a strong affinity with cancerous cells that allow both cancer imaging and therapy. One of these products is composed of a peptide ligand bio-conjugated on super-paramagnetic iron-oxide nanoparticles loaded with an anticancer drug. In order to treat the pathology, the nanoparticles are targeted to the tumor tissue and the X-ray imaging of the area is immediately well defined. Applying an alternating magnetic field, the iron nanoparticles heat up at a temperature high enough to kill the pathological cells, saving the healthy tissue. One of the advantages of such an approach is the accuracy of targeting and the preservation of healthy tissue, but one wonders what the fate will be of these nanoparticles entrapped in the tissue and impossible to remove? [1].

Another application is the synthesis of biodegradable nanoparticles to deliver drugs to the brain, an organ that, in many conditions, can be hard to reach because of the blood/brain barrier.

Multifunctional polymeric micelles or dendrimers are used to deliver imaging agents and therapeutic chemicals. Specific functionalization of nanoparticles obtained by coating their surface with a layer of proper molecules meant to interact with specific targets are being actively investigated. These innovative possibilities open new strategies for the diagnosis and treatment of cardiovascular diseases and musculo-skeletal disorders. Nanoparticles can also be used in aerosolized drugs for oral and pulmonary delivery [2].

The possibility of combinations of inorganic particles with organic substances addressed to different medical problems is impressive. Hence, great enthusiasm and expectancy on the new therapeutic nanosystems. Not only, a direct, programmed stimulation of single components of the cell with nanoparticles can answer still unanswered questions on cell metabolism and supply a huge amount of information.

Nanotechnology can also be considered science fiction: for example, with the so-called dechronification [3].

The notion, for the time being a mere theory, is that through the use of nanoparticles a limitless healthy lifespan could be achieved. Nanomedicine could combat (and, again in theory, defeat) aging, acting through two different steps: stopping the aging process, and, even more in theory, reversing the aging mechanism. The basic concept is that nanorobots like, e.g., the so-called respirocytes and the so-called microbivore devices, can act, the former, as red cells, but with a much higher efficiency in carrying oxygen, and as phagocytes the latter. A periodic removal of toxic agents and catabolites can thus assure a longer life to cells. At the same time, with similar devices, Nanotechnology proposes to eliminate or repair damaged DNA, or replace damaged chromosomes. Simultaneously, organelles like, for instance, mitochondria, with functional defects can be replaced.

At the time of this writing, all that, is no more than theory that will hopefully come true in a more or less distant future. But the problems, all that poses are many and they are not only purely technical. Accepting, and indeed, assuming that we are able to build successfully such sophisticated nanodevices and they are efficient as desired, the difficulties we could not be able to overcome are those related to biology, an aspect of nature on which we have no control. As a matter of fact, we have no idea how a cell can get rid of the "unusual" complexes made by nanoentity plus toxin. Again, as a matter of fact, we don't even know whether such a mechanism exists.

No doubt: Nanotechnology and all its apparently endless possibilities represent temptations hard to resist and much is in the pipeline but, if we don't want to fall again for the mistakes made in the past with other technologies and pay dearly for their consequences, the picture must be complete with nothing left to chance, even if the note of caution that may come is not always welcome.

And what follows is a note of caution. The core of nanomedicine is based on the right assumption that nanoparticles can negotiate virtually any physiological barrier, i.e., the filters our body opposes against the insults that come, for example, from the environment, and that they have the possibility to interact directly with proteins, enzymes, DNA, etc., due most likely (though not certainly) to their comparable size with those molecules.

Nanoparticles are "creatures" of physicists and chemists, scientists that, however excellent, in most cases have little experience with biocompatibility, i.e., the biological ability of an organism to accept exogenous objects. On the other hand, biologists, as excellent as the physicists and the chemists, have little experience of the nanoworld and its properties and, even more, little or no experience at all on how to handle nanoparticles. So, they extrapolate from their toxicological background made of atoms and molecules. But nanoparticles are neither atoms nor molecules: they are discrete bodies made of atoms arranged in a lattice and often, at least those we deal with, are not biodegradable, i.e., they are biopersistent, and in a way, not literally but just from the human life-span point of view, as already many time defined, eternal. From their part, nanotoxicologists often do not know much of the properties of matter at nanolevel and do not pay the due attention to the fate of nanoparticles and their state at the end of their experiment. So, the big problem is the gap between different cultures, all valuable in themselves but none of them sufficient when nano is at stake.

In a way, nanoparticles act as Trojan horses, since they actually do negotiate all physiological barriers finding no apparent opposition. Because of that, they can be used to introduce foreign bodies inside the cell, working as carriers for molecules that, when alone, would be recognized by the cell sensors and not allowed to enter. By breaking down the cell gates, or cheating them, we can introduce something that would otherwise be left outside.

Just as a memento of school days, in the case of the real Trojan horse, Cassandra, the Trojan "cursed prophetess," alerted her fellow citizens, warning them that accepting that gift within the city walls would have meant destruction and mourning to all of them. She was not listened to and, as is the lot of all Cassandras, not believed, and that was the last of Troy.

In the case of nanomedicine it is already evident that there is something objectionable in the basic concept. How can we inject a bolus of non-biodegradable particles into an organism and expect not to stir up a reaction to those foreign bodies? What is the fate of these nanoparticles after the end of the imaging analysis, the thermal therapy, a vaccine injection, or whatever other use? What is the fate of the nanoparticles that did hit the target and remained trapped in other parts of the body?

As repeated frequently thoughout this book, many of those particles are biopersistent and our organism cannot eliminate them or, in any case, not in an efficient way. So, once they have done their job, remaining in the organism they can interact with its most intimate parts. Drugs, at least those available so far, are ineffective at preventing the interaction with the biological substrate, being able, in the best of circumstances, to mitigate the effects acting as palliatives. It must be added that, when nanoparticles enter a cell, they occupy a volume and the intracellular trafficking can be altered just because of their physical presence [4]. The current situation is that, if we now know how to cheat the cell barrier and can introduce our nanoparticles as far as the nucleus, we don't know how to remove them.

The permanence of these foreign bodies can increase the probabilities of interaction with the DNA, and the activation of epigenetic mechanisms could contribute to non-genomic transgenerational inheritance. So, we are facing a paradox: nanoparticles could cure a cancer but later they themselves could activate cancerogenic mechanisms.

In our book *Nanopathology: The Health Effects of Nanoparticles* [5] and in this book pieces of evidence are reported of submicronic particles embedded in cancerous tissues. As already mentioned, the presence of those obviously exogenous presences bears witness to the exposure the patient underwent. It must be emphasized that those particles are not necessarily engineered but, as is by-far the most common case if only for the time being, can be originated unintentionally by many human activities. No difference is visible between engineered and unintentionally-generated nanoparticles as far as biological reactions are concerned and, in fact, there is no reason why a difference should exist. Their invasiveness is due only to their size.

Today, unintentional particulate pollution is overwhelming in comparison with that produced by Nanotechnology, but it is not impossible that engineered nanoparticles already have victims. Though unproven, it is our opinion that Professor Richard Smalley, who was awarded the Nobel prize for Chemistry in 1996 for his research work on the pyrolytic synthesis of carbon nanotubes and the so-called buckyballs, might have been one of the first victims. One of the most prominent advocates of Nanotechnology and its innumerable applications, including its potential to fight cancer, in the course of his research Professor Smalley handled a great quantity of nanoparticles and died in 2005 simultaneously of lung cancer and chronic lymphocytic leukemia, a (then) rare coincidence indeed. He himself was convinced that the dust inhaled during his work was responsible for his pathologies.

For the reasons outlined, internal dispersion of very small biopersistent foreign bodies should be strictly avoided. Our immune-system sensors do not recognize or make any particularly visible difference between an unusual cadmium selenide and titanium-dioxide. The main, though certainly not the sole, triggering factor is a very small foreign body, whatever it is made of.

When we happen to discuss the subject of prudence with nanotechnologists, the usual reaction is one of polite hostility as we are perceived as obtuse die-hards averse to any novelty. To be really fast, a racing car certainly needs a powerful engine, but without efficient brakes she will get out of control at the first demanding corner.

In short, nanoparticles offer immense possibilities but, particularly the biopersistent ones, are also very critical to handle. So, they should be treated with the utmost caution and, in any case, exposure should be avoided.

11.3 THE SUCCESS OF NANOSILVER

Although the existence of bacteria had still plenty of time to wait before being conceived as an idea, the antibacterial and preservative properties of silver have been known for over 2,000 years. For example, the ancient Greeks and Romans (the rich, to be sure) used silver vessels to keep water potable. Later, though with superstitious implications, it was customary to give a silver spoon to newborn babies that were at risk of contracting diseases from inappropriately preserved food.

Since the nineteenth-century silver-based compounds have been widely used in bactericidal applications, in burns and wounds therapy, etc, [6]. Today nanosized silver is used in hundreds, maybe thousands, of new nanoproducts that take advantage from its antibacterial property that seems to be enhanced by nanodimension [7].

Silver nanoparticles are now appearing everywhere: in deodorants, in fabrics, in toothpaste, in cosmetics, in food packaging, in spermicidal products, and even in objects somewhat unexpected such as pacifiers (a Chinese product). Thus, the exposure to those particles is growing more and more likely. Let us take, just as an example, silver-particle-treated garments. It is only natural that those who wear those clothes come in close contact with that silver both through the skin and by inhaling the particles that inevitably come off due to wear and aging. If a passage through the skin can be considered unlikely, the one through inhalation is certain. But another, less direct, contact, occurs when those garments are washed and release their particles into the sewage system. That water is treated by filters but those devices cannot retain nanosized entities that, as a consequence, re-enter the environment. One of the possibilities is through irrigation water that eventually contaminates agricultural products. A further possibility is when that water is drained in rivers and in the sea. Inevitably, many forms of aquatic life, including fish, get contaminated. The same risk is run, and in a more serious form, at that, because of the higher quantities involved, with washing machines that produce nanosilver and add it to laundry. Like all other products, nanosilver ones (e.g., some wound bandages) also have a limited life cycle and, as often happens, when the product has become waste, it is incinerated. Incineration involves the dispersion of those particles as such or in combinations of silver with other elements to form other, less predictable particles.

Alternative medicine also makes use of nanosilver in colloidal (suspension in a fluid) form as a sort of panacea, and in many cases that silver is meant for internal use. It is very probable that silver gives excellent immediate results against a number of diseases, but nobody ever evaluated or even considered the accumulation phenomena that are inevitable in chronic consumers like those who take it with the aim of strengthening their immune system. As happens in all other cases of slow accumulation, the actual negative results, if any, of protracted intake of nanosilver become visible only after a relatively long time and the responsibility of those particles may go undetected, since that form of intoxication could be different from the classical argyria well known by toxicologists. In this case silver is in particulate form with all its peculiarities and consequences.

On June 16, 2014, the European Commission issued its "Final Opinion on Nanosilver: Safety, Health and Environmental Effects and Role in Antimicrobial Resistance," which is worth reporting in full [8-10]: "The aim of this opinion is to assess whether the use of nanosilver, in particular in medical care and in consumer products, could result in additional risks compared to more traditional uses of silver and whether the use of nanosilver to control bacterial growth could result in resistance of micro-organisms."

The Scientific Committee on Emerging and Newly Identified Health Risks (SCENIHR) has also concluded that the widespread (and increasing) use of silver-containing products implicates that both consumers and the environment are exposed to new sources of that metal. Human exposure is direct (food, hand-to-mouth contact, skin) and may be life-long; while in the environment silver nanoparticles may be a particularly effective delivery system for silver to organisms in soil, water and sediment and may act as sources of ionic silver over extended periods of time. Therefore, additional effects caused by widespread and long-term use of silver nanoparticles cannot be ruled out. "Regarding the hazard associated with the dissemination of the resistance mechanism following the use of silver nanoparticles, more data are needed to better understand bacterial response to ionic silver and silver nanoparticles exposure" [11].

Our experience of direct analyses in pathological tissues suggests that submicronic or nanosized silver particles of engineered or incidental origin, like all other particles the same size, can exert a negative interaction with the human body as shown in Figure 11.1 related to a lymph node affected by Hodgkin's lymphoma where we identified numerous submicronic silver

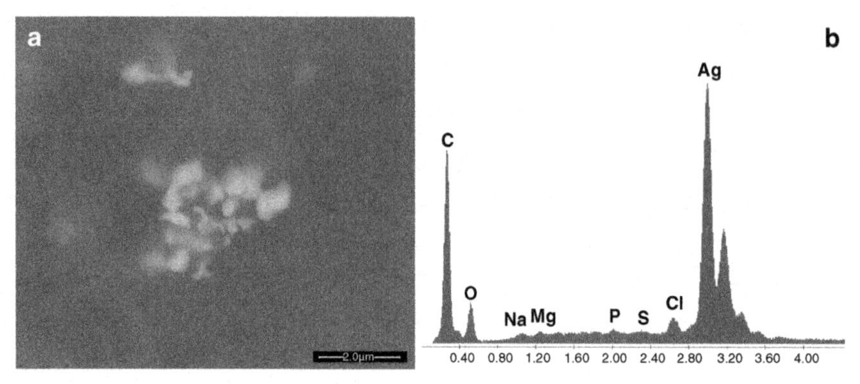

Figure 11.1 Image (a) shows silver particles (b) identified in a lymph node of a young girl affected by nodular-sclerosis Hodgkin's lymphoma.

particles together with spherical iron nanoparticles. We were unable to trace the source of this exposure.

11.4 ANALYSIS OF THE END OF THE LIFE CYCLE OF NANOPRODUCTS

We have just mentioned the obvious: all products have a life-span and sooner or later, even when recycled many times, users try to get rid of them, landfills and incinerators being their most usual lot. To add obvious to obvious, the Principle of Conservation of Mass tells us that, whatever we do, not an atom of it is actually removed from the planet. So, even if, by a sort of slight-of-hand much of what is incinerated disappears from our sight, the result is not particularly exciting. In fact, the amount of what is released into the environment as a result of waste treatment is greater than the original amount of that waste. That is because burning is a synonym of oxidation, i.e., the addition of atmospheric oxygen, which, of course, has a mass. But, for technical reasons, a substantial quantity of different materials is used to complete the incineration process: water, methane, activated carbon, ammonia, sodium bicarbonate and carbonate, etc. The result is that we roughly double the amount of waste matter that enters the environment, with the aggravating circumstance of having turned much of what was not particularly noxious into toxic gases, toxic substances (e.g., dioxins, furans, etc.), micro- and (primary and secondary) nanoparticles.

All that must be duly considered when new products are designed and, in any case, disposing of objects containing nanoparticles is an awkward matter.

The enthusiasm that took scientists and businessmen alike for nanotechnologies is not showing signs of decline even if new concerns are inevitably rising about possible, and now demonstrated, adverse effects during product manufacturing, use and disposal. Considering life cycle and trying to assess possible risks connected with use and disposal is something no one can afford to transgress. Nanoproducts are multifarious and not a day goes by without a new one being proposed on the market. As happened decades ago with plastic products, their cumbersome presence can be now hard to notice, but there is little doubt that within the near future they will become very perceivable both in variety and in quantity. As happened with plastics, nanoproducts will pose thorny problems regarding their disposal and, once again, as before, the risk is that we are caught unprepared to face them. Some of those products have a short life cycle and the problem of their disposal is growing real and urgent. The example of wound dressings just briefly mentioned above can illustrate the point: These dressings, treated with a 12-nm coating of silver nanoparticles, are now growing very popular. After 5-7 days the bandage must be removed and substituted with a new one. In almost all cases the used bandages are disposed of by incineration with all the consequences already touched upon. As already noted elsewhere in this book, the higher the temperature, the smaller the particle size is, often in the nanorange and, in the effort to reduce the quantity of pollutants such as, for instance, dioxins, modern incinerators work at higher temperatures as compared to the old ones. The filtration systems of incinerators are not sufficient to capture all the particles and a sizeable amount is freely released into the environment. Bottom ash along with what obstructs the filters is collected, and most of it is disposed of in landfills, while some is mixed with cement and asphalt without regard for what the consequences are inevitably going to be. A few studies have been published about the pathologies suffered by people living around incinerators and some concerns look more than justified [12–15].

When this pollution contaminates soil and water, it enters invariably the food chain with effects that have never been studied in-depth. This free dispersion is very hard to check and impossible to eliminate, since no effective technologies exist to that purpose, at least for the time being. While taking preventive measures to protect the people who are employed in nanotechnological factories and inside their laboratories is possible, and working to implement them is indispensable, it will be very difficult to capture nanoparticles (be they engineered or not) when they are dispersed in the air, in the water and in the soil. Figure 11.2 shows malformed chamomile

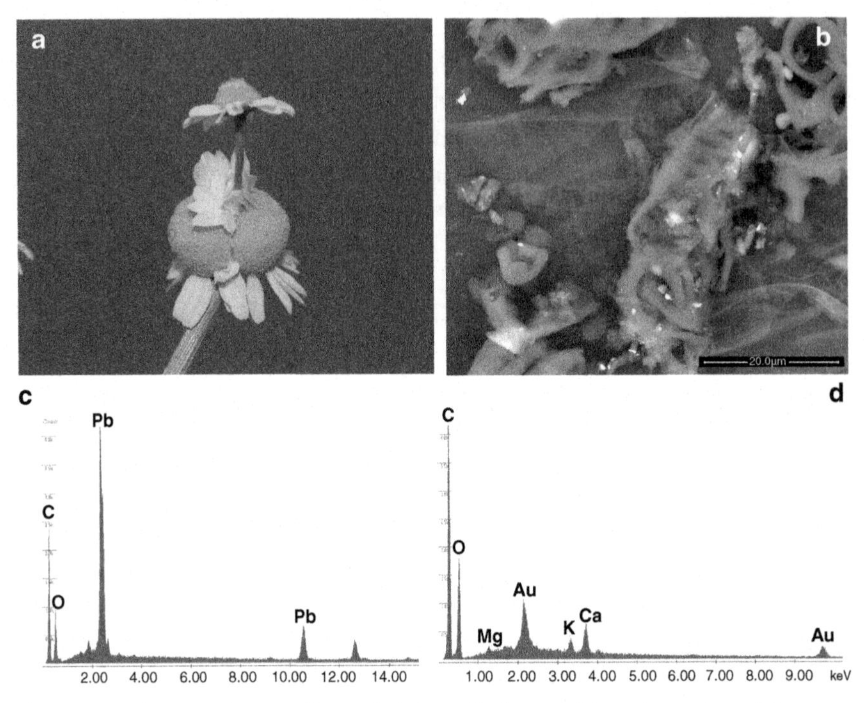

Figure 11.2 Images show malformed chamomile flowers (a) and a section of the stem observed under ESEM (b). The spectra of some of the debris identified inside the stem show the peaks of lead (c) and gold-calcium-magnesium-potassium (d).

flowers picked in a field where some fluids with inorganic pollutants were dispersed. Inside the stem, close to the soil we found many submicronic particles. Figure 11.2(c,d) show the EDS spectra of lead, gold, etc.

Besides the precautionary principle, which in many countries is the law, although it is sparingly observed, we must apply a global responsible care principle in defense of consumers, users and workers. Nanoscience and new technologies are relevant for the solution of some society problems, but occupational safety and environmental protection implications must come first. At present, there is more than one doubt on the sustainability of nanotechnologies, mainly because they are driven by myopic business and not by science and common sense. That does not mean that nanotechnologies must be hastily sacked. Certain nanoproducts, those containing degradable particles, may be considered safe and should not imply risks for humans. What we ought to be wary of are the techniques and products that can release non-biodegradable nanoparticles and, in particular, disperse them in the air, in the water, and in the soil representing a potential hazard.

Actually, we do need nanotechnologies just to cope with the possible dangers inherent in the nanoworld and to combat the nanoeffects that they can generate on the environment and on the human and animal health. Somewhat paradoxically, nano can be a problem and a solution at the same time. But "nanosafety" is the password.

11.5 PREVENTION AND SYSTEMS OF PREVENTION

In the absence of real therapeutic possibilities, the only way we currently have to defend ourselves is to resort to prevention. So, after the necessary steps of not pretending the problem does not exist and of enacting sensible laws, suitable monitoring systems should be implemented. Since the best way not to have particulate matter in the environment is not producing them, among other targets, what we call progress should be aimed at modifying old technologies and envisioning new ones for that purpose. Encouraging scientists, technicians, companies and investors to work on devices capable of capturing micro- and, especially, nanoparticles is also necessary. At the very end all this means business.

Prevention should also be applied to people who work with nanoparticles and, in particular, to students. To this aim the University of San Francisco has already released a manual containing safety procedures and recommendations to protect students during their laboratory work, and a website is open to guide them [16, 17].

Prevention measures should be adopted by people contaminated by nanopollution. They should wear proper garments at work and never wear them at home in order to avoid contaminating their families.

If already contaminated, males must avoid the risk of sexual transmission of nanopollutants. The use of a condom can work, thus preventing their female partners from contracting burning semen disease and further gynecological problems.

11.6 PRESENT AND FUTURE

Not a word of what is written in this book is in contrast with traditional medicine. We are just observing the same things, with the same aim, from a different point of view and proposing a different approach to their understanding. With the due differences and openness to scientifically-proven denial, we propose the "philosophy of everybody" in analogy to the physical "theory of everything."

What we have seen in the course of our research is something obvious: a body is not the sum of organs, tissues, cells, but is a machine, a unique machine inasmuch as is each individual, that works because there is a host of single machines that carry out functions required for the body's life. Each single machine is aimed at the harmonious functioning of the whole, but retains features that are peculiar to itself.

As to the subject of this book, due to the invasiveness of micron-sized and, even more, submicronic dust, all machines somehow come into play when a body is exposed to environmental pollution, and every one of them expresses a reaction according to the stimulus received, a stimulus that does not depend à-la-Paracelsus just on dose. Modern medicine should extend its scope, starting with not rejecting what is not immediately understandable through its received tradition. Viruses, bacteria, parasites, conventional poisons and the sheet-anchors psychology and genetics may not be enough to explain pathology or, as is growing more and more widespread, unusual, sometimes never-seen-before, syndromes. Luckily enough, often successfully, homeostasis comes to the rescue, but that is not always the case. And with particles the possibility of that kind of rescue is rather dubious. We think that an interdisciplinary approach to each case can help issue the correct diagnosis and, when available, prescribe the correct therapy. Unquestionably, with particles the cure must be based on their elimination, something that does not look likely to be feasible for the time being and that must be the goal of a branch of nanopathology research, but what can be done now and will always be of the utmost importance is remove the patient from the pathogen, be it in the air he/she breathes, in the food he/she eats or elsewhere. What must necessarily be done is ferreting out the origin, a task medicine alone as is currently conceived cannot do. As to drugs used in pathologies caused by particles, at present they can only mask the symptoms, a result that in some cases is indispensable but that may introduce elements of confusion in the attempts at diagnosis and therapy. In other circumstances drugs could depress the crucial biological reactions against the "invaders," thus weakening the patient's defenses. In any case, it must be kept in mind the basic but often forgotten pharmacological principle according to which there are no medications free from side effects and considering pros and cons objectively is indispensable for the good of the patient.

Thousands of diseases exist that are, more than once, actually mysterious in their origin. A considerable number of pathologies are tagged as cryptogenic, a definition that ultimately means, "I do not understand anything about its genesis." Now, with a growing frequency, illnesses that apparently have nothing to do with each other show up in the same patient. It is a fact that autoimmune diseases are now much more frequent than they used to be, and that means that the organism does not accept itself in its entirety anymore. It is our opinion that a few of those pathologies and syndromes could be triggered by the interaction of tissues, cells, organelles, proteins, etc., with solid particles and the consequent generation of unrecognized biopersistent new entities that cannot be eliminated by the body and against which we may have only weak defenses available or no defenses at all. We are convinced that some diseases of unknown etiology could be understood if they were investigated in a less conventional way with research aimed at detecting foreign bodies in the organism. If the search of nanoparticles were included beyond those routinely carried out correctly looking for bacteria, parasites and viruses, not only the origin of some diseases would be better understood, but we could be warned by their heralding symptoms. That could more easily lead to be more accurate in implementing prevention strategies, to find appropriate therapies and to be more timely in applying them.

One of the weapons available to medicine trying to describe pathological phenomena is epidemiology and, when correctly used, that approach allows us to be reasonably certain of the correlation between pathogen and disease. Asbestos and tobacco smoke are two good examples of the past. In both cases, no explanation of the actual pathogenic mechanisms or of the exposure threshold was given, but statistics sufficed to establish a sure-enough connection. So, after identifying a risk factor and primary prevention measures, the most cost/benefit effective ones can be implemented.

But epidemiology has a limit: like all methods based on statistics, epidemiology loses more and more reliability as the numbers to which it applies dwindle, until it becomes completely meaningless when the cohort of subjects is reduced to a few units. With illnesses and syndromes induced by particles, varieties are all but endless and, as a consequence, the numbers of those who suffer from the collection of the same symptoms is usually extremely low, rarely limited to one individual. If the approach is a purely accounting one, failure is a certainty. In those sometimes embarrassing circumstances it is far from infrequent to fall into temptation and, setting aside the primary mission of medicine, to pretend that those cases do not exist.

Nanopathological investigations could give new hope to epidemiological studies, since the direct nanodiagnostic observations on the pathological tissues allow to identify debris with their size, shape and chemical composition, thus defining, or helping to define, the actual exposure the patients underwent. Homogeneous cohorts of patients exposed to similar contaminations can be identified and, in such a way, a more accurate study can be performed, with the possibility that these direct observations can increase the reliability of epidemiology and broaden its scope. In this case, it is not the disease that is the starting point of the investigation, but the exposure.

Nanopathology can also be a tool useful for a better understanding of "well-known" pathologies such as, for example, diabetes. It is a fact that the rate of diabetes in New York showed a rise after 9/11 when many New Yorkers inhaled and ingested the dust raised by the collapse of the Twin Towers and the pulverization of two airplanes. Until now, the diagnosis is not issued through direct observation of the pancreas, but on some indirect parameters found in the blood and in the urine. Although much less easy to carry out, direct observations on pancreas specimens could show the presence of micro- and nanosized foreign bodies, already known as endocrine disruptors that impair the organ functionality.

But other not fully-explained or not-explained-at-all diseases could be better understood by the direct study of the patient's pathological tissue.

REFERENCES

[1] Brannon-Peppas L, Blanchette JO. Nanoparticle and targeted systems for cancer therapy. Advanced Drug Delivery Reviews 2012;64(Supplement):206–12.

[2] Sham JO, Zhang Y, Finlay WH, Roa WH, Löbenberg R. Formulation and characterization of spray-dried powders containing nanoparticles for aerosol delivery to the lung. Int. J. of Pharmaceutics 2004;269(2):457–67.

[3] Freitas RA. The scientific conquest of death, 2004 http://www.rfreitas.com/Nano/DesEvoForeword.htm.

[4] Gatti M, Quaglino D, Sighinolfi GL. A morphological approach to monitor the nanoparticle cell interaction, International Journal of Imaging (ISSN 0974-0627) Vol. 2 No. S09 Spring 2009 (Editorial 1) 2-21.

[5] Gatti AM, Montanari S. Nanopathology: the health effects of nanoparticles. Pan Stanford; 2008. 1-298.

[6] Klasen HJ. A historical review of the use of silver in the treatment of burns. ii. renewed interest for silver. Burns 2000;26(2):131–8.

[7] Salata OV. Application of nanoparticles in biology and medicine. J Nanobiotechnology 2004;2:1–12.

[8] http://ec.europa.eu/dgs/health_consumer/dyna/enews/enews.cfm?al_id=1494.

[9] http://ec.europa.eu/health/scientific_committees/consultations/public_consultations/scenihr_consultation_17_en.htm.

[10] http://ec.europa.eu/health/scientific_committees/emerging/docs/scenihr_o_039.pdf.

[11] http://ec.europa.eu/dgs/health_consumer/dyna/enews/enews.cfm?al_id=1494.

[12] Report of Institute de Veille Sanitaire (France) Incidence des cancers à proximité des usines d'incinération d'ordures ménagères 2006.

[13] Report of Royal Society:Review of DEFRA's health and environmental effects of waste management options, March 2004; http://www.royalsoc.ac.uk/displaypagedoc.asp?id=11459).

[14] Elliott P, Eaton N, Shaddick G, Carter R. Cancer incidence near municipal solid waste incinerators in Great Britain 2: histopathological and case note review of primary liver cancer cases. British Journal of Cancer 2000;82:1103–6.
[15] Hu SW, Shy CM. Health effects of waste incineration: a review of epidemiologic studies. Journal of the Air & Waste Management Association 2001;51(7):1100–9.
[16] Marsh W. Introduction to nanomaterials and occupational safety and health-manual for students' safety. http://www.dtsc.ca.gov/technologydevelopment/nanotechnology/index.cfm.
[17] www.goodnanoguide.com.

CHAPTER 12
Conclusions

12.1 CONCLUSIONS

Man has lived on this planet for thousands of centuries and his organism has evolved, where the concept of evolution does not necessarily mean for an indefinite, ideal better, but adapting itself to the conditions that went by. Every change took the time of generations, often many generations, but, on the other hand, also planetary conditions varied relatively slowly. In less than a couple of centuries, since the so-called Second Industrial Revolution, the environment has started to undergo important variations and the variations become faster every day, so fast that no organism can keep up with the changes. But, speed or not, the question is: can an organism, more specifically a human one, "evolve" to be able to live peacefully with the poisons we keep producing and unloading with increasing lightheadedness, many of which were unknown until very recently and, among them, many still unknown though heavily present? We are afraid that the answer is, no.

Apart from the responsibilities of each, a subject clearly outside the scope of this book, the undeniable fact is that new diseases are appearing and old ones, once relatively rare, are becoming common. Paradoxically, one of the causes of this problematic situation could also be its solution or, at least, its mitigation.

Nanotechnology, and nanopathology as part of it, offer a novel approach to medicine. It is possible that this new approach to linking the environment to man makes the understanding of many pathologies simpler. It is possible that some autoimmune diseases or cryptogenic ones can be understood, simply starting from the detection of the micro- and nanosized foreign bodies in the pathological tissues. Once micro- and nanometric foreign bodies have been found in the patient's tissues and the same particles have been detected in the environment where the patient lives or in the food he eats or in the drugs he takes, simple countermeasures can be taken: remove him from the source of particles. In most cases, in our experience, it works.

If primary prevention is possible and, indeed, mandatory, "detoxification" from those entities is surely much harder to achieve. But in the future,

Case Studies in Nanotoxicology and Particle Toxicology
http://dx.doi.org/10.1016/B978-0-12-801215-4.00012-1

nanotechnological devices could help to extract them from pathological tissues, thus eliminating those nanopollutants from the organism.

In more than one circumstance progress has required a number of deaths, but there is no reason why we should not try to dodge that toll. Nanotech medicine, if wisely used, could be a powerful weapon at our disposal.

Unlike medicine as it has been conceived so far, nanomedicine requires apparently very different skills. There, physics and chemistry are as important as biology and in some cases a few notions of biology may need to be revised, while in other cases biology must be in part rewritten. The same can be said for medicine. The Nanoworld is an immense land that we have just started to tread and, in the best of circumstances, to explore, and some of the rules we are accustomed to have no force there. That exploration demands a highly interdisciplinary background and an open mind, without which the only certainty is failure.

This book is our contribution to the understanding of problems that are either unsolved or ignored or to which a solution of convenience has been given.

We are not presumptuous enough to believe that everything we have written is free of error and we are sure that some, maybe much, of it will be corrected by better scientists. The only credits we think we deserve are having paid attention to something we are sure many researchers saw before us but erased immediately from their mind, and having observed the phenomena we are dealing with objectively and without any prejudice.

INDEX

A

Acute myeloid leukemia (AML), 56, 78, 81
Adenocarcinoma, 148
 bladder, submicronic particles in, 120
 gastric, 108
 pancreatic, 71
 pathological case, near power plant, 106
 stainless-steel spherules in tissue, 120
Aero-model, 111, 112
 environmental particles in filter of air pump, 113
 with remote control, 112
Aerosolization, bomb and target of, 71
Air pollution, 5, 208
 correlation with fetal health, 81
 environmental, 205
 events, 9
 particulate, 81
Air-pump filters
 and aero-model, 111, 113
 chemical composition of dust, 207
 debris on, 114
 and spherical nanoparticles, 104
Aluminium hydroxide, 189
Aluminosilicates, EDS spectra of, 74
Alzheimer's disease, 187
Ameloblastoma, 29, 45, 46
 case studies, 46
 chemical composition of particles found in sample, 47
 human papilloma viruses (HPVs), 45, 46
 hypotheses on origin, 45
 sample analysis, 46
Anatetall, 189, 190
Anti-inflammatory drugs, 68
Anti-wrinkle cream, 168
Archeology, pancreatic adenocarcinoma, 71
Artemia salina, 122
Arthrosis, 70
Asbestos, 33
 amphiboles, 33
 crocidolite, 33
 EDS spectrum of lung tissue, 34
 embedded in cancerous tissue, 34
 fibers, 32
 induced pathologies, 33
 serpentine, 33
Asbestosis, 33

B

Balanus amphitrite, 122
Balkan war, 50
 blood serum analysis, 136
 case studies, 136
 chemical-industry workers, 137
 cytomegalovirus infected patient samples, 138
 debris in blood samples, 137
 disease among soldiers, 135
 disease type and frequency of illnesses, 136
 Epstein-Barr, 138
 groups of patients analyzed, 137
 monoclonal gammopathy, 136, 138
 myeloma, 136
Barium sulfate, 58, 70, 101
Beta-2-microglobuline, 138
Bioenergy, 9
Biomass, 2, 9, 111, 124
Biopersistent nanoparticles, 16, 18, 139, 178, 235, 238, 244. *See also* Nanoparticles
Biopsy, 46, 70, 199
 bladder, 196, 197
  internal neoformation, 141
  nanoparticles in, 197
  particles identified, 198
 bone, 51, 87
 bone-marrow, 142
 elements found in, 76
 frontal brain, 93
 gastric mucosa, 80
 kidney, 59–61
 osteomedullary, 52, 142
 pancreas, 71
 papillary transitional neoplasm, 140
 pathological samples, 159
 plasmacytoma, 145
 prostate, 80

Biopsy *(cont.)*
samples of journalists, 149
thoracic sarcoma, 91
thoracic wall neoformation, 90
Blood vessel
ESEM images and EDS spectra of, 72
nanoparticles attached to endothelium, 210
presence of debris, 219
Bone cancer, 85
case studies, 85
chemical composition of, 86
debris found inside biopsic sample, 86
micro- and nanosized particles, SEM images of, 87
Bovine spongiform encephalopathy (BSE), 177, 180
case studies, 181
debris and nanoparticles, 184
debris in affected brain, 182
elemental compositions of debris, 186
FEG-ESEM and EDS analyses, 182
nanoparticles, 183
nanosized debris, 183
neurological symptoms in animals, 186
800nm sized particles, 184
Breast cancer, 62
benign breast lesions, 62
calcium-phosphorous spherules, 225
case studies, 62
images of debris, 62
metallic debris, 62
Bromodeoxyuridine cell proliferation test, 20
BSE. *See* Bovine spongiform encephalopathy (BSE)
Burning mouth disease, 48
ESEM image of mouth mucosa sample of, 49
tungsten carbide and stainless-steel debris, 49

C

Calcification, 39, 44, 88, 107, 149, 210, 224
and breast cancer, 62
clinical cases, 74
in liver, SEM images of, 90
reaction to foreign bodies, 48
spherular, 45
thyroid tumor sample, 43, 44
Calcitonin, 43
Calcium-phosphate crystals, SEM images of, 89, 227
Calcium-phosphorous spherules, 89
in breast cancer patient, 223
in earthworm's body, 225
in kidney cancer patient, 224
in thyroid cancer patient, 223
Carbon nanotubes, 2, 205, 237
FEG-ESEM images of, 23
with formazan dyes, 24
metallic nanoparticulate contaminants, 205
with nickel nanosized contamination, 205
Carcinoma, 147
basal-cell nodular, 115
bladder, 118, 120
breast, 62, 107
bronchioloalveolar, 118, 119
iron-phosphorous precipitates, 226
kidney, 226
iron-phosphorous precipitates, 226
lung, 147
pancreatic, 71, 72, 107, 226
squamous-cell, 219, 221
stainless-steel spherules in, 118
sulfur-silver particles, ESEM images and EDS spectra of, 72
testicular tissue
micronic particles, ESEM images of, 74
thyroid, 226
Case studies
Balkan war, 135
bovine spongiform encephalopathy (BSE) and food, 180
breast cancer, 62
ceramic-tile industry worker, 73
child with bone cancer, 85
child with prostate cancer, 75
epidemiological study on asbestos-exposure consequences, 32
Hashimoto thyroiditis, 39
incinerator
contamination around, 109
of Terni, 116

leukemia, 50
malformed child with leukemia, 78
blood examination, 78
chorionic villi sampling, 78
liver-and-spleen megaly and pallor, 78
mediastinal synovial fibrous monophasic sarcoma, 90
patient killed by repeated enemas, 88
power plant, 99
prostatic cancer, 75
spontaneous pneumothorax, 200
sudden coma, 93
toners, 199
vaccines, 187
war environmental dust, 131
Catabolites
exocytosis of, 19
inorganic-organic, 19
and toxic agents, 235
Cell-defense mechanisms, 49
Cell-nanoparticle interactions, 16, 25
Ceramic, 41, 70, 71, 213
debris, 58, 76, 200
dust, 202
materials, 34
nanometric-sized, 57
particles, 203
tile industry, 36, 73, 200
tiles, EDS spectra of, 74
tile worker, 200
worker's lung, 201
Ceratitis capitata, 122
Chronic-fatigue syndrome, 216, 217
Chronic myeloid leukemia, 216
foreign bodies attached to blood components, 218
C/N microbial ratio, 124
Coma, 93, 121
zinc-calcium phosphates precipitates, 95
zinc-phosphorus precipitates, 227
Combustion, 2, 7, 58, 75, 101, 105, 115, 122, 142, 153, 171
engines, 9
of fuel used by power station, 102
of oils, 100, 102, 185
of tobacco, 209
of toxic waste, 112
of waste, 77, 92
Congenital malformations, 54
carbon cycle, 54
internal organs of fetuses, analysis of, 55
studies on malformed babies, 55
submicronic pollutants, effect of, 56
Corrosion, 18, 66, 101, 173, 185, 211
Creutzfeldt-Jakob's disease, 180, 181
Crohn's disease, 69
Cryoglobulinemia, 56, 216
case studies, 57
ceramics, 58
frequency of elements, 59
iron-based debris, 58
kidney biopsy, SEM image of, 61
metallic materials, 58
3-micron-sized debris, SEM image of, 60
particles found in patients' blood serum, 58
silicates, 58
spherical debris, images of, 61
submicronic particle with the EDS spectrum, 60
type I, 56
type II, 56
type III, 56
Cryoglobulins, 56–58
debris in kidney biopsy, 59
gold particles, 59
Crystal germs, 93, 94
Crystalline precipitates
of calcium, 93
of phosphorus, 93
of zinc, 93
Cysts, 45
dentigerous, 45
odontogenic, 45
Cytotoxicity standard tests, 14

D

Dania rerio, 122
Dechronofication, 234
Dendrimers, 234
Dermatitis, 107
Diabetes, 216, 246
beta-cell functionality, 220
foreign-body inflammation, 220
rate in New York, 246
type-I, 220

Dust, 1, 7, 8, 10, 29, 65, 71, 92, 112, 116–118, 133–135, 138, 140, 147, 153–155, 171, 172, 204, 221, 228, 237
brick and cement, 202
ceramic, 202
chemical composition of, 207
containing uranium, 209
by depleted-uranium penetrators, 131
erupted by volcanoes, 2
inhomogeneous, 153
lung disease, 200
non-biodegradable, 203
of paint (titanium dioxide), 203
from polluting sources, 209
pollution on tobacco leaves, 215
silicates, 202
silicon-based, 73
silver, 51
submicronic, 244, 246
on surface of leaves of vegetables, 116
toner, 198
toxic, 90
volatile, 196
war environmental, 131

E

Ecography, 196
EDS. *See* Energy dispersive spectroscopy (EDS)
Electro-conductive coating, 31
Embryonic rhabdomyosarcoma, 75
Encephalitis, zinc-phosphorous precipitates in, 226
Endocytic trafficking, 16
Endocytosis, 16
Energy dispersive spectroscopy (EDS), 30
Environmental pollution, 8, 30, 39, 45, 48, 52, 54, 55, 62, 71, 79, 84, 99, 102, 105, 150, 172, 203, 208–210, 244
Environmental Protection Agency, 104
European Environment Agency, 65
Exocytosis, 223
of catabolites, 19
mechanism, 19

F

Fatal familial insomnia, 180
Fetal health, 81
Fetuin-A, 224
Fibrous flogosis, 219, 220
Food and Agriculture Organization of the United Nations (FAO), 171
Food contamination, 170
adhesive tape, debris in, 178
bacterial contamination, 170
bread, debris identified in, 177
cereals, 172
fruits and vegetables exposure to pollution, 171
hamburger analysis, 177
nanotechnological additions, 171
non-degradable compounds, 172
pollutants, 173
Food packaging, 171, 207, 238
Foreign-body reactions, 29, 69, 133, 220
Fullerene particles, 16

G

Gardasil vaccine, 189, 192
Gastric adenocarcinoma, debris images of
aluminium, 108
iron, 108
lead, 108
silicon, 108
titanium, 108
Gerstmann-Sträussler-Scheinker syndrome, 180
GLUS. *See* Granulomatous lesions of unknown significance (GLUS)
Good manufacturing practice, 173
Granulomatosis
hepatitis, 69
flogistic reaction, 71
liver, 70
reaction, 71
Granulomatous flogosis, 69
Granulomatous lesions of unknown significance (GLUS), 69
Greenockite (CdS) crystals, 205
Gulf War, 71, 129
syndrome, 130

H

Hashimoto thyroiditis, 39
affected area, SEM image of, 40
calcium compounds, 41
calcium concentration, 43

case studies, 39
ceramic materials, 41
lead-chlorine nanoparticle, SEM image of, 41
list of particles in sample, 42
metallic materials, 41
submicronic and nanosized particles, SEM image of, 41
Hazardous air pollutants (HAPs), 81
Hepatic granulomas, 69
calcium carbonate, 70
Crohn's disease, 69
granulomatous lesions of unknown significance (GLUS), 69
Hepatitis C virus (HCV), 56, 57
Hodgkin's lymphoma, 50, 107, 239, 240
HPV. *See* Human papilloma viruses (HPVs)
Human papilloma viruses (HPVs), 45, 46
high-risk types (16/18), 46
low-risk types (6/11), 46
nanoparticles trapped by tissue, as secondary effect, 49
Hydroxyapatite, 44, 56, 223
Hypotonia, 78

I

IARC. *See* International Agency for Research on Cancer (IARC)
Immunoglobulin A (IgA), 56
Immunoglobulin G (IgG), 56
Immunoglobulin M (IgM), 56
Incineration, 1, 9, 110, 118, 240, 241
dispersion of particles, 238
high-temperature treatments, 110
industrial, 109
of leaf, 209
municipal waste, 109
of organic compounds, 110
plant, 172
malformed foraminifers, 125, 126
Incinerators, 109
aero-model, 111
ashes collected in furnace, 116
ashes of waste, SEM images of, 117
biomass, 111
bronchioloalveolar carcinoma, 119
case studies, 116
contamination around, 109
debris on leaves, 112, 114, 115
dust on leaves of vegetables, 116
filtration of fumes, 111
municipal waste, 110
particulate matter, 111
pathological cases, 119
Industrial food products, 173
chocolate crust of Mars, 175
chocolate snack, 173
cold dessert, 173
micrometric debris, 174
homogenized children's food, 173
industrial hamburger, silver aggregates in, 180
metallic debris, 176
mortadella, 173, 175, 176
stainless steel debris, 175
Industrial pollution, 52, 116
Infarix, 189
Inflexall, 189, 190
In-situ hybridization, 46
Insomnia, 196
International Agency for Research on Cancer (IARC), 5, 209
Ions, 88
calcium and phosphorous, 88
diffusion of, 34, 35
iron, accumulation of, 35
metal, 185
metallic nanoparticles, corrosion of, 18, 56
nickel, 121
physiological chemical elements, 54
saturation concentration of, 94
Iron-based spherule, SEM image of, 53
Iron-phosphorus precipitates, 227
Italian biscuit, inorganic debris in, 172
Italian Institute of Technology (INESE), 164

K

Keratosis, 107

L

Lettuce leaves, 100
ESEM images of, 104
sample analysis, 102

Leukemia, 29, 50, 150, 211, 216, 221, 237
acute myeloid, 56, 78
case studies, 50
chronic myeloid, 216
iron-based spherule in bone marrow, 53
malformed child with, 78
metallic debris inside bone marrow, 53
Liver granulomatosis, 70
Lymphomas, 50, 130, 147
Hodgkin's, 50, 107, 239, 240
non-Hodgkin's, 50

M

Macrophage phagocytosis reaction, 69
Malformed children, 54, 81
case studies, 55, 81
debris in pelvis sample, 85
environmental pollution, exposure to, 55, 84
ESEM images of debris, 83
and miscarriages, 82
neural tube defects, 55, 84
Malformed foraminifers, 125, 126
Mass volcanism phenomenon, 8
MCS. *See* Multiple chemical sensitivity (MCS)
Mediastinal synovial fibrous monophasic sarcoma, 90
Membranoproliferative glomerulonephritis (MPGN), 57
Mesothelioma, 32
antimony particles
in liver, 36
in lung, 36
in pleura, 36
carbon-fluoride debris, 38
clinical cases, 36
induced by asbestos fibers, 35
peritoneal, 32, 36
pleural, 32, 36
Metabolic quotient, 124
Microbivore devices, 235
Micro-tom plants, 123
Monoclonal gammopathy, 136, 138
Mouth mucosa, 221
chemical composition of filament, 221, 222
cigarette temperature, effect of, 210
sample, ESEM image of, 49
Multifunctional polymeric micelles, 234
Multiple chemical sensitivity (MCS), 216

N

Nano-bio-interactions, 15, 16, 21, 22
NanoCare NanoSilver Beauty Soap, 164
NanoCare NanoSilver Shampoo, 165
NanoCare NanoSilver Soap, 167
NanoCare NanoSilver Toothpaste, 167
NanoCare NanoSilver Towel, 164
NanoCare shampoo, 167
Nanocontaminations
iron-rich, 166
titanium-rich, 164, 166
Nanocontent, 13, 171
Nanodevices, 235
Nanoecotoxicity, 164, 225
Nanoeffects, 3, 15
Nanoentity plus toxin, 235
Nanofood, 171
biopersistence and bioavailability, 171
nanocontent solubility, 171
Nanointeraction, 15, 127
Nanomaterials
biocompatibility of, 233
biopersistence, 233
Nanomedicine, 233, 235
Nanometric Silver, 164
Nanoparticles, 3, 9, 13, 38, 68, 163, 165, 183, 235
addition to rice plants, 124
adhesive properties, 15
in animal feed, 179
antimony oxide, 16, 17
asbestos fibers, 32
biodegradable, 234
biopersistent, 18
calcium-based, 92, 205
EDS spectrum of, 43
CeO_2, 20
FEG-ESEM images, 20
STEM images, 21
clinical cases, 29
cobalt, 123, 124
concentration, 22
dispersion of, 23
as drug-delivery agents, 13

dry, 25
effect on human health, 1
engineered, 10, 122, 170
toxicity of, 122
entrance into human body, 31
ESEM-EDS analyses, 168
ethylene-oxide process, 26
exocytosis mechanism, 19
exposure, effect of, 228
food packaging, 171
gold, 20, 70, 164
ESEM images and EDS spectra of, 72
SEM microphotograph of, 54
and 3T3 cells, 20
hematite, 19
inorganic, 25
interaction with endoplasmic reticulum, 18
interaction with hematopoietic cells, 16
intracellular trafficking, effect on, 236
in-vitro test, 24
in-vivo use of, 14
iron-oxide, 24, 123, 233
ISO standard definition, 2
metallic, 18, 86
nanosized by-products, 2
natural inorganic, 2
organic, 30
radio-labeling of, 22
sodium-pump mechanism, 19
Spanish moss, 124
titanium-rich agglomeration, 167
Trojan horses, 236
use in aerosolized drugs, 234
wet, 25
x-ray imaging, 234
Nanopesticides, 169
Nanoproducts, 51, 163, 164, 168, 238, 241, 242
chemical composition of nanoparticles, 165
end of life cycle, 240
ESEM investigation of, 166
manufacturers, 163
Nanorobots, 235
Nanosilver, 51, 169, 238, 239
antibacterial properties, 169
in colloidal form, 239
particles, 169
carbon spherules of, 169
in water filter, 169
pesticidal use, 169
Nanosized additives, 165
Nanosized debris, 48, 79, 183, 222
Nanotechnology, 4, 13, 231, 235
laboratories, working pollution in, 204
acarus entrapping environmental pollution, 208
air pollution entrapment, 208
cerium-titanium nanoparticles, 206
chemical composition of dust, 207
cross-contaminants, 206
crystals of cadmium, 206
gold-based particles, 206
iron-based spherules, 206, 207
nanometric-sized particles, 206
and medicine, 4
Nanotoxicological debate, 21
conjugate instability, 23
controversial issues, 21–27
dispersion of nanoparticles, 23
modifications in nanoparticles, 24
mutable nanoparticles, 23
nanoparticle concentration, 22
nanoparticle full-life cycle, 22
protein corona, 24
red cells of human blood, FEG-ESEM image of, 25
sterilization methods, 26
NanoUP Toothpaste Au-Ag, 164
National Organization for Rare Disorders, 216
Neu-Lexova syndrome, 82
Neural tube defects, 55, 84
Neuroblastoma, 107

O

O_2 /CO_2 exchange, 10
Organisation for Economic Co-operation and Development, 13
Orphan diseases, 68, 216
Osteosarcoma, 85

P

Paint additive, 166, 168
Palliatives, 236

Paracentrotus lividus, 122
Parkinson's disease, 187
Particulate matter (PM), 2, 8, 9, 30, 35, 54, 65, 107
  cold sources of, 9
  environmental, peach skin with, 171
  and industrial revolution, 9
  in kidneys, 79
  in liver, 79
  micro-sized, 8
  nanosized, 8
  shapes, 9
Phosphate intoxication, 88
Photochemical condensation, 7
Platinum-iridium spherules, 74, 75
Pleural coniosis, 200
Pneumoconiosis, 200
Pollakiuria, 196
Pollutants, 8, 35, 38, 42, 48, 54, 56, 58, 65, 102, 109, 111, 112, 125, 132, 149, 157, 160, 172, 173, 198, 206, 241
  biopersistent, 84
  chemical composition of, 68
  elemental composition of, 31
  exogenous, 54
  hazardous air (HAPs), 81, 84
  inorganic, 241
  metallic, 50
  nano, 243, 250
  particulate, 32, 52, 73, 137, 151, 195, 206, 216
  submicronic, 55, 56
  toxic, 214
Pollution, 4, 7–9, 26, 30, 31, 50, 55, 68, 73, 76, 81, 86, 105, 106, 109, 114, 118, 134, 151, 171, 175, 180, 195, 205, 213, 217, 241. *See also* Air pollution; Environmental pollution
  carbonaceous, 137
  dust on tobacco leaves, 215
  effective laws, 4
  electromagnetic, 150
  exogenous, 78
  industrial, 52, 116
  irreversible, 106
  nano, 204, 243
  in nanotechnology laboratories, 204
  particle, 111
  particulate, 42, 109, 214
  sources of, 5
  submicronic, 217
  tobacco, 35
  war, 50, 160
Polytetrafluoroethylene (PTFE), particles of, 30
Polyvinyl chloride (PVC), particles of, 30
Post-traumatic stress disorder (PTSD), 129, 217
Power plant, 99
  air pump, filters of, 104
  carbonaceous matrix, ESEM image of, 102
  case studies, 99
  combustion, 105
  debris deposited on lichens, 103
  dermatitis, 107
  epidemiology surveys, 109
  keratosis, 107
  lettuce leaves, ESEM images of, 104
  micro particles, 100
  nanoparticles, 100
  pancreas cancer, 107
  particles contained in oil ashes, 100
  particles in burned heavy oil ashes, 101
  pathological cases of subjects, 106
  t-shirt sample, ESEM image of, 105
Pre-natal stress, 220
Printers and nanoink, 196
  biopsic samples of patient, 196
  free dispersion, 196
  toners and inks, 196, 198
Prions, 181, 182, 185, 186
Prostatic cancer, 75. *See also* Carcinoma
  case studies, 75
  elements found in biopsy, 76
  metallic debris, 77
Prostatic neoformation, alloys in, 74
Protection masks, 123
Protein corona, 24, 186
Protein/nanoparticle interaction, 24
Pyrogens, 26

R

Renal failure, acute, 88
Respiratory distress, 78
Respyrocytes, 235
Retinal circulation, 219

S

Sarcoidosis, 69
Sarcoma affected bladder and prostate tissues, ESEM images of, 77
Scanning electron microscope, 17, 30
Seminal fluid, SEM image of, 51
Sensors, 205
  cell-membrane, 16, 18, 236
  immune-system, 237
  passive, 133
Sentinel cases, 67
Siderosis, 44
  ESEM images of, 44
  spherular calcification, 45
Silicon-based debris, 73
Silver nanoparticles, 125, 165, 167, 169, 239, 241. *See also* Nanoparticles
  carbon spherules with, 169
  nanoproducts, 165, 167, 238
  water filters, 169
Smoking, impact of, 209
  Bosnian cigarettes, 209
  cigarette, temperature of, 210
  CO_2-O_2 exchange, 211
  pathological effects, 210
  pseudoaneurysm, 210
  thrombogenicity, 210
  tobacco leaves, 209, 210
Sodium-phosphorus precipitates, 227
Solanum lycopersicum L., 122
Spherules, 61, 62, 139, 147, 148, 206
  calcification in, 90
  of calcium phosphorous, 39, 42, 43, 70, 89, 220, 223, 224, 225
  carbon, 169
  in case of pleural mesothelioma, 35
  of cerium, 35
  of heavy metals, 101
  human cancerous, 124
  of lanthanium, 35
  in lungs, 35
  nanosized, 139
  of neodymium, 35
  nickel, 205
  platinum-iridium, 74, 79
  in pleura, 35
  of praesodimium, 35
  silicon, 140
  of stainless steel, 54, 62, 120
Spontaneous pneumothorax, 200
  case studies, 200
  debris in ceramic worker's lung, 201
  debris in lung sample of mason, 201
  debris in surgeon's lung sample, 203
  debris in welder's lung, 202
  dissemination of ceramic dust in hairdresser's lung, 202, 203
  EDS spectra, 200
Squamous-cell carcinoma, 219
  stainless-steel and aluminum-titanium-phosphorus particles, 221
Stainless steel
  EDS spectra of, 74
  particles, 37
    in mesothelioma affected peritoneal tissue, 38
    micro-sized, 37
    nanosized, 37
    in pleural mesothelioma patient's pleura, 37
  seminal fluid, SEM image of, 51
Stamaril, 189, 191
Stokes' law, 23
Subependimoma, ESEM images of organic/inorganic particles in, 121
Submicronic particles, 5, 10, 41, 42, 51, 55, 69, 70, 81, 84, 120, 136, 216, 217, 241
Sulfur-silver particles, ESEM images and EDS spectra of, 72

T

3T3 cells, 19–21
Testicle cancer, 73. *See also* Carcinoma
  ceramic-tile industry workers, 73
  tungsten debris, 73
Thigmomorphogenesis, 123
Thimerosal, 189
Thoracic sarcoma
  micron-sized and submicronic debris, SEM images of, 91
  stainless-steel debris, 92
  tungsten-iron debris, 92
Thoracic wall neoformation, 90
Thyperix, 189, 191

Thyroid cancer. *See also* Carcinoma
calcium-phosphorous spherule, 223
ESEM images of, 43, 44
sodium-phosphorous precipitates, 226
Thyroid gland, 39
Thyroiditis, 107
Thyroid tissue, 39
contralateral, 42
elemental normal content of, 40
Tibia bone, image and chemical spectrum of, 53
Tillandsia usneoides, 124
Titanium oxide nanoparticles, 123, 166, 168
Tobacco
cigarettes, 211–215
chemical compositions of the debris, 212, 215
health hazards, 214
combustion of, 209
elemental composition of, 213
leaves, 35, 210, 213, 215
organic, 214
pollution, 35
smoking, 48, 211
strip, SEM image of debris found in, 48
Toners, 198
barium particles, 199
carbon particles, 199
case studies, 199
Towel thread, image of, 166
Trojan horse, 236
Tuberculosis, 69
Tumor, 39, 45, 46, 90
cases, 39
hemolymphatic, 150
of jaw and accounts, 45
maxilla-mandibular, 45
oral, 45
pediatric, 75
thyroid, 39, 43, 44
tissue, 234
yolk-sac, 84
Tungsten carbide, 48, 73, 203
anti-wear properties of, 176
ceramic tiles of, 74
debris, in mouth mucosa sample, 49
EDS spectra of, 74
wear debris, 73

U

Uranium radioactivity hypothesis, 130

V

Vaccines, 86, 87, 130, 187
Anatetall, 189, 190
case studies, 187
contamination, 187
foreign bodies in, 191
Gardasil, 189, 192
Infarix, 189
Inflexall, 189, 190
inorganic particles, 187
Stamaril, 189, 191
thimerosal, 189
Thyperix, 189, 191
Van der Waals forces, 186
Vasculitis, 216

W

War
and archeology, 71
bombing, 71
case studies, 145
environmental dust, 131
chemical compositions of dust from explosion site, 133, 134
depleted-uranium explosions, 132
depleted-uranium penetrators, 131
EDS spectra of debris from dust of explosion, 135
high-potential bombs, 131
high-temperature explosions, 132
inorganic debris, 131
nanometric droplets, 131
radioactive elements, 134
experimental activities, 52
pancreatic adenocarcinoma, 71
pollution, 50, 52, 209
theaters, 130, 135, 148
Water filters, 169
WHO. *See* World Health Organization
World Health Organization (WHO), 5, 209

X

Xanthomonas oryzae pv. oryzae (Xoo), 123
X-rays, 30, 45, 85, 141, 196, 224, 234

Z

Zirconia (zirconium oxide) particles, 37
in liver, 37
in lung, 37
in pleura, 37

Printed in the United States
By Bookmasters